The Political Economy of the Asia Pacific

Series editor
Vinod K. Aggarwal

More information about this series at http://www.springer.com/series/7840

Terry O'Callaghan • Geordan Graetz
Editors

Mining in the Asia-Pacific

Risks, Challenges and Opportunities

Springer

Editors
Terry O'Callaghan
Centre for International Risk
School of Communication, International Studies and Languages
The University of South Australia
Adelaide, SA, Australia

Geordan Graetz
Centre for International Risk
School of Communication, International Studies and Languages
The University of South Australia
Adelaide, SA, Australia

ISSN 1866-6507 ISSN 1866-6515 (electronic)
The Political Economy of the Asia Pacific
ISBN 978-3-319-61393-2 ISBN 978-3-319-61395-6 (eBook)
DOI 10.1007/978-3-319-61395-6

Library of Congress Control Number: 2017949678

Printed on acid-free paper

This Springer imprint is published by Springer Nature
The registered company is Springer International Publishing AG
The registered company address is: Gewerbestrasse 11, 6330 Cham, Switzerland

Terry dedicates this book to Margherita O'Callaghan and Cristina and Vincenza Indelicato.

Geordan dedicates this book to the many friends, colleagues and family members who have contributed to his thinking and personal and professional development over many years.

Acknowledgements

We owe a number of people an immense gratitude for their assistance in bringing this book together. Our primary thanks go to Vlado Vivoda for his significant and meaningful contribution to this project. We have been discussing aspects of, and issues in, the mining, energy and infrastructure sectors with Vlado for a number of years. We hope that we can continue our conversations and friendship with him for many years to come.

Richard Leaver has contributed the chapter on iron ore. His knowledge of the mining industry in the Asia-Pacific region is exceptional, and Richard also has had a close intellectual relationship with Vlado, Geordan and Terry. We thank him for his friendship and insights.

We have benefitted from our interactions with the mining industry. We thank Goldfields Australasia and Marathon Resources (no longer trading under this name) for their interest in, and support for, our work. We particularly thank Ed Baltis (Goldfields Australasia) and Dr Dan Tyson (ex-Marathon Resources). Our thanks also go to Phil Sutherland, former president of the South Australian Chamber of Mines and Energy, whose support was critical in the Centre for International Risk receiving Australian Government funding for our research programme. The two projects in question are an Australian Research Council (ARC) Linkage Grant (LP 0669261) and a Discovery Grant (DP 0771064). These grants provided funding to explore the social, political, economic and environmental aspects of mining and infrastructure in the Asia-Pacific region. The book is best viewed as the culmination of the ARC's investment in the Centre for International Risk.

The book also would not have been possible without the significant contribution of Professor Sumi Odono of Shiga University, Hikone, Japan. He was important in securing industry funding from Nomura Securities, Japan's largest finance corporation. This book honours Professor Odono, the professional staff at Shiga University, and Nomura Securities for their significant contribution to our work.

We thank, too, the University of South Australia. The Centre for International Risk received two grants to undertake specific projects associated with our research. The first was a grant to examine the Australian mining industry's performance in relation to the resource-rich countries in the Asia-Pacific region. The second was a grant to examine the development implications of a lack of infrastructure in the Asia-Pacific region, especially in relation to the problem of energy poverty.

We thank Jon Gurstelle and Kevin Halligan from Springer International, New York, for their consideration of, and assistance with, our manuscript. Finally, we thank the commissioning editor, Professor Vinod Aggarwal, who leads the International Political Economy of the Asia-Pacific Region series at Springer International, for his favourable endorsement of this project.

This book is fortunate to include contributions from some of the most esteemed thinkers and practitioners in the mining industry. We thank those contributors for their interest and willingness to participate. We hope, as editors, that we have done justice to their ideas.

The book is meant to be a reflection of the 'state of the art' of the mining industry and a forecast of its future. If the sector is to flourish in the coming decades, we believe that the lessons contained in this book will need to be heeded and the prescriptions adopted.

Contents

1 **Introduction** 1
Terry O'Callaghan and Geordan Graetz

Part I Theorising Risk in the Mining Industry

2 **Determinants of Foreign Direct Investment in the Mining Industry** 19
Vlado Vivoda

3 **Regulatory Regimes, Foreign Mining Investment, and Risk in the Asia-Pacific Region: Comparative Evaluation and Policy Implications** 35
Terry O'Callaghan and Vlado Vivoda

4 **Property, Politics, and Power: Theorising Political Risk in the Mining Industry** 49
Daniel Feher

5 **Social Risk and Business Risk in Mining Projects** 65
Geordan Graetz and Daniel M. Franks

6 **Security of Resource Supply and the Obsolescing Bargain** 85
Tom Johnson

7 **Mining and Human Rights** 105
Eboni Tiller

8 **Mergers and Acquisitions: The Politics of Megadeals in the Mining Industry** 121
Vlado Vivoda and Geordan Graetz

Part II The Major Resources

9 Iron Ore 141
Richard Leaver

10 Coal 155
Philip Crowson

11 Rare Earth Elements 177
Ming Hwa Ting

12 Natural Gas 189
Vlado Vivoda

13 Copper 217
Philip Crowson

14 Coal Seam Gas 237
Cristelle Maurin

Part III The Major Issues

15 Mining and Sustainable Development 253
Glen Corder

16 Artisanal and Small-Scale Mining 271
Tamsin Scholz

17 Mining, Corporate Social Responsibility, and Corporate Reputation 285
Terry O'Callaghan and Belinda Spagnoletti

18 Mining and Climate Change 301
Vigya Sharma

19 Mining and Infrastructure 321
Tom Johnson

20 Legacy Issues and Abandoned Mines 333
Corinne Unger

Part IV Mitigating Risk

21 Investment Decision-Making Tools for the Mining, Oil, Gas, and Infrastructure Sectors 373
Terry O'Callaghan and Vlado Vivoda

Abbreviations

4QA	Four-quadrant analysis
ABS	Australian Bureau of Statistics
ACARP	The Australian Coal Industry's Research Program
ADB	Asian Development Bank
AFC	Asian financial crisis
AMD	Acid mine drainage
AMWG	Abandoned Mines Working Group
ANCOLD	Australian National Committee on Large Dams
ANZECC	Australian and New Zealand Environment and Conservation Council
ANZMEC	Australian and New Zealand Minerals and Energy Council
AONSW	Audit Office of New South Wales
APEC	Asia-Pacific Economic Cooperation
ARC	Australian Research Council
ARD	Acid rock drainage
ARF	ASEAN Regional Forum
ARMCANZ	Agriculture and Resource Management Council of Australia and New Zealand
ASEAN	Association of Southeast Asian Nations
ASM	Artisanal and small-scale mining
AUSIMM	Australasian Institute of Mining and Metallurgy
BCCCSP	British Columbia Crown Contaminated Sites Program
BIT	Bilateral investment treaty
BNE	Business and the natural environment
BOM	Bureau of Meteorology
BOO	Build-operate-own
BOP	Bottom of the pyramid
BOT	Build-operate-transfer
BTO	Build-transfer-operate
BTU	British thermal unit
CASM	Communities and Small-Scale Mining
CBM	Coal bed methane

CCD	Corporate community development
CCP	Chinese Communist Party
CCR	Corporate community relations
CCSP	Crown Contaminated Sites Program
CE	Creeping expropriation
CEO	Chief executive officer
CHP	Combined heat and power plants
CLM	Contaminated land management
CMLR	Centre for Mined Land Rehabilitation
CNOOC	China National Offshore Oil Corporation
CNPC	China National Petroleum Corporation
CO_2	Carbon dioxide
COAG	Council of Australian Governments
COCHILCO	Chilean Copper Commission
COW	Contract of work
CPI	Corruption Perceptions Index
CSG	Coal seam gas
CSR	Corporate social responsibility
DE	Direct expropriation
DEHP	Department of Environment and Heritage Protection, Queensland
DMC	Domestic material consumption
DME	Department of Mines and Energy, Western Australia
DMP	Department of Mines and Petroleum, Western Australia
DNRM	Department of Natural Resources and Mines, Queensland
DRC	Democratic Republic of the Congo
DRET	Department of Resources, Energy and Tourism
DSB	Dispute Settlement Body
DSTP	Deep-sea mine tailings placement
DTIRIS	Department of Trade and Investment, Regional Infrastructure and Services
EAAU	East Asia Analytical Unit
EAS	East Asia Summit
EIA	Energy Information Administration
EIA	Environmental impact assessment
EITI	Extractive Industries Transparency Initiative
EMP	Environmental management plan
EMOS	Environmental Management Overview Strategy
ENI	Ente Nazionale Idrocarburi, Italy
EPA	Environmental Protection Agency
EPBC	Environment Protection and Biodiversity Conservation Act (Cth) 1999
EPI	Environmental Performance Index
ERA	Energy Resources of Australia
EU	European Union
FA	Financial assurance
FDI	Foreign direct investment

FIFO	Fly-in, fly-out
FMG	Fortescue Metals Group
FOB	Free on board
G7	Group of 7
GAIL	Gas Authority of India Limited
GATT	General Agreement on Tariffs and Trade
GAZPROM	Gazovaya Promyshlennost
GDP	Gross domestic product
GFC	Global financial crisis
GFCF	Gross fixed capital formation
GHG	Greenhouse gases
GIIGNL	International Group of Liquefied Natural Gas Importers
GIS	Geographical information system
GMI	Global Mining Initiative
GMS	Grammes
GRI	Global Reporting Initiative
GSP	Gross state product
GUCAS	Guided Uniform Criteria Assessment System
GW	Gigawatt
H&S	Health and safety
HH	Henry Hub
HHI	Herfindahl-Hirschman Index
HRC	Human Rights Council
HSEC	Health, safety, environment and community
IAEA	International Atomic Energy Agency
IBA-SEE	Internationale Bauausstellung Fürst-Pückler-Land, Germany
ICEL	Indonesian Center for Environmental Law
ICMM	International Council on Mining and Metals
ICSG	International Copper Study Group
ICSID	International Centre for Settlement of Investment Disputes
IEA	International Energy Agency
IEEJ	Institute of Energy Economics, Japan
IESC	Independent Expert Scientific Committee on Coal Seam Gas and Large Coal Mining Development
IFC	International Finance Corporation
IFOE	International Friends of the Earth
IIED	International Institute for Environment and Development
IISI	International Iron and Steel Institute
IMF	International Monetary Fund
INAP	International Network for Acid Prevention
IOC	International Oil Company
IPCC	Intergovernmental Panel on Climate Change
IPE	International political economy
IRR	Internal rate of return
ISO	International Organization for Standardization

IUCN	World Conservation Union
JBIC	Japan Bank for International Cooperation
JCC	Japan Crude Cocktail
JKM	Japan/Korea Marker
JOGMEC	Japan Oil, Gas and Metals National Corporation
JATAM	Jaringan Advokasi Tambang, Indonesia
KOGAS	Korea Gas Corporation
LMU	Legacy Mine Unit
LNG	Liquefied natural gas
LSM	Large-scale mining
LTO	Licence to operate
M&A	Mergers and acquisitions
MCA	Minerals Council of Australia
MCMPR	Ministerial Council on Mineral and Petroleum Resources, Australia
MEG	SNL Metals Economics Group
METI	Ministry of Economy, Trade and Industry, Japan
MIGA	Multilateral Investment Guarantee Agency
MIRECO	Mine Reclamation Corporation
MMBTU	Million British thermal units
MMSD	Mining, Minerals and Sustainable Development
MMSS	Mining and Metals Sector Supplement
MNC	Multinational corporation
MOF	Ministry of Finance, Japan
MPMD	Mines and Petroleum Management Division, Northern Territory
MRF	Mining Rehabilitation Fund, Western Australia
MRRT	Minerals resource rent tax
MT	Million tonnes
MTCE	Million tonnes of coal equivalent
MTOE	Million tonnes of oil equivalent
MTPA	Million tonnes per annum
NAO	North Atlantic Oscillation
NGO	Nongovernment organisation
NOAMI	National Orphaned/Abandoned Mines Initiative
NOC	National oil company
NPV	Net present value
NSW	New South Wales
NT	Northern Territory
NTMC	Northern Territory Minerals Council (Inc.)
OAGBC	Office of the Auditor General of British Columbia
OBM	Obsolescing bargain model
OECD	Organisation for Economic Co-operation and Development
OHCHR	Office of the United Nations High Commissioner for Human Rights
OPEC	Organization of the Petroleum Exporting Countries
OT	Oyu Tolgoi, Mongolia
PJ	Petajoules

PMSEIC	Prime Minister's Science, Engineering and Innovation Council
PNG	Papua New Guinea
POSCO	Pohang Iron and Steel Company
PPI	Policy Perception Index
PPP	Public-private partnership
PPP	Purchasing power parity
Q2	Second quarter
QAO	Queensland Audit Office
QCA	Queensland Commission of Audit
QFCI	Queensland Floods Commission of Inquiry
RN	Resource nationalism
SA	South Australia
SCER	Standing Council on Energy and Resources
SCO	Shanghai Cooperation Organisation
SD	Sustainable development
SLO	Social licence to operate
SMCRA	Surface Mining Control and Reclamation Act, 1977
SO	Southern Oscillation
SOE	State-owned enterprise
SOHO	Sustainability Opportunities and Hazards Overview
SOI	Southern Oscillation Index
SPeAR	Sustainable Project Appraisal Routine
SWF	Sovereign wealth fund
SX-EW	Solvent extraction and electrowinning
TEPCO	Tokyo Electric Power Company
TNC	Transnational corporation
TRIP	Trade-related intellectual property rights
TWH	Terawatt hour
UCG	Underground coal gasification
UDHR	Universal Declaration of Human Rights
UK	United Kingdom
UN	United Nations
UNCTAD	United Nations Conference on Trade and Development
UNEP	United Nations Environment Programme
US	United States of America
USD	US dollar
USGS	United States Geological Survey
VAGO	Victorian Auditor-General's Office
WA	Western Australia
WBCSD	World Business Council for Sustainable Development
WBMS	World Bureau of Metal Statistics
WHO	World Health Organization
WOAG	Western Australian Office of the Auditor General
WTO	World Trade Organization

List of Figures

Fig. 10.1 Coal's percentage share of primary energy supply in selected countries, 1990–2011 159
Fig. 10.2 Supplies of coal and coal products, 1990–2011 (mtce) 160
Fig. 10.3 Domestic supplies of coking coal, 1980–2012 (million tonnes) 162
Fig. 10.4 Domestic supplies of steam coal, 1980–2012 (million tonnes) 163
Fig. 10.5 EIA projections of Chinese coal demand to 2040. Index numbers based on consumption expressed in BTUs (2010 = 100) 168
Fig. 10.6 Asia-Pacific coal production and consumption, 1981–2012, excluding China and Indonesia 169
Fig. 10.7 China's coal production and consumption, 1981–2012 170
Fig. 10.8 Annual prices of Australian steam coal in money and in 2013 terms since 1980 172
Fig. 10.9 Japanese coal prices and an Asian marker price, 1987–2012 173

Fig. 12.1 LNG price in the three basins, 2002–2013 (US$ per million Btu) 191
Fig. 12.2 Japan's long-term LNG contracts in Australia, 2009–2036 (mtpa). 207

Fig. 13.1 Index numbers of the usage of refined copper 221
Fig. 13.2 Association of South East Asian Nations (ASEAN) and Asia-Pacific Economic Cooperation (APEC) per cent share of global usage of refined copper, 1973–2012 222
Fig. 13.3 Asia-Pacific sources of supply to meet usage of refined copper, 2003–2012 227
Fig. 13.4 Sources of Chinese supplies of refined copper metal, 2003–2012 229
Fig. 13.5 Mine output of copper in the Asia-Pacific Region, 2003–2012 231
Fig. 13.6 ICSG forecasts of global copper mine capacity, 2010–2013 ('000 tonnes contained copper) 234

Fig. 18.1 Australia's Southern Oscillation Index from 2000–2012 307

Fig. 20.1 Leading practice elements of abandoned mine regeneration 352
Fig. 20.2 Improving level of maturity model 359
Fig. 20.3 Maturity of abandoned mine program in Western Australia 360
Fig. 20.4 Maturity of crown contaminated sites program in British Columbia, Canada 360

List of Tables

Table 2.1 Key foreign investment criteria in the mining sector 23

Table 5.1 Types of social risk and examples........................... 71
Table 5.2 Social change processes and examples 72
Table 5.3 Examples of social impacts arising from private sector developments... 72
Table 5.4 Types of business risk and examples......................... 73

Table 6.1 Summary of raw material exports and their value............... 93
Table 6.2 Asia-Pacific countries introducing RN policies, 2004–2014 94
Table 6.3 National oil companies in the Asia-Pacific Region.............. 96

Table 7.1 Nussbaum's list of central capabilities 115

Table 8.1 M&A megadeals in the mining industry (Asia-Pacific Region) 2005–2014.. 129

Table 10.1 Top ten coal producers, 2014 (million tonnes) 157
Table 10.2 Total global coal production in 2013, 2014 and 2015, compared to 1990 158
Table 10.3 Coal consumption in the Asia-Pacific Region, 2010–2015....... 158
Table 10.4 Domestic coal supplies in the Asia-Pacific Region, 1980–2011 (million tonnes).. 160
Table 10.5 Domestic supplies of coal by type, 1980–2011 (million tonnes) .. 161
Table 10.6 Usage of coal in electricity generation in Asia-Pacific countries, 1980–2011 (million tonnes) 166
Table 10.7 Production and consumption of various coal types in selected countries and years, 1990–2012 170
Table 10.8 Estimated Asia-Pacific coal reserves, end 2011 (million tonnes) .. 171

Table 11.1 OPEC basket price, 2005–2014............................. 179

Table 12.1 Basic natural gas data 199
Table 12.2 Key market participants in the LNG trade 199

Table 13.1 Global Uses of copper in 2013 219
Table 13.2 Usage of refined copper 2002–2012, Asia-Pacific and World. 221
Table 13.3 Annual average rates of change of real gross domestic product . . . 223
Table 13.4 Shares of gross fixed capital formation in GDP and copper intensities 223
Table 13.5 Usage and sources of refined copper in 2012 ('000 tonnes) 228
Table 13.6 Asia-Pacific production and trade in primary copper raw materials in 2012 230

Table 15.1 The ICMM's 10 guiding principles 257
Table 15.2 Relevant sustainability principles, frameworks, and standards 257
Table 15.3 Features of the Five Capitals 258
Table 15.4 A sample of sustainability tools utilised by industry 261

Table 17.1 An adaptation of Arnstein's ladder and its applicability to the mining industry 297

Table 18.1 Global mining hotspots vulnerable to climate change 303
Table 18.2 Impacts of climate change on various economic sectors. 305
Table 18.3 The mining-herding-climate nexus in Mongolia. 312

Table 19.1 Common investment structure for infrastructure projects. 324

Table 20.1 Estimated numbers of orphaned, abandoned or derelict mine sites (outside Australia) 339
Table 20.2 Towards a sustainable future for the mining and minerals sector 341
Table 20.3 Elements of a mature abandoned mine program applied to Australia's strategic framework 359

Chapter 1
Introduction

Terry O'Callaghan and Geordan Graetz

Abbreviations

ARD	Acid Rock Drainage
CEO	Chief Executive Officer
CSG	Coal Seam Gas
CSR	Corporate Social Responsibility
DRC	Democratic Republic of Congo
DSTP	Deep Sea Mine Tailings Placement
ICMM	International Council on Mining and Metals
IFOE	International Friends of The Earth
RN	Resource Nationalism
SLO	Social Licence to Operate
SOE	State-Owned Enterprises

Late on the afternoon of 5 November 2015, the Fundão iron ore tailings dam collapsed. An estimated 60 billion litres of tailings flowed down the Santarém Valley, located in the south-eastern region of Brazil. The 'mudslide' destroyed Bento Rodrigues, a small town located below the dam wall.[1] It also washed away nearby roads and bridges, and flowed into the Doce River. The waste plume flowed downstream for some 600 km, draining into the Atlantic Ocean. The plume then traversed the coastline of Espirito Santo, a neighbouring state, before gradually dispersing (Australian Broadcasting Commission 2015). Nineteen people died in the incident; while many others seriously were injured. This was a calamity, not only for those directly affected by the dam collapse, but also for the Brazilian economy.

The impact of the waste plume on fish stocks, wildlife, swamplands, and coastal areas was significant. There were reports that a large number of dead fish washed up

[1] Bento Rodriguez is in the state of Minas Gerais, in the south-eastern part of Brazil. Other seriously affected towns in the region include Barra Longa.

T. O'Callaghan (✉) • G. Graetz
Centre for International Risk, School of Communication, International Studies and Languages, The University of South Australia, Adelaide, SA 5001, Australia
e-mail: Terry.OCallaghan@unisa.edu.au; g.graetz@uq.edu.au

T. O'Callaghan, G. Graetz (eds.), *Mining in the Asia-Pacific*, The Political Economy of the Asia Pacific, DOI 10.1007/978-3-319-61395-6_1

along the Atlantic coast. Local authorities warned people not to touch or enter the contaminated water. There also were concerns that the waste plume could disrupt the breeding cycle of fish that spawned in the mouth of the river. In addition, the contamination of the river system led to a suspension of water supplies to towns and cities in close proximity.[2]

There was no alarm system in place to warn villagers of the impending danger. Indeed, it was locals who rushed through the streets of Bento Rodrigues, alerting their neighbours of the mudslide that was heading their way. Reports have suggested that the plume was travelling about 40 km an hour (Fitzgerald 2016). It was fortunate that the failure of the tailings dam occurred during the day. Had it occurred in the middle of the night, the death toll likely would have been much higher, as most of the villages in the plume's path would have been asleep (Knight 2016).

Some commentators have described the tailings collapse as the worst environmental disaster in Brazil's history (Gallas 2016).[3] The Brazilian environment minister even went as far as to suggest that it could take 30 years to restore the Doce River to health (Timson 2015). Without doubt, it ranks alongside the Marcopper and Ok Tedi disasters in the Asia-Pacific Region for the level of devastation to the environment. Unfortunately, after two decades of efforts to demonstrate to communities that mining companies are responsible to—and for— the host communities—with significant gains in community support for mining projects, the Fundão incident undermines those gains. Consequently, the impact of the disaster will be felt across the industry for many years to come.[4]

Fundão is owned by Samarco Mineração S.A., a 50:50 joint venture partnership between Vale S.A. and BHP Billiton Brazil. Samarco's principal line of business is the export of iron ore pellets. In 2014, this amounted to 25.075 million tonnes. The net income generated by Samarco amounted to R$2805.5.

Samarco had been warned of problems in the dam wall many months before its collapse. However, it is alleged that the company did not act on these warnings (Keirman 2016). According to engineer Joaquim Pimenta de Avila, the company was warned that there were potential problems with the dam wall in September 2014 (Sonawane 2016).[5] Knight notes that the dam wall also had changed shape (Knight 2016). Notwithstanding the warnings, the most likely cause of the disaster was liquefaction—that is, waste water building up behind the dam wall, weakening its structural integrity. Samarco has acknowledged that there had been drainage

[2] According to Samarco, Fundão's operator, 29 cities/towns along the Doce River were affected by the collapse of the dam wall.

[3] Given the degree of destruction of the Amazon jungle over the last decade, it is difficult to accept this assessment.

[4] This sometimes is referred to as a 'contagion effect' or 'reputational interdependence'. When something bad happens to one company, it tends to taint the sector/industry as a whole.

[5] Samarco disputes this view.

problems in the past. There also is a suggestion that a small earthquake the previous day may have triggered the collapse.

The cost of the incident is likely to be enormous. Samarco already has agreed to pay US$2.3 billion over 6 years (Mehra 2016). The agreement covers two programs: one is to repair the damage to local communities and the environment; the second aims to ensure that, where satisfactory repairs are not possible, individuals and communities receive fair compensation. However, the programs do not prevent the levying of any future court-sanctioned fines. For example, the Federal Public Prosecution Service of Brazil is seeking a US$43 billion penalty against Samarco, Vale, and BHP Billiton. Moreover, as a result of the disaster, the mine's operating licence has been suspended and twenty one of Samarco's employees have been arrested and charged with homicide, including Ricardo Vescovi, the company's president (Fitzgerald 2016).

Despite the charges, the company has been proactive in dealing with the disaster. On 19 May 2016, Samarco released *To Do What Must be Done: This is Our Commitment* (Samarco 2016), which sets out the company's current and future commitments, including:

- Provision of temporary accommodation for families that have lost their homes
- The resiting and reconstruction of Bento Rodrigues and other affected towns (in the Paracatu de Baixo district)
- Emergency reforestation along the Doce River to stop further sedimentation
- Monitoring of water quality along the length of the Doce River and the coastal areas near its mouth
- Rebuilding damaged bridges
- Issuing 7542 financial assistance cards
- Stabilising the tailings dam, with real-time monitoring
- Ensuring that the affected children are able to finish their school year with as little disruption as possible.

In addition to these 'first-order' issues, the company also has renovated houses in the nearby township of Barra Longa, refurbished its public spaces, built retaining walls, renewed 145,000 m of fencing, rescued and assisted nearly 7000 animals, and hired a large number of health professionals to attend to the physical and psychological needs of affected residents.

Quite rightly, the focus of attention has been on those directly affected by the disaster. However, mention also needs to be made of its broader impact. Samarco does not believe it will have the necessary approvals in place to restart operations in 2016–2017. This means that there will be redundancies within the company. In addition, there will be a negative impact on the Brazilian economy. According to Samarco, the company has generated R$6.3 billion in taxes over the last 5 years, with R$1.5 billion in tax paid in 2014. Between 2011 and 2015, the company invested R$9.1 billion in Brazil. Finally, Samarco's export representation is about 1% of Brazil's total exports (Samarco 2016). This revenue stream has been disrupted by the disaster, and is likely to continue to be disrupted for some years to come.

This case is a stark reminder of the challenges and risks associated with mining operations. Mining often occurs in remote areas, far from cities and large towns. Dedicated infrastructure has to be built, equipment and supplies have to be transported—frequently over long distances, and skilled workers are required. Often, the terrain is rough, the weather inhospitable, and death and lost time injuries in the workplace are an ever-present risk. It is not hard, then, to admire companies that are able to overcome such challenges in their search for, and exploitation of, a deposit. In addition to the logistical challenges, mining companies have to negotiate with numerous stakeholders, including national and regional governments, local communities, and increasingly sophisticated activist groups. This takes skill, patience, and a commitment to negotiation. It is not unheard of for such negotiations to take many years. It is understandable, then, that mining is characterised as a high risk, but also, a high reward activity.[6]

By risk, we mean those issues that may arise during the various stages of a mining project (from planning to closure) that have the capacity to undermine a project's financial and operational feasibility and viability. These risks are diverse, but pertain to political, economic, social, business, and environmental issues. The key characteristic of such risks is that they cannot be predicted with any certainty. Scenario plans or models, the development of complex matrices, and other similar strategies cannot predict outcomes with full certainty. Understanding these risks is not a scientific endeavour; instead, it relies on intuition, good fortune, and the application of knowledge and experience. For example, a country may have a stable, investment-oriented government for many years. It is very difficult to predict and scenario plan for an overnight coup that leads to the installation of a government that nationalises the assets of foreign mining companies. A mining company cannot divest itself of its assets in the way an equity investor can. This, then, is just one reason why mining companies generally prefer to operate in jurisdictions that are less prone to risk. Operating in Australia, New Zealand, or the United States generally is less 'risky' than operating in the Democratic Republic of the Congo (DRC), the highlands of Papua New Guinea, or Pakistan—despite the attractive resource endowments of these countries.[7]

[6] Samarco's revenue in 2015 was approximately US$600 million (R$1,921,460). In 2015, the 5 largest mining companies in the world had combined revenues of US$306.2 billion. These are Glencore (US$170.497 bn), BHP Billiton (US$52.267 bn), Rio Tinto Group (US$34.829 bn), Vale (US$25.609 bn), and Anglo American (US$23.003 bn). See: Fortune (2016) Global 500. Anglo American sits just outside the Fortune global 500. See: Anglo American 2016 Annual Report.

[7] This is not always the case. Freeport McMoRan is seeking permission to develop a copper/cobalt mine in the DRC—the Kisanfu project. This is despite the fact that, in 2015, the DRC was ranked the fifth most fragile country in the world. Transparency International also found it to be one of the most corrupt countries in the organisation's annual survey. In addition, the Fraser Institute has ranked the DRC as the second most politically unstable country in the world (Fraser Institute 2015; Fund for Peace 2015).

Challenges

The chapters of this book do not just draw attention to the high risk profile of the mining industry, but also to the significant challenges the industry faces. Rather than list every challenge confronting the industry, here, we draw readers' attention to five issues that we foresee will be at the forefront of corporate thinking in the next few years.

The Challenge of Resource Nationalism

When countries begin to suffer financial distress, whether as a consequence of poor domestic policy or a global crisis, it is natural that new revenue streams are considered. This can lead to higher income taxes for individuals, increased regulatory costs to small and medium enterprises, or attempts to place a higher fiscal burden on large, multinational corporations. If the latter, it generally means an increase in the cost of doing business in host countries. The outcome of this might be relatively benign, such as a 0.5% tax on all goods leaving a particular port. It also might mean an increase in the tax on plant and machinery entering the country. These costs generally can be absorbed by a company and not threaten its overall profitability. The same cannot be said for resource nationalism (RN).

RN occurs when national economic priorities take precedence over those of the global economy. That economy, generally speaking, is characterised by four key assumptions: First, capital should be able to move freely between countries; second, international trade should be free and unfettered by tariffs and other regulatory instruments; third, multinational corporations are engines of growth and should be allowed to invest in destinations of their choosing; and fourth, multinational investment creates jobs, drives prosperity, and is a global civilising force.

RN rejects some, or all, of these policies. It is a political (and economic) position, which rejects internationalism and globalisation, and is hostile to multinational corporations. It is a doctrine that puts the 'national interest' first. In a sense, it is a heretical position, not swayed by the dominant neo-liberal rhetoric of the United Nations, World Trade Organization, regional trade associations, central banks, and internationalist politicians.

Johnson argues that we are seeing a return of RN policies in a number of countries across the Asia-Pacific and around the world more broadly. This is a consequence of the recent mining 'super cycle'. With the high commodity prices of the early twenty-first century, some governments have sought a greater share of the wealth generated by mining operations in their jurisdictions. The other reason, of course, may be ideological. A number of leaders have come to power over the course of the last decade or so who generally are anti-corporate, anti-globalisation, and anti-American in their thinking. Nationalising foreign-owned assets is seen as one way to protest against the dominance and imposition of neo-liberal forces on their economies and to derive populist electoral support at the same time.

Given this trend likely will continue for some time to come, especially as commodity prices have begun to rise again, the challenge for mining companies is to determine how to shield their operations and businesses from RN. Johnson suggests that one strategy might be for mining companies to become part of a local industrial cluster, which could facilitate their integration into the local economies in which they operate. BHP Billiton's cluster program in Chile is one such example (Chilean Economic Development Agency. n.d.). The key point is that what counts now is 'creating shared value'.[8] Companies that can demonstrate such a value proposition may be able to avoid the risk of RN becoming an impact. A related, but alternative, strategy is put forward by O'Callaghan and Spagnoletti in Part III. They argue that the pursuit of a traditional corporate social responsibility agenda may not pay dividends to mining companies in the coming years. Instead, the authors look at the potential of *meaningful* development partnerships. They employ Arnstein's Ladder of Citizen Participation to highlight the difference between meaningful partnerships and those that employ the language of partnerships, but that only are instrumental in their approach to communities. The challenge for mining companies, in the view of O'Callaghan and Spagnoletti, is to deepen the quality of their interactions with local communities to create shared value.

The Challenge of the Environment

Mining and the environment inextricably are linked. Mining companies scar the environment as they develop the infrastructure that facilitates their operations. They also use water to process ore and large amounts of energy, usually coal and diesel, to run their plant and allied machinery. Arguably, then, it is fair to say that without environmental impacts, mining could not occur.

Historically, there was little acknowledgement of the intimate connection between mining and the environment. The environment was an externality that was exploited in order for mining companies to profit. Disposing of tailings into a watercourse, changing the direction of rivers, and clearing large tracts of land simply was viewed as the price that had to be paid for the extraction of mineral resources. One of the best illustrations of this attitude appears in Clint Eastwood's film, *Pale Rider*.

Set in the late 18 hundreds, a ruthless mining company—owned by Coy LaHood—sought to drive a group of struggling artisanal gold miners from their legitimate claims. LaHood's tactics were violent and inhumane. The context of the plot is about the different ways in which people relate to, and interact with, the environment. The 'tin panners' sought to make a living from the environment, but with little negative impact on it. They simply panned for gold in the fast-flowing stream. In contrast, LaHood's upstream operation used an extraction method called hydraulic mining, or hydraulicking. A high-pressure stream of water was used to

[8] See: Shared Value Initiative, 'About Shared Value', Shared Value Initiative, https://www.sharedvalue.org/about-shared-value. Accessed 6 April 2017.

dislodge soil and rock from the cliff face. The dislodged material was then guided into a sluice to separate the gold from the waste, with the waste flowing downstream, polluting the river. The blasting process destroyed any flora and fauna on the rock face, and the operation required enormous amounts of water. Just as LaHood did not care about the tin panners, he took an instrumental approach to the environment—it was there to be exploited in order to derive a profit.

Hydraulic extraction was common during the Californian gold rush, particularly in the Sacramento Valley. The consequences for the environment were profound and, by the turn of the century, the practice was stopped. While hydraulic mining methods still are used in parts of the world today, it no longer is considered viable by most Western domiciled mining companies. Yet, the environment remains a fundamental challenge for the industry.

A number of chapters in this book address aspects of the relationship between mining and the environment. Corder and Sharma contribute chapters on sustainable development and climate change, respectively, while Scholz examines the environmental impacts of artisanal mining. The clear message from these chapters is that mining companies cannot ignore environmental risks and impacts. They are mainstream, in the public eye, and therefore present a critical challenge for mining companies to address. More than any chapter in this book, however, Unger draws attention to the environmental impacts of poor mine closure programs and regulation. Closure programs often are underfunded, negatively affect local communities, and, through their poor design, can leave negative environmental legacies. Unger's case study on the Mount Morgan mine in Queensland is illuminating in this regard. That mine was a copper and gold mine, and ceased operations in 1990. Since closure, acid rock drainage (ARD) problems have plagued the site. In 2013, the tailings dam overflowed and contaminated the Dee River for some 20 km. Unger concludes with policy recommendations for national, state, and territory governments in Australia, as well as suggestions for industry and communities about how better to manage mines after their closure

The Challenge of Social Activism

The mining industry is no stranger to opposition. Social activist organisations, such as International Friends of the Earth (IFOE) and Mines and Communities, have been vocal critics of mining operations for many years. Similarly, national groups frequently are critical of mining activities in their countries. Groups such as Jaringan Advokasi Tambang (JATAM) and Friends of the Earth Indonesia in Indonesia, Save the Palawan Movement in the Philippines, and GetUp! in Australia are representative of the many activist groups who focus their attention on what they perceive to be the ills of the mining industry. First Peoples also voice their concerns about the construction of new mines on their lands. For example, the Mirarr Traditional Owners in Australia's Northern Australia stridently have opposed uranium mining on their lands for more than 30 years.

Opponents of mining do not fit easily into a single category. Some are single issue interest groups and individuals, such as farmers who oppose coal seam gas developments on their properties or anti-uranium mining activists. Others oppose mining on environmental grounds. GetUp! opposes coal mining in the vicinity of the Great Barrier Reef. There also are those for whom opposition to mining is ideological. Korten (2002) calls mining companies 'predatory' organisations. Moody (2005, 2007) suggests the claim that mining is a sustainable activity is an oxymoron. Evans, Goodman, and Lansbury's (2002) edited collection, *Moving Mountains*, critiques the mining industry from a green-ideological perspective. What these ideological critiques of mining activities have in common is the link they draw between mining, globalisation, and capitalism, which they view as symptomatic of the problems the modern world faces.

Social activism poses a serious challenge to mining companies. Protests outside of head offices, barricading of work sites, awareness-raising on social media, political lobbying, and legal challenges are some of the mechanisms through which activists are able negatively to affect existing and new mining ventures. Business advocate Peters (1999) likens social activist groups to the ancient Velociraptor—a predatory dinosaur. Peters argues that they are capable of 'shredding' the reputations of companies of which they disapprove.

There are three main challenges facing mining companies in this regard. First, while mining companies are highly skilled and technically proficient operators, they often are not socially and politically astute. When challenged, particularly in the media, corporate leaders can be flippant in their responses to journalists' or community members' questions, can misreport events/actions/data, and can appear distant and unconcerned about the damage their businesses cause. This criticism has been levelled at Freeport McMoRan's former Chief Executive Officer (CEO), James 'Jim Bob' Moffett. The challenge, then, is to develop strong engagement, communications, and crisis management programs and processes.

The second challenge is to have meaningful engagement with First Peoples and to ensure that mining companies negotiate fairly and in good faith. There is a natural power differential at work in many negotiation scenarios, and it is important for mining companies not to abuse their power (Graetz and Franks). Should they act improperly, companies will be at the mercy of community and activist campaigns. O'Callaghan and Spagnoletti offer some guidance on this matter.

The third, related challenge is to manage the development issues that arise during mining activity. While mining companies are profit-making entities, many civil society groups have called for companies to play a more useful role in realising social and economic development objectives. Many of the people that live near mining operations are poor. The fly-in-fly-out camps, in particular, can introduce sexually transmitted diseases, prostitution, and violence to communities, affecting women and children disproportionately. The International Council on Mining and Metals (ICMM) has produced two useful 'toolkits' aimed at addressing the development challenges mining companies face. Those toolkits are not perfect, as O'Callaghan and Spagnoletti show, but they are an important reference point if mining companies are to reduce the possibility of an activist backlash against them and to manage the associated reputational risks.

The Challenge of Non-Western Mining Companies in Matters of Social and Environmental Corporate Policy

Western mining companies are subject to relatively stringent regulations that cover the life-cycle of a mine. The regulatory regimes cover four broad areas deemed critical to ensuring that mines are operated in accordance with community expectations. These areas cover finance and tax, environmental, operational, and social regulations. Each regime is in place to ensure that successful mining ventures can occur without there being negative externalities, which so often are associated with mining. Current corporate social responsibility principles offer guidance for companies that wish to be viewed as good corporate citizens.

Some critics argue that these regulations are not strict enough. For example, deep sea mine tailings placement (DSTP) still is allowed in some parts of the world, even though the practice causes immense damage to the marine environment (Moody 2005). Coal seam gas operations are believed by landholders and activist groups to pollute underground water tables, even though there are strict regulations governing such operations. And the protections available to Traditional Owners often are viewed as inadequate and as having failed to translate international principles on the rights of First Peoples into domestic rights frameworks.

Yet, despite the frequent criticism that Western mining companies inadequately are regulated, current rules governing the industry arguably are sufficiently robust and, in most jurisdictions, are seen as adequate both to facilitate mining and to prevent negative environmental and social impacts.

However, the practices of non-Western mining companies—including private firms, state-owned enterprises (SOEs), and companies operating in jurisdictions with lax regulations—present a challenge. First, they are not subject to the kinds of regulation that companies operating in Australia, Canada, or the United States (and elsewhere) are. Second, their CSR initiatives are not tied to shareholder expectations in the same way as they are in non-Western countries. Corporate governance requirements and accountability are also far less stringent. Third, the ability of civil society organisations to hold these companies to account often is weak. Fourth, they can source labour and material inputs cheaper. This means that the cost of production is lower. This makes them more competitive internationally. Fifth, most Western mining companies are subject to anti-corruption laws; acts of corruption committed overseas are an offence in their domicile country and are punishable by fines and other sanctions. However, non-Western mining companies often are not constrained by such laws. Consequently, their cost of doing business can be lower, increasing their profitability. For example, Chinese mining companies operating in Sudan are less encumbered by regulation or monitoring by civil society organisations. Human rights groups regularly assert that they cannot gain access to such countries; therefore, monitoring to ensure adherence to global human rights principles and other regulatory norms is difficult. This means that companies are free to act with relative impunity, because regulation in both the domicile and host countries is poor. Companies that seek to operate to a higher ethical and regulatory standard are at a commercial disadvantage then. The challenge, therefore, for Western mining companies is to demonstrate to stakeholders that acting to a higher ethical standard is valuable and has social and environmental benefits. This also may help to counter the perception that mining companies act according to their own interests alone, that they are unimpeded by regulation, and that they have scant regard for the communities, environments, and economies in which they operate.

The Challenge of Reputation Risk

There are a number of factors that contribute to corporate success. These include strong economic performance, which usually hinges on the sale of high quality products/services that are demanded by customers, good customer service, sound decision-making by management and boards, and the provision of good working conditions for employees. One way to think about the sum of these factors is to view them as the building blocks of a strong corporate reputation. By strong corporate reputation we do not mean, simply, the way in which a company is viewed by its customers; a strong corporate reputation is more than a company's image or its ranking on various commercial league tables. Instead, a strong corporate reputation is built on the behaviour of companies in the conduct of their business activities. Companies that are corrupt, lie to clients and/or regulators, defraud investors, or effect negative environment impacts will never gain a strong corporate reputation. In fact, such behaviours may result in a company's insolvency or, at the very least, a plummeting share price. In extreme cases, detrimental behaviours may result in complete corporate collapse. Enron, Arthur Anderson, and A.H. Robins are good examples of companies that were not able to recover from their misdemeanours. Even in less severe cases, the financial cost of poor corporate behaviour can be staggering. BP's Deepwater Horizon disaster and Volkswagen's emissions scandal have cost those companies billions of dollars in fines, rectification costs, and brand damage.

Mining companies tend not to have strong corporate reputations. Historically, mining companies have been faulted for their poor environmental impact management and for their abuses—on their own or in concert with host governments of militia forces—of human rights. Protests around mine sites have been commonplace and companies regularly have been found to be in breach of national and sub-national regulations.

But, it would be wrong to suggest that all mining companies flagrantly flout regulations. Mining is a complex activity; in some cases, companies face contradictory regulations. Mining laws can overlap with forestry regulations, such that a company finds itself not knowing to which competing regulation it should adhere. In other cases, a company might follow national regulations to the letter of the law, only to find itself in contravention of sub-national regulations. Mining companies sometimes also are unfairly or incorrectly blamed for impacts/events that are beyond their control. For example, in 2004, mining executives from Newmont Mining Corporation were arrested in Indonesia on suspicion of poisoning Buyat Bay with arsenic and mercury. Government testing later showed that there was no pollution in the bay and the charges were dropped. Unfortunately, Newmont's executives spent 5 days in jail.

Over the last two decades, reputation risk has emerged as a major challenge for the mining industry. This largely is why mining companies have begun to embrace CSR programs. Such programs are seen an important to successful business operations, as they demonstrate to stakeholders that mining activities can be undertaken responsibly and add value to the societies in which they operate. Companies, in the main, do not adopt CSR programs for altruistic reasons, but rather for instrumental reasons. Indeed, they see their CSR agenda as a bulwark against criticism from social activist groups and other groups or individuals that may be sceptical of the industry's claims to be socially and environmentally responsible. A CSR policy says to stakeholders that a company is a good corporate citizen and can be trusted to operate with integrity and according to leading practice standards.

Nevertheless, it is likely that mining companies will come under increased pressure around their social and environmental performance in the coming years. There are a number of reasons for this: First, responding to changing community expectations, Western governments are developing new and more robust regulation to govern the mining industry. Failure to adhere to community expectations and new regulatory standards could result in sever penalties being levied against companies and corporate managers. Second, in this new digital world in which social media reigns supreme, real time sharing of news and views means that all 'crises' now can be witnessed by a global audience. This changed media environment—in which reputations can be damaged in seconds—will necessitate significant adjustments by the industry in terms of its engagement with the media, civil society organisations, communities, and regulators. Third, social activism is an ever-more potent force in global politics. Activists can bring an operation to a stop, disseminate 'fake news' to communities/landholders, and, by damaging a company's reputation, can affect market share and/or profitability. Finally, the number and influence of shareholder activists is growing. Such shareholders want to see profitable companies, but they also want the companies in which they invest to act ethically. Companies that do not deliver on both these expectations likely will suffer reputational risks.

While the notion of the corporate reputation often is referred to as an intangible asset, reputational risk can undermine a company's commercial value very quickly. We suggest that maintaining a strong corporate reputation will be a top three risk confronting mining companies in the years ahead. The challenge, then, is how to manage reputational risk. In their respective chapters, Graetz and Franks and O'Callaghan and Spagnoletti chart different paths for companies to follow in managing reputational risk.

Opportunities

There is growing recognition among some industry leaders that the current approach to doing business is inadequate. First, it can be wasteful and use resources inefficiently. Second, the continual depletion of mineral, metals, and energy resources is not sustainable. Third, the dominant approach seeks short-term profits at the expense of longer-term, mutually beneficial outcomes for all stakeholders. Fourth, it fails to recognise that there is a good business case for doing things differently, including opportunities to develop new and innovative technologies to drive efficiencies, reduce waste and pollution, and increase yields, thereby growing market share and, potentially, improving interactions with stakeholders. Ray Anderson, the former CEO of Interface Inc. referred to this as the 'take-make-waste' model of business activity.

A new industrial model turns these four points on their head. It begins by reducing waste, improving resource efficiency, and identifying new methods to produce minerals and metals that are sustainable. Reducing waste is a vital first step, as the derived savings can be reinvested in the company to drive competitiveness and innovation in sustainability practices. Finding new ways to recycle and reuse minerals and metals can reduce the use of other natural resources, such as water, and the energy consumed by companies in their operations. Not only could this militate against environmental challenges such as climate change, but it also will reduce a company's environmental footprint, with attendant benefits for a company's social and environmental performance, profitability, and reputation.

The best example of the success of this model comes from the manufacturing sector. In 1994, Ray Anderson realised that his carpet manufacturing company was having a detrimental impact on the environment. The company was sending thousands of tonnes of carpet waste to landfill each year, was using large amounts of petroleum in its manufacturing process, and was wasting thousands of litres of water annually. At the time of his realisation, Interface Inc. was a relatively small company. However, the changes that Anderson implemented over a number of years has resulted in his company becoming the world's largest manufacturer of carpet tiles, with sales of more than one billion dollars annually. The company reduced the amount of its process waste to negligible amounts and now recycles all water used in the dying process. The company also generates a large proportion of its electricity from renewable sources. Interface Inc. also now has product lines that are made from 100% recycled materials. The company estimates that it has saved almost US$500 million dollars since implementing its new industrial model; savings that have helped it pay for the innovations in its manufacturing process.

The application of a new industrial model resulting in increased profits in the mining industry, then, is not as fanciful as it first may appear. Lavery and Pennell (2014) suggest the net profit increase from improvements in material efficiency, energy efficiency, and renewable energy could be as much as 100 billion Euros across the 51 European countries annually. In addition, they forecast that as many as 168,000 new jobs could be created. It is clear, therefore, that the kinds of arguments that reject the adoption of socially and environmentally responsible business practices are flawed. How each company incorporates more sustainable processes and practices will differ depending on the focus of a company's activities. Of course, manufacturing and mining are very different activities; however, we believe that the principles underpinning the new industrial model readily can be applied to the mining industry. Indeed, the chapters of this book offer many insights into how such a model might be applied to the industry, as do the Investment Decision-Making Tools in Part IV.

The Sections of the Book

The book is divided into four parts. The division is important. Part I theorises risk in the mining industry, which has not been done previously in the literature. The section examines the determinants of foreign direct investment, the role of regulation in the industry, the concepts of social and business risk, the obsolescing bargain and resource nationalism, and merger and acquisition activity in recent years. The centrepiece of Part I is the first academic attempt (that we know of) to theorise the concept of political risk with regard to the mining industry. For many students of the subject, political risk is considered pre-theoretical. It often is defined either as a list of risks, a matrix to determine the severity of a particular risk, or some version of a Delphi approach. In Chap. 4, Feher demonstrates how theoretically void such approaches are. By this, we mean that political risk as a concept generally is considered to be self-evident—a political risk is defined by the concepts that fall within it. So, hitherto, the definition of political risk has been defined as anything that includes factors such as political instability, crisis, regulatory change, and war, among others. By contrast, Feher shows that the concept of political risk can be theorised—with benefits for the industry.

Parts II and III are, in a sense, self-evident. The two key areas in which mining companies are interested are the economics of the particular minerals/metals they want to exploit and the socio-political and environmental challenges that hamper (or facilitate) their extraction. Part IV presents three decision-making tools to aid investors in their decision-making about where to invest based on the risks and challenges they may face as they seek to pursue opportunities. We hope that these tools provide a practical guide to help mining companies navigate the risks to their operations and, potentially, to mitigate them.

A Word on Nomenclature and the Book's Scope

There probably is no other industry that has as many alternative names as the mining industry. Other names include the extractive industry/ies, the natural resources sector, the resource sector, the minerals industry or sector, the mining sector, and the mining and metals industry. These various terms refer to a broad range of activities that involve the extraction of minerals, metals, petroleum, and gas. For the reader's ease of navigation and because some of the other names refer to a wider range of activities than mining, such as forestry, fishing, agriculture, water harvesting, and oil and gas extraction, we have settled on the use of 'mining industry' here. We also have adopted the term, mining industry, because many of the subsequent chapters of the book address issues related to the mining of minerals and metals—not those other activities. This is why Part II is titled, The Main Metals.

We concede an anomaly in Part II, we have included chapters on natural gas and coal seam gas. While these commodities are neither minerals nor metals, they critical commodities in the global extractives marketplace. Unconventional gas resources almost entirely have reduced the United States' dependence on foreign oil and gas, with the effect of significantly improving America's energy security. Gas also is seen as a key plank in the transition to a lower emissions global energy system, as its emissions footprint is far lower than that of coal and oil. The gas chapters give significant insight into recent developments in these sectors.

Finally, a note on the book's scope: there are other minerals and metals equally deserving of a hearing, including silver and gold, precious and semi-precious stones, and metals such as bauxite, zinc, tin, and nickel. These are significant commodities that contribute to growth in the global economy. We would have liked to include chapters on these commodities, but limitations on the size of the book and willing and able contributors prevented us from doing so. In Part III, we also would have liked to include chapters on the nexus between mining, gender, health, military conflict, water, geopolitical issues, and resource depletion. The last of these issues runs counter to the narrative that mining is a sustainable activity in the longer term. These subjects deserve greater examination in the literature, and it is our hope that a future volume brings attention to the risks, challenges, and opportunities facing the mining industry in these areas in the coming decades.

We note that the book focuses on the mineral rich countries of the Asia-Pacific Region. We broadly exclude from consideration developments in South America, Africa, and Europe. Perhaps future publications might address this deficit. We also believe that there is need for a manuscript that is country focused. Indeed, a valuable addition in this regard would be a country-by-country account of the performance of the industry against a number of criteria, notably, those presented in the decision-making tools in Part IV.

References

Anglo American. (2016). Annual Report. http://www.angloamerican.com/investors/annual-reporting/reports-archive/report-2016. Accessed 25 August 2016.

Australian Broadcasting Commission (2015). Samarco mine tragedy: Toxic mud from Brazil mine spill reaches Atlantic Ocean. http://www.abc.net.au/news/2015-11-22/brazil-mine-spill-mud-to-hit-atlantic-ocean/6962040. Accessed 23 August 2016.

Chilean Economic Development Agency. (n.d). http://www.english.corfo.cl/. Assessed 4 April 2017.

Evans, G. Goodman, J. and Lansbury, N. (2002). *Moving mountains: mining communities confront globalisation*. London: Zed Books.
Fitzgerald, B. (2016). BHP face dam criminal charges. *The Australian*, October 22–23.
Fortune (2016). Global 500. http://beta.fortune.com/global500/list. Accessed 25 August 2016.
Fraser Institute. (2015). Survey of Mining Companies. https://www.fraserinstitute.org/sites/default/files/survey-of-mining-companies-2015.pdf. Accessed 25 August 2016.
Fund for Peace (2015). The Fragile State Index 2015. http://fsi.fundforpeace.org/rankings-2015. Accessed 25 August 2016.
Gallas, D. (2016). Seven charged over Samarco dam disaster. http://www.bbc.com/news/business-35645960. Accessed 15 June 2016.
Keirman, P. (2016). Samarco warned of problems at dam, engineer says: Joaquim Pimenta de Ávila consulted for mining company, inspected a crack 14 months before collapse.
Knight, B. (2016). BHP's deadly dam collapse linked to ramping up production. http://www.abc.net.au/news/2016-02-29/bhp-samarco-dam-collapse-brazil-linked-to-ramping-up-production/7201022. Accessed 16 June 2016.
Korten, D. (2002). Predatory Corporations. In Evans, G., Goodman, J., Lansbury, N. Eds. *Moving mountains: communities confront mining globalisation* (pp.1-18). London: Zed Books.
Lavery/Pennell. (2014). The new industrial model: greater profits, more jobs and reduced environmental impact. Stoke on Trent: United Kingdom.
Mehra, P. (2016). Brazil court ratifies Samarco Settlement. http://www.smh.com.au/breaking-news-business/brazil-court-ratifies-samarco-settlement-20160506-4el8m.html. Accessed 23 August 2016.
Moody, R. (2007). *Rocks and hard places: the globalization of mining*. London: Zed Books.
Moody, R. (2005). *The risks we run: mining communities and political risk insurance*. Dublin: International Books.
Peters, G. (1999). *Waltzing with raptors: a practical roadmap to protection your company's reputation*. New York: John Wiley and Sons.
Samarco. (2016). *To do what must be done: this is our commitment*.
Shared Value Initiative. (n.d.). About shared value. https://www.sharedvalue.org/about-shared-value. Accessed 6 April 2017.
Sonawane, V. (2016). BHP-Vale's Samarco was warned about crack in dam responsible for deadly burst in Brazil, engineer says: report. http://www.ibtimes.com/. Accessed 23 August 2016.
Timson, L. (2015). BHP-Vale Samarco dam: mud reaches Atlantic in Brazil's worst environmental disaster. http://www.smh.com.au/world/bhpvale-samarco-dam-mud-reaches-atlantic-ocean-in-brazils-worst-environmental-disaster-20151124-gl6ma5.html. Accessed 23 August 2016.

Part I
Theorising Risk in the Mining Industry

Chapter 2
Determinants of Foreign Direct Investment in the Mining Industry

Vlado Vivoda

Abstract A range of social, political, and economic factors determine where mining companies invest their scarce capital. This chapter identifies nine areas of risk that investors need to consider before investing in the sector. These include high sunk costs, the finite life of a deposit, and the long period to achieve a positive financial return. Add to this, legacy issues and it is clear that mining is a high-risk venture. This chapter argues that increased attention to the nine areas of risk may benefit mining companies in the future.

Abbreviations

BIT	Bilateral investment treaty
EITI	Extractive Industries Transparency Initiative
EPI	Environmental Performance Index
FDI	Foreign Direct Investment
FIFO	Fly-in-fly-out
GIS	Geographical Information System
GUCAS	Guided Uniform Criteria Assessment System
ICSID	International Centre for the Settlement of Investment Disputes
IRR	Internal Rate of Return
MEG	SNL Metals Economics Group
NGO	Non-governmental Organisation
NVP	Net Present Value
PMSEIC	Prime Minister's Science, Engineering and Innovation Council
UNCTAD	United Nations Conference on Trade and Development

V. Vivoda (✉)
Centre for Social Responsibility in Mining, Sustainable Minerals Institute, The University of Queensland, St Lucia, QLD, Australia
e-mail: v.vivoda@uq.edu.au

T. O'Callaghan, G. Graetz (eds.), *Mining in the Asia-Pacific*, The Political Economy of the Asia Pacific, DOI 10.1007/978-3-319-61395-6_2

Introduction

The mining industry has a certain combination of characteristics that are not found in other industries. Although some determinants of foreign direct investment (FDI), such as the quality of formal institutions or tax policies, are important in influencing the location of manufacturing and services, as well as extractive industry (mining and oil & gas) investment, most are sector specific (Imbun 2006). According to Andrews-Speed (1996) and Saidu (2007), these include:

- High capital intensity
- Low labour intensity
- Long lead times
- High risk
- Non-renewable resources
- Finite life
- Volatile markets
- Failures
- Late payback

Mining companies have numerous jurisdictions from which to choose when determining the targets of their exploration and development budgets. In an ideal world, investment would flow to countries that have the most abundant and highest-grade deposits. However, in practice, many other factors besides geological endowment influence investment decisions. Prior to investment, mining companies require as much assurance as possible as to the security of the investment. This particularly is so given that mining investments involve large sunk costs that are irreversible (Barham and Coomes 2005). Sunk costs refer to investments that, once undertaken, cannot fully be recovered through their transfer or sale due to industry-specific characteristics (Dixit 1980; Spence 1977). For example, the task of extracting, separating, and transporting large volumes of raw materials requires secure facilities with high installation costs. Moreover, required investments in the mining industry are significant in order to achieve a minimum efficient scale (Barham and Coomes 2005). Due to the nature of the industry, mining companies are immobile for periods of time and, as such, their investments typically are associated with high levels of risk (Naito et al. 1998). It therefore is in the commercial interests of any mining investor to undertake due diligence when considering any new major investment, and risk assessment should be part of any due diligence effort.

One of the most convenient ways to measure geological potential and country-specific risk is by survey. The three best-known annual surveys of exploration investment are those conducted by SNL Metals Economics Group (MEG), Behre Dolbear, and the Fraser Institute. The MEG survey, published since 1991, provides general information on the sources and destinations of investment by various commodities, but does not provide much information on the reasons behind the direction of investment flows (SNL MEG 2014). Behre Dolbear's 'Ranking of Countries for Mining Investment', compiled annually since 1999, ranks 25 countries that host major exploration or extractive development efforts and/or mining operations, based on seven criteria. The rankings are based on qualitative opinions gathered from company professionals and research from various public and confidential sources. While the survey undoubtedly has some value, it provides no detailed explanatory notes on why surveyed countries have been assigned a particular score for each of the criteria. The scores are based on the 'collective experiences' gained by company professionals, who, according to Behre Dolbear (2014), 'have had the unique opportunity to travel widely and experience many different cultures.' While the staff's expertise is extensive, and the survey has merit, the country ranking methodology arguably lacks rigour.

The Fraser Institute survey, which commenced in 1997, is of slightly more use with regard to the determinants of FDI in the mining industry. It ranks, among other things, policy potential, mineral potential, and the investment attractiveness of a growing number of jurisdictions globally. The latest survey, which evaluates 109 jurisdictions, was sent to approximately 3,800 mining-related stakeholders worldwide, receiving responses from 690 organisations (The Fraser Institute 2015). Respondents include both junior and major mining companies, but also regulators, government officials, non-government organisations (NGOs), and many other groups and individuals involved in the mining industry. While liaising with government departments and NGOs involved in the sector, the present author was informed that many of these organisations are asked to fill out the Fraser Institute survey on an annual basis and informants confirmed that they regularly respond to it. It therefore is unclear whether the survey represents responses only from mining companies as the Fraser Institute claims. Regardless, allowing regulatory agencies to respond to a survey that ranks mining jurisdictions adds a degree of bias to the results and makes the findings less objective.

Whereas most mining companies will be interested in surveys like those prepared by Behre Dolbear and the Fraser Institute, in-house assessments are the norm for guiding actual investment decisions. These vary from *ad hoc* decision-making in junior companies, to a classic Delphi-type approach or a guided uniform criteria assessment system (GUCAS) in major companies (Otto 2006). Some companies, particularly risk-taking juniors that aim to establish a foothold in the industry, will target jurisdictions with good prospectivity regardless of risk. However, most mining companies will balance prospectivity against risk criteria when making investment decisions. Analysing the main criteria to which mining companies refer when making investment decisions, Morgan (2002) found that a significant proportion reflect perceptions of administrative procedures and regulatory agency functions. He argues that the perception of *security of tenure* is of paramount importance to foreign mining investment. Morisset (1999) examines determinants of FDI in Africa's resource-rich states. He argues that the implementation of a few visible actions is essential in the strategy of attracting FDI, which involves opening the economy through trade liberalisation reform; launching an attractive privatisation program; modernising mining and investment codes; adopting international agreements related to FDI; developing several priority projects that have a multiplier effect on other investment projects; and mounting an image-building effort with the participation of high political figures, including the president or prime minister.

A study by Kasatuka and Minnitt (2006) finds that non-commercial risks deter foreign investment in resource-rich jurisdictions. Such risks include government instability, poor socio-economic conditions, conflict, corruption, political terrorism, civil war, the quality of bureaucracy, racial, national, and religious tensions, and other issues. Tole and Koop (2011) examine firm location decisions by the world's major gold mining companies using a dataset of political, economic, regulatory, infrastructure, and investment risk variables observed since 1975. They find that firms prefer not to venture far away from their home offices, are strongly attracted to countries that have low levels of corruption, and are attracted to fairly developed economies that provide a good business environment characterised by predictability, efficient institutions, transparent laws, and advantageous tax codes. They also found, perhaps counter intuitively, that stringent environmental regulations attract firms.

The most comprehensive study to date on determinants of FDI in the mining industry is that by Otto (1992a). He identified over 60 factors that may influence a company's decision to invest in one jurisdiction over another. He divided the investment criteria into nine categories: geological, political, regulatory, marketing, fiscal, monetary, environmental and social, operational, and profit. In an effort to establish the relative importance of the criteria, Otto (1992b) surveyed 39 internationally active junior and major mining companies. Surveys by Johnson (1990) and Naito et al. (2001) have yielded similar results. The criteria ranking results for the most important determinants of investment are shown in Table 2.1 and are listed under Otto's nine principal categories. Table 2.1 also includes additional criteria identified by O'Neill (1993), Morisset (1999), Morgan (2002), Kasatuka and Minnitt (2006), Penney et al. (2007), and Tole and Koop (2011). The following section elaborates on the key criteria that influence FDI in the mining sector.

Sources: Johnson (1990), Otto (1992b), O'Neill (1993), Morisset (1999), Naito et al. (2001), Morgan (2002), Kasatuka and Minnitt (2006), Tienhaara (2006), Penney et al. (2007), Tole and Koop (2011)

Geology

Geological prospectivity has a major influence on the level of exploration activity, as it determines the likelihood that a mining company will discover an economic deposit (O'Neill 1993). The prospectivity of a region is determined by its geology and known resource endowment. The availability of a credible and publicly accessible geological database is an essential element to increasing the transparency of mining-related information. A lack of adequate geological information can result in a downgrading of assessments of a country's potential to host exploration and mining development. Transparent information promotes and facilitates exploration by reducing the costs and associated risks (PMSEIC 2001). If the Geographical Information System (GIS) is not linked to geological databases, it may be a daunting task for an investor to find out which area is open to investment. Inadequate geological database information (including quality and scale of maps, ease of access, etc.) is likely to be a significant investment deterrent.

Table 2.1 Key foreign investment criteria in the mining sector

Category	Specific criteria
Geological	Geological potential for target minerals
	Ability to apply geological assessment techniques
	Quality of mining titles system
Political	Consistent and constant mining policy
	National security and political stability
	Internal and external conflicts
	High level of transparency
	Low level of corruption
	Adoption of international agreements related to mining
Investment promotion	Trade liberalisation
	Privatisation program
	Image-building effort to attract investment
	Import-export policies
	Existence of previous priority projects
Regulatory	Stability of exploration/mining terms (security of tenure)
	Modern mining legislation
	Efficient regulatory institutions/administrative procedures
Fiscal	Ability to predetermine tax liability (predictability)
	Stability of fiscal regime
	Method and level of taxes and levies
Financial	Ability to repatriate profits
	Realistic foreign exchange regulations
	Ability to raise external financing
	Permitted external accounts
Environmental and social	Ability to predetermine environment-related obligations
	Stringent environmental regulations
	Ability to gain the support of local stakeholders
Operational	Majority management control held by investor
	Right to transfer ownership
	Quality of infrastructure
Profit	Projected measures of profitability (IRR, NPV, pay-back)

Political Stability

Political stability and the absence of conflict and tension also are important determinants of FDI in the mining sector (O'Neill 1993). Mining companies are unlikely to invest in a country if they perceive that the government may be destabilised or overthrown by unconstitutional or violent means, or if there is a high degree of political or ethnic violence. A stable political environment reduces the risk of regulatory changes and licences being revoked without warning. In unstable and/or conflict-riven jurisdictions, companies will be concerned with the safety of their employees, equipment, and tenements. Where violence is common, companies increase their expenditure on security measures for land holdings, mining equipment, and staff. It also can precipitate conflicts with host communities. These factors inevitably raise operating costs (Penney et al. 2007).

Regardless of a government's policies and regulations, it is important that information on the regulatory system is transparent. Transparency reduces uncertainty and increases commercial confidence (O'Neill 1993). Increased transparency allows companies to make an informed judgment when considering or conducting exploration. In particular, the ability to access relevant information makes investors more willing to undertake exploration. If foreign companies are prevented from fully understanding the relevant regulations, or are denied access to necessary geological and other information, they are unlikely to invest in exploration activities. The lack of procedural/administrative transparency in the distribution of licences/leases, environmental clearances, exit-policies, incentives, and/or access to geological information is a significant obstacle to attracting foreign mining investment. If a potential host jurisdiction has implemented or shown support for the Extractive Industries Transparency Initiative (EITI) (2014), this may make it more attractive to potential mining investors. The EITI is a global coalition of governments, companies, and civil society organisations that works collaboratively to improve openness and accountability in the management of mining industry revenues (Kemp 2010). Investing in a jurisdiction that is EITI compliant reduces potential reputational risk to the investor. As of February 2016, 51 countries were classified as EITI compliant.

Mining involves risks, and investors seek to reduce these risks by obtaining legal protection for their investments (Naito et al. 1999). Resource-rich countries, which adopt international agreements related to mining, are likely to attract FDI in their mining sectors (Tienhaara 2006). These international agreements include bilateral investment treaties (BITs) between the home and the host jurisdiction, which give mining companies from domiciled jurisdictions access to international arbitration and prohibit expropriation without compensation from host countries (UNCTAD 2014). It also is important to consider whether a potential host jurisdiction is one of the 150 states that have ratified the Convention on the Settlement of Investment Disputes between States and Nationals of other States. The International Centre for Settlement of Investment Disputes (ICSID) provides facilities for conciliation and arbitration of international investment disputes and is the leading international arbitration institution devoted to investor-state dispute settlement (ICSID 2014). If a host country does not have a BIT with the domiciled state and if it has not ratified the ICSID Convention, this may act as a deterrent to FDI in the mining industry.

Investment Promotion

An important task of a jurisdiction's mining policy is to communicate investment conditions to potential investors and thereby improve competitiveness. In order to attract foreign mining investment, a jurisdiction's mining policy should be consistent, clear, and concise (Otto 2006). The methods available to promote investment are both indirect, such as reforms of the underlying legal system and institutions, and direct, such as advertising and promotion of pro-investment policies. Naito and Remy (2001) argue that long-term success in attracting private investment in mining exploration is affected not only by a favourable natural endowment, but also by effective implementation and promotion of policies. According to Naito et al. (2001), national governments focus their mining industry policies on how they can attract investments. Increasingly, governments are assuming the role of investment promoter. Otto (2006) argues that the key to successful promotion is to, first, bring the mining potential to the attention of investors and, second, to assure investors that risks are low or manageable. In this context, a resource-rich country or jurisdiction that wishes to attract mining FDI should publicise its mining policy and the mining-related international agreements it has signed. More importantly, the mining policy should be combined with broader sectoral reforms, which include the establishment of a competitive investment climate for private sector participation, such as privatisation and trade liberalisation. The mining industry is trade focused and, as such, general trade restrictions that affect the industry will impede investment in exploration and project development. For example, any trade restrictions on the import of capital equipment required for mining exploration, or taxes on the export of certain commodities, are likely to affect investment decisions (Davis 2010). Finally, if there is evidence of prior successful priority projects in the jurisdiction by large international mining companies, this also may attract potential investors.

Regulatory Regime

Foreign investment likely will expand when regulations and procedures are clear, efficient, and transparent. This only will occur when all levels of government are devoted to delivering good outcomes. Overall government effectiveness is determined by the quality of public provision and the regulatory agencies, the competence of civil servants and their independence, and the credibility of government decisions (Jalilian et al. 2007). Ineffective governance, in particular, excessively complex administrative and/or regulatory procedures required to establish and/or operate a business, discourage inflows of foreign mining investment (Globerman and Shapiro 2002). Regulatory overlap between various government departments (often Energy/Resources/Mining and Environment) may result in unclear lines of authority regarding the regulation of FDI (Vivoda 2008). Regulatory overlap or duplication is a significant investment deterrent.

Conversely, a high level of cooperation between government departments or the existence of a 'one stop shop' to manage the regulatory process can reduce the time required for, and cost of, obtaining approvals, which may encourage mining exploration in a jurisdiction (Penney et al. 2007). In addition, mining laws in most jurisdictions provide a two-step process in which a company gains, first, a licence to explore and, second, should an economic deposit be discovered, a licence to mine. The strength of the linkage between the right to explore and the right to mine is a measure of a company's security of tenure. If a company perceives that it will have high security of tenure, the more likely it is that it will invest in a particular jurisdiction.

In some federal systems of government, state or provincial governments may be empowered to design and regulate their own FDI policies. They also may have a different regulatory and policy focus from that of the national government. Sub-national governments may create their own qualifying conditions and hold discretionary powers to terminate a mining lease. If the authority to grant exploration and mining licences is conditional upon sub-national government approval, this can add duplication and complexity to the regulatory regime. Division of mining project approval mechanisms between national and sub-national governments therefore can undermine FDI promotion efforts by the national government and ultimately raise project costs (O'Callaghan 2010). If sub-national governments establish their own standards for the acquisition and cancellation of mining licences, it is essential that national governments provide mining licence holders with sufficient safeguards against the intrusion of lower levels of government. However, this often does not occur in practice. Consequently, as a project moves through the various phases of development (exploration, feasibility, construction, development, operation, and closure), investors may be required to comply with standards that overlap, are unclear, and are subject to a degree of administrative discretion at the sub-national level (Vivoda 2011). This can undermine investor confidence in the security of tenure and add to investment uncertainty.

Inadequate financial and human resources can be a problem in developing economies across the Asia-Pacific Region. Regulatory agencies may not have the adequate administrative machinery to deal with their responsibilities (O'Callaghan and Vivoda in this volume). This problem can be exacerbated at the sub-national level following decentralisation efforts that transfer more powers to lower-level governments without simultaneously providing them with additional human and/or financial resources. A further challenge may arise from a shortage of adequately trained compliance and enforcement officers, many of whom find employment in private industry more rewarding (Bridge 1999; Haselip and Hilson 2005). This point is addressed in the following chapter.

Fiscal Regime

Exploration activity involves high costs, and the profitability of a project can be influenced by a government's fiscal regime. Generally, foreign mining investment will be enhanced by resource taxation regimes that are simple, stable, predictable, transparent, equitable, efficient, and competitive (Saidu 2007). For mining companies, the overall level of tax, including royalties and tax incentives (for example, tax holidays), influences decisions to explore and develop new projects. All other factors being equal, companies prefer to invest in low-tax jurisdictions (Otto et al. 2006).

Moreover, when deciding where to invest, mining companies consider the perceived stability of a fiscal regime over time, and this requirement probably is the most important for companies, due to the long life-cycle of most mining projects (Andrews-Speed 1996). Mining investors look for predictable fiscal regimes, as transparent and clear tax regulations allow companies to predetermine their tax liability (O'Neill 1993).

Different types and levels of taxes imposed on mining companies also have a direct bearing on the rates of return, and thus influence investment behaviour (Saidu 2007). For example, taxes or royalties based on units of production, irrespective of profitability, may create economic inefficiencies by discouraging the exploitation of lower grade ore and shortening the operational life of some mines. Such a regressive tax instrument can contribute to inefficient resource exploitation and premature mine closures with attendant negative impacts on investments (Andrews-Speed 1996; Otto et al. 2006). Taxes on corporate profits, and to a lesser degree, income, are more efficient, and recognise the inherent risks in mining operations, particularly the wide fluctuations in international commodity prices and the difficulty in anticipating all geological, technical, financial, and political factors over a mine's life-cycle (Mitchell 2009). Finally, federal systems often allow taxes and royalties to be taken at multiple levels of government, potentially resulting in complicated fiscal regimes in which multiple levels of government compete for their share of revenue. This can lead to excessive administrative costs for potential investors.

Financial Conditions

The ability to repatriate profits and capital to the domiciled jurisdiction also has an impact on the attractiveness of that jurisdiction to foreign investors (O'Neill 1993). If foreign owned companies are able to convert profits earned in local currency to US dollars or another major international currency, this may enable them to remit their profits to the parent company. However, foreign-invested entities, either structured as subsidiaries or joint ventures with domestic mining companies, often face restrictions on profit repatriation (UNCTAD 2007). Constraints may be imposed on the timing and magnitude of transactions to repatriate foreign investors' profits. This subsequently may place a discriminatory burden on foreign investors and discourage investment (Penney et al. 2007).

Exploration is a risky and capital-intensive activity. Large capital investments usually are required to identify recoverable ore deposits. However, the risk that an economically viable deposit will not be found is high enough that companies often encounter difficulties in raising funds through debt financing (Penney et al. 2007). The majority of funds used for mining exploration are raised on the Sydney, London, Toronto, and Johannesburg stock exchanges. In order to improve investment attractiveness, it is important that countries do not limit the ability of mining companies to raise external finance.

Environmental and Social Regulation

In 2004 Hilson and Haselip found that the desperation of many developing country governments to promote foreign mining investment provided little incentive for mining multinationals to engage in environmental best practice. Consequently, they asserted that few of the mining companies operating in the developing world had embraced sound environmental management practices. This may have been the case in the 1980s and 1990s, with the Ok Tedi, Grasberg, and Marcopper environmental disasters in Papua New Guinea, Indonesia, and the Philippines, respectively, providing excellent examples of past industry practice. However, a more recent empirical study has shown a positive change in environmental management practices among multinational mining companies operating in the developing world. Tole and Koop (2011) found that stringent environmental regulations have no effect in influencing investment location decisions. Indeed, it is often more the case that stringent environmental regulations attract mining investment. Regardless of any short-term cost savings from lower environmental standards, most multinational mining companies now view their presence in environmentally 'dirty' parts of the world as potentially damaging to their reputation, even though they implement international best practice environmental standards. Complex and inconsistent environmental regulations likely will deter FDI, with mining companies unable to predetermine environment-related obligations.

Because of the nature of mining activities, the demand for socially and environmentally responsible performance from multinational mining companies has become the norm (Kepore and Imbun 2011; Graetz and Franks 2013). Constructive approaches to stakeholder engagement and collaboration in line with a host community's long-term development agenda is becoming standard practice for mining companies (Owen and Kemp 2013). There are numerous historical examples where local stakeholders have rejected mining and, consequently, deposits have not been brought into production. Local communities have the power to influence the security of tenure between exploration and mining. The key to a community accepting or rejecting exploration and mining often depends on the extent to which the community and its members will directly or indirectly benefit from a development (for example, through taxation/royalty payments, employment, infrastructure, joint ownership, etc.), balanced against the perceived harm (for example, environmental and social impacts) a development may cause. In many resource-rich developing countries, mines usually are located in remote areas, populated by the rural poor and with inadequate social and manufactured capital resources, including political and regulatory institutions and infrastructure. Under these circumstances, foreign mining investors often are *de facto* social infrastructure providers. To obtain local government approval and a 'social licence to operate', companies often are obliged to build hospitals, schools, and other social infrastructure. The importance of gaining the social licence is critical to the ultimate success of a mining project (Imbun 2007; KPMG 2013). Hence, the ease with which host community support can be obtained is one of the key determinants of mining investment (Otto 2006). Jurisdictions that recognise that communities have roles to play in the mining investment process and that actively address community concerns will be more attractive to potential investors than jurisdictions in which problems are left solely for mining companies to address (Campbell 2012).

Operational Environment

Foreign mining investment likely will grow when jurisdictions allow investors operational flexibility so that they can form corporate structures to suit the operating environment. The ability of mining investors to invest in exploration activities is impeded when they cannot alter the operational structure of their business to suit their changing needs (for example, joint ventures to share risk). Equity requirements or requirements on foreign investors to take a local joint venture partner can act as an impediment to potential investment, and can limit the level of interest from foreign companies in establishing operations (O'Neill 1993).

The existing infrastructure of a particular economy, such as access to water and electricity, and the quality of roads, influences the level of foreign investment. Resources often are found in remote or regional locations, which makes it difficult to attract employees. These areas generally have limited access to hospitals, schools, roads, airstrips, and other infrastructure. Inadequate infrastructure typically will increase the cost of developing a deposit and, therefore, will reduce the expected profitability of the development, particularly if the full costs of transport, water, power, and housing are borne by the company. Similarly, a lack of basic infrastructure can impede access to exploration areas and increase the cost of exploration activities (Penney et al. 2007). Fly-in-fly-out (FIFO) operations require unique infrastructure investment and may affect operational expenditure based on the circumstances of the development, host community, regulatory expectations, and environmental impacts.

Potential Profitability

Otto (2006) argues that not only are determinants of FDI industry specific, but that they are also company specific. The criteria that any one company will apply when deciding whether to invest in a particular resource-rich jurisdiction will be unique to that company and time (Bhappu and Guzman 1995). Decisions are based on projected internal measures of profitability (for example, the IRR, NPV, and pay-back), which are commercially sensitive. These measures represent an estimated percentage of an effective net cash flow of all amounts payable by the company to the government (Feibel 2003). Investment decisions also are couched within broader industry trends. For example, in 2013, responding to lower metals prices, uncertain demand, and poor market conditions, mining companies across various commodity sectors cut their exploration activity sharply, with budgets 29% lower than in 2012 (SNL MEG 2014).

Conclusion

The existence and extractability of minerals is the most important determinant of where mining companies invest in exploration and extraction. While the presence of deposits is a necessary requirement to attract investment, it is not a sufficient condition. Despite the high geological potential, a combination of various challenges often puts resource-rich jurisdictions in an unfavourable position in the global market for investment. As a consequence, many countries that are endowed with resources traditionally have been unable to attract and retain FDI.

Governments of resource-rich jurisdictions increasingly are aware that mining companies are selective in their choices and that, in order to attract investors, they need to implement reforms to establish more effective regulatory structures. Jurisdictions can improve the likelihood of mining investment by taking steps to satisfy investor decision criteria through informed policies and regulations. However, while over the past three decades more than one hundred countries have amended their mining and related rules and regulations, policy and regulatory changes have not led automatically to a reduction in risks for investors. Improved regulation *per se* does not automatically attract more foreign investment.

The low level of FDI in the mining industry may result from inconsistent and unclear policy towards investment, inadequate and ineffective fiscal, regulatory, and environmental regimes, a lack of geological information, an unfavourable political environment, unresolved social issues and the inability to secure a social licence to operate, inadequate infrastructure, and various other challenges. Levels of FDI also may be influenced by a strong and competitive domestic mining industry, which may be shielded from the competition of foreign majors. Moreover, domestic policy-makers may view participation by international companies in their mining sectors with suspicion, due to beliefs about exploitation of the national heritage or patrimony. Consequently, it may be a national duty to have direct control of the industry (Johnson in this volume).

If resource-rich jurisdictions aim to increase the inflow of foreign capital into the mining industry, the performance of regulatory regimes governing foreign mining investment is essential. Attracting foreign mining investment requires favourable and consistent government policies and effective regulatory and fiscal systems. This needs to be supplemented with active investment promotion programs and holistic stakeholder engagement processes that assure potential investors, local communities, and other stakeholders that mining activities can be sustainable and mutually beneficial.

Many systemic and geological risks remain that are beyond the control of government. However, governments can improve the likelihood of mining sector investment by taking steps to satisfy investor decision-making criteria by establishing regulatory structures that are clear and by guaranteeing 'security of tenure' (Otto 2006). The challenge for countries in the Asia-Pacific Region is to regulate multinational mining companies' activities in line with national socio-economic development priorities, while making it attractive for these companies to invest.

References

Andrews-Speed P. (1996). Fiscal systems for mining in countries with mineral resources and their effect on the commercialisation of minerals with special reference to Brazil. Paper presented at the Mining and petroleum: the coming years international conference, Brasilia, 22–23 October.

Barham B., & Coomes, O. (2005). Sunk costs, resource mining industry, and development outcomes. In P. Ciccantell, D. Smith & G. Seidman (Eds.) *Nature, raw materials, and political economy*. (pp. 159-186). London: Emerald Group Publishing Limited.

Behre Dolbear. (2014). 2014 Ranking of countries for mining investment: where 'not to Invest'. http://www.dolbear.com/_literature_209687/2014_Ranking_of_Countries_for_Mining_Investment. Accessed 13 Aug 2014.

Bhappu, R., Guzman, J. (1995). Mineral investment decision making: a study of mining company practices. *Engineering and Mining Journal,* 196(7), 36-38.

Bridge, G. (1999). Harnessing the bonanza: economic liberalization and capacity building in the mineral sector. *Natural Resources Forum,* 23(1), 43-55.

Campbell, B. (2012). Corporate social responsibility and development in Africa: redefining the roles and responsibilities of public and private actors in the mining sector. *Resources Policy,* 37(2), 138-143.

Davis, G. (2010). Trade in mineral resources. *World Trade Organization*, Economic Research and Statistics Division, Staff Working Paper ERSD-2010-01. https://www.wto.org/english/res_e/reser_e/ersd201001_e.htm. Accessed 17 May 2014.

Dixit, A. (1980). The role of investment in entry deterrence. *Economic Journal,* 90(357), 95-106.

Extractive Industries Transparency Initiative. (2014). What is the EITI? http://eiti.org/eiti. Accessed 13 Aug 2014.

Feibel, B.J. (2003). *Investment performance measurement*. New York: Wiley.

Fraser Institute. (2015). Survey of mining companies 2015. https://www.fraserinstitute.org/sites/default/files/survey-of-mining-companies-2015.pdf. Accessed 13 Aug 2015.

Globerman, S., Shapiro, D. (2002). Global foreign direct investment flows: the role of governance infrastructure. *World Development*, 30(11), 1899-1919.

Graetz, G., Franks, D. (2013). Incorporating human rights into the corporate domain: due diligence, impact assessment and integrated risk management. *Impact Assessment and Project Appraisal*, 31(2), 97-106.

Haselip, J. and Hilson, G. (2005). Winners and losers from industry reforms in the developing world: experiences from the electricity and mining sectors. *Resources Policy*, 30(2), 87-100

Hilson, G., Haselip, J. (2004). The environmental and socioeconomic performance of multinational mining companies in the developing world economy. *Minerals & Energy-Raw Materials Report*, 19(3), 25-47.

International Centre for Settlement of Investment Disputes (ICSID). (2014). List of contracting states and other signatories of the convention. https://icsid.worldbank.org/ICSID/FrontServlet?requestType=ICSIDDocRH&actionVal=ContractingStates&ReqFrom=Main. Accessed 13 Aug 2014.

Imbun, B. (2007). Cannot manage without the 'significant other': mining, corporate social responsibility and local communities in Papua New Guinea. *Journal of Business Ethics*, 72(2),177-192.

Imbun, B. (2006). Multinational mining and petroleum companies' perceptions of the policy framework in Papua New Guinea. *Pacific Economic Bulletin*, 21(1), 225-242.

Jalilian, H. Kirkpatrick, C., Parker, D. (2007). The impact of regulation on economic growth in developing countries: a cross-country analysis. *World Development*, 35(1), 87-103.

Johnson, C. (1990). Ranking countries for mineral exploration. *Natural Resources Forum*, 14(3), 178-186.

Kasatuka, C. and Minnitt, R. (2006). Investment and non-commercial risks in developing countries. *The Journal of the Southern African Institute of Mining and Metallurgy,* 106(12), 849-856.

Kemp D. (2010). Mining and community development: problems and possibilities of local-level practice. *Community Development Journal*, 45(2), 198-218.

Kepore, K., Imbun, B. (2011). Mining and stakeholder engagement discourse in a Papua New Guinea mine. *Corporate social responsibility and environmental management*, 18(4), 220-233.

KPMG. (2013). The community investment dividend: measuring the value of community investment to support your social licence to operate. http://www.kpmg.com/AU/en/IssuesAndInsights/ArticlesPublications/documents/community-investment-dividend-social-licence-resources-nov-2013.pdf. Accessed 13 Aug 2014.

Mitchell, P. (2009). Taxation and investment issues in mining. In C. P. Mitchell, & F. Paris (Eds.), *Advancing the EITI in the mining sector: a consultation with stakeholders* (pp. 27-31). Oslo: EITI Publishing.

Morgan, P.G. (2002). Mineral title management - the key to attracting foreign mining investment in developing countries? *Applied Earth Science: IMM Transactions Section B,* 111(3), 165-170.

Morisset, J. (1999). Foreign direct investment in Africa: policies also matter. *World Bank Policy Research Working Paper*, No. 2481, November. http://papers.ssrn.com/sol3/papers.cfm?abstract_id=632551 Accessed 3 March 2015.

Naito, K. Remy, F. (2001a). Mining sector reform and investment: results of a global survey. *Mining Journal Books*. London: World Bank Group.

Naito, K., Remy, F. and Williams, J. (2001b). Review of legal and fiscal frameworks for exploration and mining. *Mining Journal Books.* London: World Bank Group.

Naito, K., Otto, J., Smith, D., Myoi, H. (1999). Legal aspects of exploration and mining: a comparative study of mining law in Asia. *Journal of Energy and Natural Resources Law*, 17(1), 1-12.

Naito, K., Otto, J., Eggert, R. (1998). Mineral investment risk and opportunities in Asia. *Resources Policy*, 24(2), 77-78

O'Callaghan, T. (2010). Patience is a virtue: regulation and governance in the Indonesian mining sector. *Resources Policy*, 35(3), 218-225.

O'Neill, D. (1993). Mining investment in Africa: an Australian perspective. *Natural Resources Forum*, 17(4), 262-273.

Otto, J. (2006a). The competitive position of countries seeking exploration and development investment. *Society of Economic Geologists*, Special Publication, 12, 109-125.

Otto, J., Andrews, C., Cawood, F., et al. (2006b). Mining royalties: a global study of their impact on investors, government, and civil society. New York: The World Bank. http://siteresources.worldbank.org/INTOGMC/Resources/336099-1156955107170/miningroyaltiespublication.pdf. Accessed 22 March 1915.

Otto, J. (1992a). Criteria for assessing mineral investment conditions in mineral investment conditions in selected countries of the Asia-Pacific Region (pp. 6-34). New York: United Nations.

Otto, J. (1992b). A global survey of mineral company investment preferences. in mineral investment conditions in selected countries of the Asia-Pacific Region (330-342). New York: United Nations.

Owen, J., Kemp, D. (2013). Social licence and mining: a critical perspective. *Resources Policy*, 38(1), 29-35.

Penney, K., McCallum, R., Schultz, A., Ball, A. (2007). Mineral exploration in APEC economies: a framework for investment. Australian Bureau of Agricultural and Resource Economics (ABARE) research report, December. http://publications.apec.org/publication-detail.php?pub_id=244. Accessed 24 June 2014.

Prime Minister's Science, Engineering and Innovation Council (PMSEIC). (2001). *Australia's mineral exploration*. Canberra, 28 June. http://www.industry.gov.au/science/PMSEIC/Documents/AustraliasMineralExploration.pdf. Accessed 4 July 2014.

Saidu, B. (2007). How taxes, royalties, and fiscal regime stability affect mining investment: a comparison of Niger and Indonesia. *The Journal of Structured Finance*, 13(3), 105-111.

SNL Metals Economics Group (MEG). (2014). Worldwide exploration trends. a special report from SNL Metals Economics Group for the PDAC International Convention. http://go.snl.com/rs/snlfinanciallc/images/WETReport_0114.pdf. Accessed 13 Sept 2014.

Spence, A. (1977). Entry capacity investment and oligopolistic pricing. *Bell Journal of Economics*, 8(2), 534-544.

Tienhaara, K. (2006). Mineral investment and the regulation of the environment in developing countries: lessons from Ghana. *International Environmental Agreements*, 6(4), 371-394.
Tole, L., Koop, G. (2011). Do environmental regulations affect the location decisions of multinational gold mining firms? *Journal of Economic Geography*, 11(1), 151-177.
United Nations Conference on Trade and Development (UNCTAD). (2007). World investment report: transnational corporations, mining industry and development. New York: United Nations. http://unctad.org/en/docs/iteiit20074a5_en.pdf. Accessed 14 Jan 2015.
United Nations Conference on Trade and Development (UNCTAD). (2014). Country-specific lists of bilateral investment treaties. http://unctad.org/en/Pages/DIAE/Internationalper cent20Investmentper cent20Agreementsper cent20(IIA)/Country-specific-Lists-of-BITs.aspx. Accessed 13 Aug 2014.
Vivoda, V. (2011). Determinants of foreign direct investment in the mining sector in Asia: a comparison between China and India. *Resources Policy*, 36(1), 49-59.
Vivoda, V. (2008). Assessment of the governance performance of the regulatory regime governing foreign mining investment in the Philippines. *Minerals and Energy - Raw Materials Report*, 23(3), 127-143.

Chapter 3
Regulatory Regimes, Foreign Mining Investment, and Risk in the Asia-Pacific Region: Comparative Evaluation and Policy Implications

Terry O'Callaghan and Vlado Vivoda

Abstract This chapter analyses the major issues associated with regulatory governance of foreign mining investment in the Asia-Pacific Region. After surveying relevant literature on regulatory regimes, the chapter draws on the seminal work of Stern and Holder to evaluate the major regulatory governance issues in China, India, Indonesia, Papua New Guinea, and the Philippines. Our analyses indicate that there are six key regional challenges: regulatory overlap, regulatory capture and a lack of independence from government, a lack of impartiality in decision-making, a lack of transparency in decision-making, inadequate stakeholder engagement and access to regulators, and a lack of institutional capacity. These challenges are more pronounced in some countries than in others. Policy implications that may aide regional governments to improve governance infrastructure in their mining industries are outlined.

Abbreviations

ADB	Asian Development Bank
CoW	Contract of Work
CPI	Corruption Perceptions Index
EIA	Environmental Impact Assessment
FDI	Foreign Direct Investment
IFC	International Finance Corporation
PNG	Papua New Guinea

T. O'Callaghan (✉)
Centre for International Risk, School of Communication, International Studies and Languages, The University of South Australia, Adelaide, SA 5001, Australia
e-mail: Terry.OCallaghan@unisa.edu.au

V. Vivoda
Centre for Social Responsibility in Mining, Sustainable Minerals Institute, The University of Queensland, St Lucia, QLD, Australia
e-mail: v.vivoda@uq.edu.au

T. O'Callaghan, G. Graetz (eds.), *Mining in the Asia-Pacific*, The Political Economy of the Asia Pacific, DOI 10.1007/978-3-319-61395-6_3

Introduction

The Asia-Pacific Region has substantial reserves of mineral and energy resources, including coal, copper, iron ore, uranium, gas, and petroleum. The most abundant resources are found in China, India, Indonesia, Papua New Guinea (PNG), the Philippines, and Australia. Since the early 1990s, the Region's major non-fuel mineral producers have passed new regulations aimed at attracting more foreign investment with a view to increasing production. The trend towards regulatory reform, to a large extent, had been fuelled by the recent minerals 'super cycle', which saw increased commodity prices due to an upswing in demand. India and Indonesia have completely redrafted their mining laws, while China, PNG, and the Philippines have undertaken more piecemeal changes to their mining legislation. Indonesia currently is undertaking a further review of its mining regulations.

Despite more favourable regulation, poor performance (in terms of application and enforcement of rules) of the regulatory regimes continues to create a high level of regulatory risk for foreign investors in the China, India, the Philippines, and PNG. In this chapter, we suggest that in order to attract more investment into the industry, the performance of regulatory regimes governing foreign mining investment also must be improved.

The comparative political economy of the mining industry in the Asia-Pacific Region is an under-studied area. Existing studies either are dated (Naito et al. 1998; Naito et al. 1999; Otto 1992a, 1992b; Otto and Cordes 2002) or focus on a single country (Holden and Jacobson 2007; Vivoda 2008, O'Callaghan 2009 for the Philippines; Resosudarmo et al. 2009, O'Callaghan 2010 for Indonesia; Imbun 2006 for Papua New Guinea; Andrews-Speed et al. 2003, Tse 2003, Suxun and Chenjunnan 2008 for China; Jhingran 1997, Singh and Kalirajan 2003, Sames 2006 for India). The existing literature also fails to establish a sophisticated set of evaluative criteria for comparative analysis and assessment of the performance of regulatory regimes in the Asia-Pacific Region.

Here, we develop a set of evaluative criteria that enables comparative analysis and assessment of the performance of the regulatory regimes governing the mining industry across the Region. This study provides investors with a clearer picture of the various regimes governing exploration and mining activity, and provides governments with the means to assess the performance of their regulatory regime relative to their regional neighbours.

The chapter is organised as follows. In section "Regulatory Regimes, Governance Infrastructure, and Foreign Direct Investment", we survey relevant literature on evaluating the performance of regulatory regimes governing FDI in the mining industry. In section "Analysing the Performance of Regulatory Regimes", we analyse comparatively the performance of regulatory regimes in five countries in the Region. The findings of our study provide the first comprehensive regional assessment of the performance of regulatory regimes governing mining investment. In the final section, we outline major policy implications. Understanding these implications better may assist the ability of regional policy-makers to assess and improve regulatory practice and increase future investment.

Regulatory Regimes, Governance Infrastructure, and Foreign Direct Investment

Regulations are the administrative laws that governments establish to provide guidance to individuals, groups, and corporations about what is considered possible in a well-governed society. In other words, regulation is a set of standards or principles. Failure to adhere to them generally leads to some form of sanction, such as a fine or imprisonment.

In the context of attracting FDI, regulation acts as a framework to help companies know what they can and cannot do within their particular industry. For example, some jurisdictions have regulations that prevent mining companies from forcibly removing Indigenous Peoples from their tribal lands in order to mine. However, governments must be mindful that if they place too onerous restrictions on investors, then investors may look for opportunities elsewhere. Regulation, then, is a balancing act between the need to promote certain behaviours and to control others.

There are two challenges that mining companies face with respect to regulation. The first is that no two regulatory environments are the same. This is a particular challenge for companies that operate across multiple jurisdictions. Environmental regulation and tax regimes, for example, will vary from country to country, and even within a country. This heightens the chances that a company might find itself breaking the law in China, but be compliant in Vietnam. The second challenge is how to assess the quality of a regulatory regime. Indeed, the question that arises is: How do we compare mining regulation in different jurisdictions, and how do we ascertain whether one regulatory system is performing better than another?

There have been few attempts in the literature to develop a systematic method to answer these questions. Mining companies have had to rely on information and analysis from political risk analysts and consultants, which often is provided on a country-to-country basis. Companies often utilise the findings of the Fraser Institute's annual survey of preferred mining destinations. That survey asks respondents to rate their preferred mining destinations according to 18 criteria. These include environmental concerns, regulatory uncertainty, and mineral potential. There are three major strengths to the survey. First, it is comprehensive, in that it compares all the countries in which mining companies operate. Second, the annual nature of the survey means that changes in the perceptions of the mining community can be tracked fairly well. Finally, the survey is cumulative, providing annual survey data for a number of years. However, as noted in the previous chapter by Vivoda, the survey is not without flaws.

In contrast to the subjective judgement of mining companies used in the Fraser Institute survey, a more objective approach is to employ a range of quantifiable criteria to assess the performance of regulatory regimes (Stern and Holder 1999; Brown and De Paula 2002; Gutiérrez 2003; Cubbin and Stern 2004; Kaufmann et al. 2004; Kurtzman et al. 2004; World Resources Institute 2005). These studies identify general 'appraisal criteria', which are used to assess the quality and performance of regulatory regimes governing the infrastructure sector in the Asia-Pacific Region.

In this chapter, we employ criteria developed by Stern and Holder (1999). As they note, 'our assessment framework provides a useful basis for appraising and discussing the effectiveness of regulatory frameworks in supporting private investment in infrastructure industries. We have demonstrated its applicability for developing Asian economies and we look forward to seeing how it may be applied and developed in other contexts' (Stern and Holder 1999). Their criteria are: (1) clarity of roles and objectives; (2) autonomy; (3) participation; (4) accountability; (5) transparency; and (6) predictability. We use these criteria to assess the performance of mining regulatory regimes across the Region.

Analysing the Performance of Regulatory Regimes

In this section, we explore cross-country similarities and differences, focusing on the negative characteristics of respective regulatory regimes. In the following section, we analyse some of the positive characteristics as a basis for improved regulatory performance throughout the Asia-Pacific Region. A common feature across the Region is that there is significant overlap between various regulatory bodies, which results in a lack of clarity about roles and objectives. There are two types of regulatory overlap: (1) among various levels of government (for example, national, sub-national, and local); and (2) among various agencies at the same level of government. The problem is particularly pronounced in China, India, Indonesia, and the Philippines, where decentralisation processes have diluted national government authority and empowered local government units. As part of the process of decentralisation, sub-national government agencies have been endowed with a high degree of decision-making power. When they have had conflicting policies with, or objectives to, a national government, they have not been timid in exercising power. Generally speaking, in these four countries, sub-national governments have substantial power over decisions regarding mining investment. The reluctance of certain local governments to consent to mining projects has seen some come into conflict with the mining-based development paradigm pursued by their respective national governments.

Despite the move by the Philippines Government towards reengagement with the mining industry, several provinces, including Aurora, Capiz, Mindoro Oriental, and Eastern Samar, have passed moratoriums on mining (Vivoda 2008). This is due to strong opposition to mining activities following the Marcopper mining disaster in 1996.

A recent expression of sub-national authority is the veto of the approval of the Sagittarius mine in Mindanao, forcing parent company Xtrata to mothball the project. In India, where state governments are empowered to design and regulate their own FDI policies, the division of authority regarding project approval applications between the national and state governments often undermines the FDI promotion efforts of the national government (Singh and Kalirajan 2003; Bloodgood 2007). The regulatory burden on foreign investors tends to be higher at the state level, where application and approval procedures can vary widely across states. FDI projects already approved at the national government level tend to encounter obstacles as they proceed at the state level, since nearly 70% of the approvals needed for FDI projects are issued by state governments. State-level impediments to FDI can be severe to the point that multinational corporations have been known to abandon FDI projects before project completion (Planning Commission, the Government of India 2006).

In Indonesia, as new rule-making powers have been ceded to provincial and local governments, regulatory contradictions have emerged. For example, new taxes have been implemented that conflict with the terms of some earlier Contracts of Work (CoW), the main mine licensing agreements issued by the national government. Indonesian Regional Autonomy Watch claimed that more than 30% of 693 regional regulations showed a 'lack of sensitivity with respect to the creation of a conducive business atmosphere' (Rabasa and Chalk 2001). In addition, the new mining law, which was introduced to replace the CoW system, has not been favourably received, with increased 'red tape' being cited as an impediment (Santosa 2013). In China, local governments also have been given greater autonomy in terms of mining project approvals, with investors being required to obtain approvals from four tiers of government. There is a high degree of duplication and complexity in the approval process. For example, two levels of government can issue exploration licences and four levels of government can issue mining licences (Ward et al. 2003).

Regulatory overlap and confusion is prevalent not only between national and local governments, but also among various agencies at the same level of government. In Indonesia, the inconsistency between the 2009 Mining Law and the 1999 Forestry Law has given rise to significant regulatory overlap between the Ministry of Energy and Mineral Resources and the Ministry of Environment and Forestry. In China, there is significant regulatory confusion and overlap between various departments, including the Ministry of Land and Resources, the Ministry of Environmental Protection, and the National Development and Reform Commission with regard to the mining licence approval process. In India, there is a jurisdictional overlap between the Ministry of Environment and Forests and the Ministry of Mines regarding the approval of environmental permits for mining operations (Government of India, National Mineral Policy 2006). A long-standing discrepancy between the need to conserve forest resources and a desire to exploit the country's mineral resources has created obstacles to the timely development of mineral resources. In PNG, the lines of responsibility between regulatory agencies are not well defined or coordinated. For example, there is no clarity about the respective roles of the Mineral Resources Authority, the Department of Mining, and the Department of Planning and Rural Development in the management of mineral wealth.

Various studies have found that regulators and the judiciary across the Region often lack autonomy. A conference organised by Ernst & Young and Frontier Strategy Group, which was attended by senior mining company executives, found that despite the liberalisation of India's investment restrictions, the Indian Government continues to have a strong preference for domestic mining companies (Bailey 2007). Similarly, in China, regulators and the judiciary tend to favour domestic over foreign interests. In contrast, rulings on dispute cases in the Philippines often favour mining companies to the detriment of the local communities whose consent to mine legally is required, which causes significant social risks for the communities that lack a voice during the social licensing process (Christian Aid 2004).

A related issue is that, across the Region, regulatory agencies lack independence from sectional influence and capture. In India's mining industry, the payment of bribes by mining companies to avoid bureaucratic red tape is commonplace (Government of India, National Mineral Policy 2006). Corruption continues to be a problem in India, though it is not exclusive to the sub-continent. The 2015 Corruption Perception Index (CPI), compiled by Transparency International (2014), neatly highlights this challenge. Australia ranks 13/167, China 83/167, India 76/167, Indonesia 88/167, the Philippines 95/167, and PNG 139/167. The Fraser Institute survey tells a similar story, with respondents acknowledging that China, India, Indonesia, and PNG have problems with corruption (Fraser Institute 2015).

Another challenge is the independence of the regulatory agencies. In India, the government is both the 'development or project implementation agency', as well as the regulatory authority (Subramaniam and Ashwin 2006). This does not mean that a conflict of interest will arise in all cases. However, in India, regulatory uncertainty has been a major challenge for companies seeking mining permits. Similarly, in PNG, the Government participates in the development of mineral resources as a joint venture partner. In Indonesia, the Government can engage in business deals and sign contracts with third parties.

Although the picture with regard to access to the regulator and information, or participation in the regulatory process generally is positive, a number of problems have been reported. In the Philippines, access to adequate mining information is limited. For example, it is difficult for Indigenous Peoples potentially affected by proposed mining activities to obtain access to relevant information prior to project approval (Christian Aid 2004). Moreover, the Department of Environment and Natural Resources, the Mines and Geosciences Bureau, and the Environmental Management Bureau in particular have been found to be 'averse to disclosing information to the public' (Aguilar 2008). In PNG, the government's capacity to disseminate information about the costs, impacts, and benefits of individual mining projects often is found to be inadequate. Government agencies are unable to share information with each other, let alone with stakeholders outside the public sector. These problems persist, as a Fraser Institute mining survey attests (Dinnen 2009; Fraser Institute 2012/2013). One of the main reasons that many mining companies decide not to invest in Indonesia is that they do not understand the revenue system. Since important information on the revenues paid by mining companies to government is not publicly available, sourcing this information adds cost and uncertainty for potential new investors (Bhasin and Venkataramany 2007; Laodengkowe 2008).

Problems associated with the independence, transparency, and accountability of the recruitment processes for the regulatory agencies are most pronounced in Indonesia and the Philippines. In these countries, positions in regulatory agencies often are bought or secured via family and/or clan networks, and there is no comprehensive evidence of promotion or recruitment through a meritocratic and open selection process. Often, this practice is tied up with the high levels of corruption.

The problem is not a new one. Indeed, in 1988, Wurfel noted that 'by the early years of independence the pattern had been set; bureau directors and division chiefs received appropriations from the legislature in exchange for appointing friends, relatives, and needy constituents of congressmen.'

Inadequate financial and human resources also pose a challenge throughout the Region. Most regulatory agencies do not have the adequate administrative capacity to undertake their responsibilities. This challenge has been exacerbated at the provincial/state level following the move towards decentralisation, which has seen powers transferred to lower-level governments without the simultaneous provision of additional resources. In the Philippines, studies have found that the Mines and Geosciences Bureau and the National Commission on Indigenous Peoples have failed effectively to apply the Indigenous Peoples Rights Act as the agencies have limited human and financial resources to enforce the legal provisions. The large number of applications from mining companies makes their task particularly difficult. Commission officials have reported that they do not have sufficient monies in their budget to be able to inform communities about proposed plans and to monitor consultation processes (Christian Aid 2004).

A similar problem has been reported in Indonesia. In particular, local governments generally lack financial and human resources to provide services to mining companies and other stakeholders (Resosudarmo et al. 2009). In China, when the government promoted the National Environmental Protection Administration to the ministry level and renamed the agency the State Environmental Protection Administration, it reduced the number of staff from 600 to 300. For a country with the size and complexity of China, the number of staff seems low, especially when compared with, for example, the United States Environmental Protection Agency, which has 6,000 staff (Vivoda 2011).

Behre Dolbear undertakes an annual mining survey with the foreboding subtitle, 'Where Not to Invest'. In 2014, PNG was ranked 21/25, with only Mozambique, Mongolia, Russia, and the Democratic Republic of the Congo performing worse. The reasons for this low ranking include: (1) the country was not favourable to FDI; (2) it rated poorly on social issues; and (3) it faced on-going challenges with corruption.

As a result of the lack of resources and regulatory overlap discussed above, regulators and other government agencies across the region are inefficient and slow in approving mining project applications, which makes the processes unpredictable. Since an application process may require up to 100 different approvals from different agencies at various levels of government, it may take a number of years from the initial application to the start of mining operations, as in the cases of India and Indonesia. Even in states within Australia, it can take many years to obtain a licence to mine. In Queensland, for example, some mining companies have waited over seven years for a coal permit to be approved.

The lack of human and financial resources is not only the cause of delays in mining approvals, but also behind the problem of ineffectiveness of enforcement and compliance mechanisms during mining operations. This is unsurprising given the chronic shortage across the Region of adequately trained compliance and enforcement officers, for whom employment in private industry may be more rewarding. Indeed, attracting and retaining skilled and professional staff is a challenge across the Region.

In addition, changes in FDI policy can make mining investment unpredictable, as in the case of China. In what has been a change since the early 2000s, recent evidence from China indicates that foreign investment in the country's mining industry is not welcome. International companies continue to feel stymied by an inconsistent and convoluted mining policy and their inability to create relationships of trust with local stakeholders. China no longer courts foreign mining companies, because it already has benefited from the desired influx of foreign capital, technology, and management techniques. For example, it has adopted environmentally friendly technologies that improve coal liquefaction and coal bed methane production, and management techniques, such as the reorganisation of companies via mergers and acquisitions. Risk-averse Chinese companies seem to be satisfied with the technology and management techniques they have acquired, and are not actively seeking joint ventures with foreign partners. These reactions express a form of Chinese protectionism that frustrates the aspirations of foreign investors attracted by China's abundant mineral resources (Caprioni 2013).

What adds to the unpredictability for foreign mining investors is the ineffectiveness and uncertainty regarding enforcement of regulations by regulatory agencies. In China, the State Environmental Protection Administration sends compliance officers to mines and often issues operators with fines for non-compliance. However, the fines are small, or not enforced, and violators often face few consequences (Economy 2007). In Indonesia and the Philippines, where companies are required to undertake an environmental impact assessment (EIA) prior to the establishment of their operations, EIAs frequently only are an on-paper exercise and even best practice EIAs do not necessarily lead to effective enforcement (Tan 1998; Resosudarmo et al. 2009).

Conclusions and Policy Implications

In this chapter, we have assessed the performance of regulatory regimes governing foreign mining investment in the Asia-Pacific Region. Various regulatory challenges that plague the mining industry across the Region contribute to the high level of risk for investors and consequently, the low levels of foreign investment in the countries we surveyed. Our analysis points to six key challenges affecting mining industry governance in the region: (1) regulatory overlap; (2) regulatory capture and a lack of independence from government; (3) a lack of impartiality in decision-making; (4) a lack of transparency in decision-making; (5) inadequate stakeholder engagement and access to the regulator; and (6) a lack of institutional capacity. Below, we address a number of weaknesses in the various regimes; our policy implications serve to indicate where improvement is needed.

However, before we turn to discuss the policy implications of our empirical findings, it is important to note various positive developments across the Region.

First, a raft of international investment agreements has helped to return some confidence to the regional economy. One indicator of this is that governments in the Region are beginning to speak about the need for new infrastructure projects to foster domestic growth. Tahilyani et al. (2011) estimates that around US$8 trillion 'will be committed to infrastructure projects over the next decade to remedy historical underinvestment and accommodate the explosion in demand.' This view has been echoed by the Australian Trade Commission (2013). Second, in China, India, and Indonesia, some local governments have made considerable efforts to improve the regulatory environment for foreign investors. In the Indonesian province of Riau, the government actively is promoting the jurisdiction to investors. A few other regions, such as Balikpapan in East Kalimantan, have pledged to guarantee the security of both domestic and foreign investors. Others, such as Yogjakarta, have reduced the red tape burden (Brodjonegoro 2004; Fox et al. 2005; Wahyuni 2010). Not only has the provincial mayor been awarded for his efforts in this regard, but Yogjakarta was named the country's most investor-friendly city by the International Finance Corporation (IFC). The local government of Rajasthan (India) has developed and implemented regulations that are attractive to foreign mining investors, which have led to economic growth (Singh and Kalirajan 2003).

Moreover, since the introduction of the new Mining Law in Indonesia in early 2009, the Indonesian mining sector has seen continued investment growth despite what some have perceived as a lack of regulatory certainty and the need for further clarification of the law. While this is in large part due to high commodity prices and sustained demand for key products, the fact that the mining regulations continue to evolve within a clear framework appears to be providing sufficient certainty for investors to view the Indonesian mining industry favourably. Developments such as benchmark pricing and domestic market obligation rules have been in place for some time, and do not appear to have disturbed the investment outlook (PwC 2012). These positive examples may serve as a guide for other governments across the Region to further improve both their mining regulation and its implementation if foreign investment is desired.

Empirical analyses in the previous section indicate that there is a high degree of overlap in the regulatory regimes governing mining investment across the Asia-Pacific Region. Given the significance of this problem, it may be necessary for some governments to review their regulatory architecture in order better to delineate responsibilities among regulatory agencies and ministries. One of the greatest challenges for mining investors is navigating both mining and forestry regulatory and environmental requirements in order to complete necessary approvals. Often, mining companies receive conflicting messages from various regulatory agencies, which unnecessarily slows down and/or impedes the approvals process. One possible solution to this problem is the creation of an inter-departmental coordinating agency with the capacity to resolve jurisdictional issues and demarcate clear boundaries over respective areas of control. This most likely will result in the reduction of regulatory overlap and red tape in the exploration and mining licence application process.

Most regulators across the Region are not independent from government influence and/or capture by other stakeholders. Of all the policy challenges that plague developing countries, regulatory capture is potentially the most difficult to solve because of the power and intransigence of entrenched interests. However, as a first step, governments should appoint impartial bodies to oversee regulatory agencies. While there is no way to ensure that such bodies themselves would not fall victim to capture, well-resourced and independent bodies charged with regulatory oversight would go some distance to improving the performance of bureaucracies in the Region. At the very least, formal and informal accountability mechanisms are essential. In addition, it is crucial that regulatory decisions and the reasons for their decisions are made public and are open to challenge by stakeholders. This would ensure a higher level of communication and transparency in the mining industry, and go some distance towards preventing capture.

We acknowledge the importance of incorporating the interests of landowners and host communities into mining projects. However, one of the emerging challenges for foreign investors is stakeholder engagement. There are two principal challenges for foreign investors in this regard. The first is that legislation seeking to protect the interests of host communities often is exceedingly complex and unhelpful to miners. Approvals take too long and the process often can be biased against mining companies. In PNG, for example, dealing with landowners over land access is a difficult process during all stages of resource development. Miners seeking to invest in the Philippines have experienced similar problems (Chase and Lugue 2006). In fact, it has become common practice that no project can be finalised without acknowledging and incorporating the interests of landowners and local communities in the project area. Obtaining a social licence to operate is a crucial step in ensuring that mineral exploration and production can begin and proceed without social and business risks for both the local community and the mining company (Graetz and Franks in this volume). However, the procedures for obtaining a social licence to operate need to be streamlined and made more transparent, particularly in the cases of PNG and the Philippines.

Although not a widespread problem, the tactics of anti-mining activist groups also present a challenge to mining investors. Many of these groups interfere in the legislative process and artificially fuel tensions between communities and mining companies. Newmont Mining Corporation employees were arrested by Indonesian police and charged with damaging the environment. Allegedly, anti-mining activists were behind the claims, and the basis upon which the Indonesian police acted. However, upon closer examination, the claims were found to be false and the local environment had sustained no damage (outside of what was permitted) as a consequence of Newmont's activities. This also has been a problem in Australia, where anti-mining activists have attempted to 'wedge' local communities to stop projects moving forward.

In response to stakeholder engagement problems, all stakeholders, including governments, should address the following issues. First, national governments should formulate consistent standards and regulation, and insist on consistent implementation and monitoring. Second, governments and mining companies should be more transparent and accountable in providing socio-economic and environmental information about their operations. Third, all stakeholders, particularly companies, should take shared responsibility for the socio-economic and environmental impacts of mining activities. Fourth, the distribution of revenues from mining companies among various stakeholders should take into account equity and justice considerations from the perspectives of the various stakeholders. Appropriate socio-cultural considerations increasingly have become central to successful mining operations. While this is not an easy task, governments and private industry should move in this direction. Stakeholder engagement legislation should be efficient, fair to all parties, and immune to the tactics of anti-mining activists.

A lack of institutional capacity is a key challenge for regional governments. Regulators suffer from a lack of human and financial resources and, as a result, undermine the effectiveness of compliance and enforcement mechanisms. This especially is a challenge for lower levels of government. Institutional capacity building requires high levels of expenditure for training and development, and mining revenues can be used to address this problem. However, the continued low levels of investment across the Region mean that institutional capacity building will be difficult to achieve. Consequently, governments across the Region need to redouble their efforts to build institutional capacity by seeking funding from international agencies. There have been some attempts by regional multilateral organisations, such as the Asian Development Bank (ADB), to promote regulatory capacity building. Governments should seek further assistance from the ADB and other regional and global bodies and make institutional capacity building a priority. The World Bank's capacity building work in PNG has had some success in this regard, with funds being used to establish the Mineral Resources Authority (Mineral Resources Authority 2008).

Conclusion

All regulatory regimes are a work in progress. Sometimes it is a question of two steps forward, one step backward. Events such as a change of government, political coup, civil unrest, or even an environmental disaster can be challenging for regulators. The US is reputed to have some of the most effective regulatory regimes in the world. But this did not stop the economic chaos that accompanied the Global Financial Crisis. It also is important to understand that regulation is influenced by local factors as well. Such factors form and develop within unique circumstances, and include language and culture, history, norms, taboos, conventions, forms of government, and institutional structures. North (1990) refers to this as a country's 'institutional endowment'. The key point is that institutions and regulatory frameworks evolve according to the particular institutional endowment of a country and this ultimately determines the efficiency of its regulatory regime, its capacity for reform, and its ability to attract foreign investment. As Levy and Spiller (1994) note, 'the credibility and effectiveness of a regulatory framework—and hence its ability to facilitate private investment—varies with a country's political and social institutions.' This insight is an important one as it helps to temper the force of a critique of a developing economy by ensuring we understand a country's institutional capacity. The criticisms and policy recommendations serve to highlight just how important good regulation is to a well-functioning economy and to attracting foreign mining investment.

References

Aguilar, J. (2008). Data mining. *Newsbreak*, special issue, July-September.

Andrews-Speed, P., Yang, M., Shen, L., Cao, S. (2003). The regulation of China's township and village coal mines: a study of complexity. *Journal of Cleaner Production*, 11(2), 185-196.

Australian Trade Commission. (2013). Infrastructure trade opportunities in the Asian Century. www.austrade.gov.au. Accessed 25 May 2015.

Bailey, S. (12 May 2007). India: attractive destination for western mining Investment. *Resource Investor*. http://www.resourceinvestor.com/2007/05/11/india-attractive-destination-western-mining-investment. Accessed 2 February 2015

Behre Dolbear. (2014). 2014 Ranking of countries for mining investment: where not to invest. http://www.dolbear.com/resources/. Accessed 14 November 2014.

Bhasin, B., Venkataramany, S. (2007). Mining law and policy: replacing the 'Contract of Work' system in Indonesia. Paper presented at the 10th International Conference of the Society for Global Business and Economic Development (SGBED), Kyoto, Japan, 8-11 August. www.dundee.ac.uk/.../files.php?file=CEPMLPIJ-Mar08-Indonesia.pdf. Accessed 6 July 2014.

Bloodgood, L. (2007). Competitive conditions for foreign direct investment in India. Office of Industries, U.S. International Trade Commission, Publication 3931, July. http://digitalcommons.ilr.cornell.edu/cgi/viewcontent.cgi?article=1394&context=key_workplace&sei-redir=1&referer=httppercent3Aper cent2Fper cent2Fscholar.google.com.auper cent2Fscholarpercent3Fqper cent3DLaurapercent2BBloodgoodper cent2BCompetitivepercent2BConditionsper cent2Bforper cent2BForeignpercent2BDirectper cent26btnG-percent3Dper cent26hlpercent3Denpercent26as_sdtper cent3D0percent252C5#search=per cent22Laurapercent20Bloodgoodpercent20Competitivepercent20Conditionspercent20Foreignpercent20Directpercent22. Accessed 8 May 2014.

Brodjonegoro, B. (2004). The effects of decentralisation on business in Indonesia. In M. Chatib Basri, & P. van der Eng (Eds.), *Business in Indonesia: new challenges, old problems* (pp. 125-140). Singapore: Institute of Southeast Asian Studies.

Brown, A., De Paula, E. (2002). Strengthening of the institutional and regulatory structure of the Brazilian power sector. Washington, DC: World Bank, Report on the PPIAF Project for Brazil Power Sector, Task 4.

Caprioni, E. (2013). Foreign mining companies flee China. OilPrice.com. http://oilprice.com/Metals/Gold/Foreign-Mining-Companies-Flee-China.html. Accessed 10 Dec 2015.

Chase, R., Lugue, F. (2006), October 18. The Philippines: reviving fortunes. *Ambrian Economic Updater*.

Christian Aid. (2004). Breaking promises, making profits: mining in the Philippines. London: A Christian Aid and PIPlinks report, December. http://cpcabrisbane.org/Kasama/2004/V18n4/BreakingPromises.htm. Accessed 12 May 2015.

Cubbin, J., Stern, J. (2004). Regulatory effectiveness: the impact of good regulatory governance on electricity industry capacity and efficiency in developing countries. London: City University, Department of Economics Discussion Paper Series 04/04. http://facultyresearch.london.edu/docs/paper57.pdf. Accessed 5 September 2014.

Dinnen, S. (2009). Thirty years of law and order policy and practice: Trying to do 'too much, too badly, with too little? In R.J. May (Ed.), *Policy making and implementation: studies from Papua New Guinea* (pp. 233-260). Canberra: ANU ePress.

Economy, E. (2007). The great leap backward? the costs of China's environmental crisis. *Foreign Affairs*, 86(5), 38-59.
Fox, J., Adhuri, D., Resosudarmo, I. (2005). Unfinished edifice or Pandora's box? decentralisation and resource management in Indonesia. In B. Resosudarmo (Ed.), *The politics and economics of Indonesia's natural resources* (pp. 92-108). Singapore: Institute of Southeast Asian Studies.
Fraser Institute. (2015). Fraser Institute annual survey of mining companies. https://www.fraserinstitute.org/sites/default/files/survey-of-mining-companies-2015.pdf. Accessed 3 November 2015.
Fraser Institute. (2012/2013). Fraser Institute annual survey of mining companies 2012/2013. http://www.fraserinstitute.org/. Accessed 3 November 2014.
Gutiérrez, L. (2003). Regulatory governance in the Latin American telecommunications sector. *Utilities Policy*, 11(4), 225-240.
Holden, W., Jacobson, D. (2007). Mining amid armed conflict: nonferrous metals mining in the Philippines. *The Canadian Geographer*, 51(4), 475-500.
Imbun, B. (2006). Multinational mining and petroleum companies' perceptions of the policy framework in Papua New Guinea. *Pacific Economic Bulletin*, 21(1), 225-242.
Jhingran, I. (1997). National mineral policy of India: an overview. *Resources Policy*, 23(1-2), 91-96.
Kaufmann, D., Kraay, A., Mastruzzi, M. (2004). *Governance matters III: governance indicators for 1996–2002*. New York: World Bank. http://siteresources.worldbank.org/INTWBIGOVANTCOR/Resources/govmatters3_wber.pdf. 6 July 2014.
Kurtzman, J., Yago, G., Phumiwasana, T. (2004). The global costs of opacity. *MIT Sloan Management Review*, 46(1), 38-44.
Laodengkowe, R. (2008, May 26). Extractive sector transparency initiative. *The Jakarta Post*.
Levy, B., Spiller, P. (1994). The institutional foundations of regulatory commitment: a comparative analysis of telecommunications regulation. *Journal of Law, Economics and Organisation*, 10(2), 201-246.
Mineral Resources Authority, Papua New Guinea. (2008). Mineral Resources Authority: corporate profile. http://www.mra.gov.pg/AboutUs.aspx. Accessed 2 February 2014.
Ministry of Mines, Government of India. (2006). National mineral policy. http://www.mines.nic.in/writereaddata/Filelinks/46ff58f0_rep_nmp.pdf. Accessed 24 January 2015.
Naito, K., Otto, J., Smith, D., Myoi, H. (1999). Legal aspects of exploration and mining: a comparative study of mining law in Asia. *Journal of Energy and Natural Resources Law*, 17(1), 1-12.
Naito, K., Otto, J., Eggert, R. (1998). Mineral investment risk and opportunities in Asia. *Resources Policy*, 24(2), 77-78.
North, D. (1990). *Institutions, institutional change and economic performance*. Cambridge: Cambridge University Press.
O'Callaghan, T. (2010). Patience is a virtue: regulation and governance in the Indonesian mining sector', *Resources Policy*, 35(3), 218-225.
O'Callaghan, T. (2009). Regulation and governance in the Philippines mining sector. *The Asia Pacific Journal of Public Administration*, 31(1), 91-114.
Otto, J., Cordes, J. (2002). The regulation of mineral enterprises: a global perspective on economics, law and policy. Institute for Global Resources Policy and Management. Rocky Mountain Mineral Law Foundation. Colorado School of Mines.
Otto, J. (1992a). *Criteria for assessing mineral investment conditions: mineral investment conditions in selected countries of the Asia-Pacific Region*. New York: The United Nations.
Otto, J. (1992b). *A global survey of mineral company investment preferences: mineral investment conditions in selected countries of the Asia-Pacific Region*. New York: The United Nations.
Planning Commission, Government of India, . (2006). National mineral policy: report of the High Level Committee, December. http://www.planningcommission.nic.in/reports/genrep/rep_nmp.pdf. Accessed 14 February 2015.
PwC (2012), Mining in Indonesia: investment and taxation guide. April. https://www.pwc.com/id/en/publications/assets/mininginindonesia-investmentandtaxation.pdf. Accessed 15 January 2015.

Rabasa, A., Chalk, P. (2001). Reinventing Indonesia: the challenge of decentralisation. In A. Rabasa & P. Chalk (Eds), *Indonesia's transformation and the stability of Southeast Asia* (pp. 47-51). Jakarta: The Rand Corporation.
Resosudarmo, B., Resosudarmo, I., Sarosa, W., Subiman, N. (2009). Socioeconomic conflicts in Indonesia's mining industry. In R. Cronin & A. Pandya (Eds.), *Exploiting natural resources: growth, instability, and conflict in the Middle East and Asia* (pp. 33-46). Washington, DC: Stimson Center.
Sames, C. (2006). The mineral industry of India. *Minerals & Energy - Raw Materials Report*, 21(3-4), 19-34.
Santosa, F. (2013). New developments for the Indonesian mining industry. SSEK: Indonesian Legal Consultants. http://blog.ssek.com/index.php/2013/02/new-developments-for-the-indonesian-mining-industry/. Accessed 24 November 2014.
Singh, K., Kalirajan, K. (2003). A decade of economic reforms in India: the mining sector. *Resources Policy*, 29(3-4), 139-151.
Stern, J., Holder, S. (1999) Regulatory governance: criteria for assessing the performance of regulatory systems. *Utilities Policy*, 8(1), 33-50.
Subramaniam, V., Ashwin, M. (2006, 20 November). Ministries must stop being regulators. *India Together*. http://indiatogether.org/reglpower-editorials. Accessed 15 March 2015.
Suxun, Chenjunnan. (2008). Private capital: what impedes its entry into China's minerals. *Resources Policy*, 33(1), 23-28.
Tan, A. (1998). Preliminary assessment of Philippines' environmental law. APCEL Report. Singapore: Asia-Pacific Centre for Environmental Law.
Tahilyani, N., Tamhane, T., Tan, J. (2011). Asia's $1 trillion infrastructure opportunity. *McKinsey Quarterly*, March. http://www.mckinseyquarterly.com/Asias_1_trillion_infrastructure_opportunity_2765. Accessed May 23 2015.
Transparency International. (2014). The Corruption Perceptions Index. http://www.transparency.org/cpi2014/results. Accessed December 10 2014.
Tse, P. (2003). The mineral industry of China. U.S. Geological Survey Minerals Yearbook.
Vivoda, V. (2011). Determinants of foreign direct investment in the mining sector in Asia: a comparative analysis between China and India. *Resources Policy*, 36(1), 49-59.
Vivoda, V. (2008). Assessment of the governance performance of the regulatory regime governing foreign mining investment in the Philippines. *Minerals & Energy - Raw Materials Report*, 23(3), 127-143.
Ward, M., Izzard, C., Cornelius, S. (2003). Focus: mining in China. Allens Arthur Robinson, August. http://www.allens.com.au/pubs/res/fomcaug03.htm. Accessed 23 May 2015.
Wahyuni, S. (2010, 28 October). Herry Zudianto: for the good of Yogyakarta. *The Jakarta Post*. http://www.thejakartapost.com/news/2010/10/28/herry-zudianto-for-good-yogyakarta.html. Accessed 24 June 2015.
Wurfel, D. (1988). *Filipino politics: development and decay*. Ithaca: Cornell University Press.
World Resources Institute. (2005). *Electricity governance toolkit: benchmarking best practices and promoting accountability in the electricity sector.* Prayas-Prune: National Institute of Public Finance and Policy. http://www.wri.org/publication/electricity-governance-toolkit. Accessed 5 March 2014.

Chapter 4
Property, Politics, and Power: Theorising Political Risk in the Mining Industry

Daniel Feher

Abstract Political risks are critical variables for investors to consider when undertaking business in a globalised economy. Yet the concept itself remains under-theorised and, consequently, suffers from a lack of conceptual clarity. Attempts to define political risk thus far have failed to consider and analyse how the concept is theoretically grounded. This has led to a significant knowledge deficit and to a lack of consensus as to what constitutes the idea of political risk. This chapter addresses this deficit by exploring the theoretical nature of political risk. It examines the inter-related elements of property, politics, and power to develop a new understanding of how these elements underpin the idea of political risk. This chapter is, therefore, theoretical in nature. As such, it does not directly address risks, challenges, and opportunities facing the mining industry in the Asia-Pacific Region. Instead, it examines key ideas that inform the idea of political risk. It will be up to individual mining companies to adapt these abstract ideas to their individual circumstances.

Abbreviations

BIT	Bilateral Investment Treaty
TRIP	Trade Related Intellectual Property Rights
US	United States

Introduction

Mining companies operate in political societies and therefore are subject to political risk. This is because they generate enormous wealth, polarise communities, and have substantial impacts on the environment. These, and other factors, make the mining industry controversial. Mining companies have sought to overcome these

D. Feher (✉)
Centre for International Risk, School of Communication, International Studies and Languages, The University of South Australia, Adelaide, SA 5001, Australia
e-mail: feher.daniel@gmail.com

T. O'Callaghan, G. Graetz (eds.), *Mining in the Asia-Pacific*, The Political Economy of the Asia Pacific, DOI 10.1007/978-3-319-61395-6_4

challenges by developing a range of political risk mitigation strategies. These include joint venture partnerships, developing risk matrices that rate potential political risks and their severity, and using commercial rating systems, such as those discussed in the previous chapters.

While these strategies are of some value to mining companies, the concept of political risk is not well defined in these strategies; indeed, its meaning largely is taken for granted. This is not just a problem for the mining industry; it also is a problem that is present in much of the academic literature on political risk. Currently, there not only is a lack of consensus as to the definition of political risk, but there also is no adequate direction as to how political risk might be conceptualised. This is a 'political' concept, and there should be adequate tools within the field of political science to assist in developing a more robust understanding of this important concept.

In this chapter, I argue that there are tools within political science to progress our understanding of political risk. However, in order to make progress, it is necessary to look below the surface of the concept and examine its foundational elements. I contend that the foundational elements are property, politics, and power. I examine each of these ideas individually, as well as how they interact with each other to become political risks for mining companies. It is hoped that this new conceptualisation will lead to a more nuanced understanding of political risk and, additionally, that it will inform future research on the topic.

Political Risk Research: A Dead End?

Political risks are events, phenomena, or occurrences that obstruct, or have the potential to obstruct, commercial outcomes. Political risks arise out of the unique social, cultural, religious, and political landscapes of operating environments, and are most notable for their fundamentally unpredictable nature. Political risk is a convenient catch-all term used to classify hard-to-define, yet financially, devastating risks. The term's convenience and broad applicability has guaranteed its survival and continued application in an evolving global marketplace. However, an inability to be adequately conceptualised has relegated it to the status of what Meldrum (2000) calls a 'soft art', rather than a hard science.

The lack of conceptual consensus on what political risk is and is not leaves it without a foundation on which any sort of consistent definition can be achieved. This problem means that political risk most accurately is characterised as a form of subjective, or 'Knightian', uncertainty (Knight 1927). It is for this reason that political risks thus far have resisted easy quantification. Knight suggested that subjective uncertainty exists when neither the nature nor the number of outcomes definitively can be ascertained. Therefore, a concrete scenario cannot be established, probability cannot be associated with outcomes, and 'risk proper' cannot be determined (Knight 1927). Although measures can be taken to reduce subjective uncertainty *a posteriori* by empirical means, the unique and, therefore, unpredictable nature of most political risk events cannot be modelled actuarially or scientifically.

This does not mean that the study of political risk should be neglected. Instead, by focusing on the three key components of political risk, we can begin to build a foundation for this conceptualisation. In the following sections of this chapter, I provide such a foundation. I highlight why existing definitions have failed and why conceptual progress on political risk has been slow. What I demonstrate is that the debate about political risk needs to be taken in a new direction based around the foundational elements noted above.

Where Did the Research Go Astray?

A review of the literature on political risk provides valuable insights into how political risk came to its present state of development, and how it lost conceptual focus, with scholars becoming divided and disillusioned by the 1980s.

The literature on political risk begins in the 1920s with debates about the legality of expropriation and rights to compensation. This phase was stimulated by the emergence of state socialism, the two world wars, and decolonisation. These factors contributed to the gradual unravelling of the great power empires that had spawned and nurtured early multinational enterprise, and meant that empire-backed corporations were no longer immune from host country interference in their operations (Bauzon 2000). The discussions in this phase moved logically to the causes and prevention of these events and potential forms of protection, showing promising and fertile ground for conceptual progress. Authors such as Boublikoff (1935) began to speculate about potential turmoil in Soviet Russia and the feasibility of taking out a protective insurance policy through brokers such as Lloyds of London. The most promising and influential scholars during this period were Halperin (1940), who questioned the conditions under which expropriation occurred, and Herz (1941), who argued that expropriation was one form of an undefined area of risk: 'strange as it may seem, a coherent theory of expropriation as a separate institution has not been developed.'

Between the late 1940s and 1960s, the growing spectre of the Cold War and the subsequent formation of the Point Four Program in 1949, by United States (US) President Harry S. Truman, marked a turning point in research on political risk. In a speech to Congress, Truman acknowledged the problem of expropriation and similar risks faced by corporations investing abroad. To counteract these concerns, he allocated $45 million in funding to kick-start a program to attract private investment, authorising the Export-Import Bank of America to 'guarantee private capital … in productive enterprise abroad … against the risks peculiar to those investments' (Truman 1949). This program led to the birth of political risk insurance and the Export-Import Bank became the first provider of 'financial guarantees', as they were then termed. This was in addition to its original role as a facilitator of US exports and trade (Greene 1965). More significantly, Truman's program became a catalyst for focused research into the risks this new insurance was designed to protect against and, in the early 1960s, the term 'political risk' became formally established in the literature. The formal cementing of the term's semantic association can be pinpointed to an announcement in 1962 by the Export-Import Bank and the US commercial insurance sector of a joint venture to offer 'political risk insurance' (Greene 1965; Becker and McClenahan 2003). It also was around this time that research lost its focus on developing a theory of political risk, and began focusing on the nature of events and phenomena *a posteriori*.

The primary catalyst for the shift in focus was the need to identify and predict political risks in order to insure against them. From here, the first serious efforts to define political risk emerged in the literature, as scholars sought to contextualise their research on the basis of this new form of insurance and the risks it was intended to mitigate. Contributors to the literature depended on the analysis of previously categorised political risk events and scenarios experienced *a posteriori*, while failing to consider political risk as a concept in its own right. Consequently, a new point of focus was set for scholars that would change the direction of the literature, from developing an understanding of political risk as a theoretical concept, as authors such as Herz (1940) and Halperin (1940) had suggested, to the identification and analysis of its perceived phenomena. Adding to this was the increasingly crises-driven nature of research into political risk. The 'political crises' of the 1960s and 1970s, the 'debt crisis' of the 1980s, and 'financial crises' of the 1990s became the focus of attention for political risk scholars (Simon 1985; Bouchet, Clark and Groslambert 2003; Bremmer and Keats 2009). These events stimulated volumes of research on their nature and cause within various contexts, but offered little practical or theoretical clarity to the concept of political risk.

This is not to say that the research of the past 50 years on political risk has been fruitless. On the contrary, it has provided significant insight into the nature of certain political risks, with much of this research now incorporated into mainstream political risk management practice. What this change in direction failed to do, however, was to conceptualise political risk. This is reflected in the disillusionment in the literature from the late 1970s onwards, as scholars gradually realised that the key to conceptualising political risk did not lie in the study of its phenomena. For example, Kobrin (1979) argued that 'while the literature reflects substantial progress in a relatively short period of time, it still does not provide an analytic framework which can adequately contribute—in either a taxonomic or an operational sense—to improved practice', adding that political risk events were still 'not defined in a manner that allows for unambiguous classification of environmental events.' He went on to note that the conceptual and empirical void in political risk analysis needs to be filled with 'better definitions of the phenomena, a conceptual structure relating politics to the corporation, and a great deal of information about the impact of the political environment' (Kobrin 1979). Two years later, Brewer (1981) described the political risk assessment process as 'too impressionistic and intuitive' and 'conceptually and methodologically constrained.' Simon (1985) attributed the growing stagnation of the field to scepticism in industry about the concepts, theories, and methods of political risk analysis, suggesting dryly that the 'lack of a hot crisis' also could be to blame. The ultimate outcome of these insights has been that research into political risk has since avoided the problem of its conceptualisation, leading to a weakness in this aspect of research. Both academic and industry researchers instead have continued to focus on the analysis of known political risk phenomena, building on *a posteriori* knowledge of past events to provide assessments and predictions of their occurrence through various methods, and with varying degrees of success. Observing the growing political risk analysis and insurance sector, this 'work-around' solution is no doubt sufficient for general application. However, it nonetheless means that the concept of political risk continues to lack clarity.

This stagnation is yet to be overcome and, accordingly, research into political risk remains conceptually weak. In the sections that follow, I sketch out an argument that I hope will advance the understanding of political risk, or at least push debate about the concept in a new direction. I suggest that we are able to gain a better understanding of political risk by looking at three interrelated elements: property, politics, and power. I call these the three pillars of political risk.

The Three Pillars of Political Risk

I use the term 'pillars' purposely, as the conceptualisation of political risk offered here illuminates the foundational ideas that give rise to political risk events. By focusing on property, politics, and power, it is possible to ground the concept of political risk and open a pathway for new research.

Property

When we think of property we think of ownership and control over objects that have important value to individuals. These may include houses, cars and other personal possessions. I call this the 'everyday understanding' of property. However, property is much more than this. It is an institution, a system of rights and obligations backed by law, an economic system that underpins everything that is transacted in a capitalist society and, most fundamentally, it is the core tenet of Western liberal political and economic thought.

In some ways, the Cold War can be interpreted as a contest over differing interpretations of the meaning of property. For the Soviet Union, property was a communal asset that supposedly was shared by all. In the West, ownership of private property is an individual right. This is because the West equates ownership of property with 'independence and self-reliance' (Heywood, 1994). Ownership provides individuals with choices, opportunities, and status. It cements their place in society's hierarchy, and is regarded as a measure of their well-being. Unfortunately, however, those without property generally are the poor.

Just as individuals in liberal societies have benefitted from the right to own property, so too have corporations. This partly is because, legally, corporations have the same rights of ownership as individuals. Indeed, the right of corporations to own property has contributed to their success. A century and a half ago, there were very few corporations in existence; those that did exist were 'chartered' by governments to serve the public interest. This may have been building transport corridors, providing potable water, or the provision of electricity. The mandate was quite narrow and specific. The shift from this view of a corporation to the modern corporation occurred in the 1890s. According to Bakan (2004), three things precipitated the change. First, the court repealed rules that required businesses to incorporate for narrowly defined purposes and for relatively short periods of time. Second, it loosened the controls governing mergers and acquisitions. Third, it abolished rules that prohibited one company from owning shares in another company. These changes allowed companies to buy shares in other companies and provided them with the rights to own capital and property and hold assets indefinitely. The development of subsequent mechanisms, such as international treaties, has helped corporations to consolidate their control over property. The provision of political risk insurance (Spagnoletti and O'Callaghan, 2011), Trade Related Aspects of Intellectual Property Rights (TRIP) and Bilateral Investment Treaties (BIT) are directed at protecting the right of corporations to own property, and to use property to their economic advantage. It is at this point that property and political risk intersect.

I define political risks as non-market events, characteristics, or phenomena in an operating environment, or the threat of such features, which may cause adverse commercial outcomes or obstruct otherwise tangible or achievable commercial potential. This complex and unwieldy description can be simplified by factoring in the concept of property. Political risk obstructs, or threatens to obstruct, successful commercial outcomes, whether they are actual or potential. At the root of 'successful commercial outcomes' is the acquisition, maintenance, and accumulation of *property* in the form of resources and value, through their prosperous utilisation. Therefore, in a strict commercial sense, the threat of political risk lies in its potential to affect adversely a company's *property rights* or the *value* of those rights. The first conceptual pillar in my framework of political risk therefore is the commercially relevant ownership of property and the risk of loss posed to this ownership by an operating environment's non-market events, phenomena, or circumstances. From a strictly commercial viewpoint, property and value can be seen to encompass all tangible and intangible sources of value, from plant equipment and other physical resources, through to intellectual property rights and the contractual rights to the skills and labour of a corporation's human resources.

Politics

In its most rudimentary form, politics is an activity that arises out of human or actor interaction as a result of our diversity of needs and wants (Heywood 1994). Politics is underpinned by four essential elements:

- Differing (and sometimes clashing) goals or interests
- The resources needed to facilitate these interests

- The capacity to obtain those resources
- The structure, setting, or environment in which these elements come together to form political interaction, and the constraints, rules, and boundaries of such structures, or the absence thereof.

While there are a number of definitions of politics, there are two views that are relevant to our understanding of political risk. The first view emphasises the unstructured nature of politics. The second view emphasises a more structured notion of politics.

Politics: An 'Unstructured' Approach

Politics primarily is about the pursuit of interests, and the power and resources needed to realise those interests (Heywood 1994). The broad nature of this approach to politics is described by Leftwich (2004):

> politics comprises all the activities of co-operation, negotiation and conflict, within and between societies, whereby people go about organising the use, production or distribution of human, natural and other resources in the course of the production and reproduction of their biological and social life.

According to Leftwich (2004), the interactive ingredients of politics are people, commonly with different interests, preferences and ideas, scarce resources, and power. The divergence of interests described by Heywood and Leftwich, and the resources and power needed to fulfil them, are what drive political activity, while also giving rise to the process (or indeed absence) of decision and rule making through which conflict can be managed. These two factors form the basis of politics, and are the most important to note here. This approach to politics conceptualises it as a ubiquitous and pervasive feature of all human interaction, based on the pursuit of interests and the resources necessary to fulfil them. According to Leftwich (2004), 'politics consists of *all* the activities of conflict (peaceful or not), negotiation and cooperation over the use and distribution of resources, wherever they may be found, within or beyond formal institutions … between two or more people.'

In this model, political activity is seen to take place on a number of levels across a range of human interaction, with the key elements being competing interests, scarce resources, and the means of how they are distributed and used to meet competing demands. The strengths of this model are that it is not constrained by parameters regarding the nature of interaction between actors, nor on the origin or allegiance of the actors themselves. It acknowledges that the pursuit of interests often can be chaotic and incorporate a spectrum of powers exercised by various actors, from exertion of influence and persuasion to the use of coercion. Furthermore, this model usefully can be applied to political interactions that are not confined to recognised state or territorial boundaries.

It has been observed that the broad and all-encompassing nature of this interpretation puts it at risk of rendering the term meaningless with arguably any form of interaction considered as politics (Heywood 1994). However, this shortcoming also is this interpretation's strength and the source of its robustness. By focusing broadly on interests, resources, and the use of various forms of power, this conceptualisation of politics better accommodates scenarios in which formal structures are absent, lack effectiveness, or are not applicable, as is commonly found in commercial operating environments where political risks feature prominently. Furthermore, this interpretation particularly is relevant to interpreting risks that arise outside of the institutional structures of organised and enforceable governance, such as state failure, cross-border conflict, and terrorism.

Politics as Structured Governance

The second approach to politics is structural. Politics in this sense is most commonly used to describe the apparatus of government. This includes policy-making, regulation, and the maintenance of social harmony. Politics involves the making and enforcement of collective decisions on behalf of the community. In a liberal democracy, elected representatives are conferred with the right to make laws as a consequence of regular elections. The decisions made by these elected representatives then form the basis for the community's *policy*—that is, the guiding principles that provide the community with a strategy to achieve collectively beneficial action. For example, Lasswell (1936) observed that politics is how a governing body determines 'who gets what, when, where and how' within the community. In this sense, the concept of politics lies in the machinery of government and is distinguished from the public. A key dimension of this view of politics is the role of government as an institution, which sets the 'rules of the game' for a community (North 1995). These structures are comprised of the formal and informal rules encompassing the various norms, behaviours, and expectations that shape a functioning society, and the social and economic conditions that arise as a result of these rules. In addition, the rules determine the extent and nature of cooperation between actors subject to them, while providing the foundation for the exercise of legitimate power and authority. As Crick (1964) suggests, 'politics is the way in which free societies are governed.' He interprets politics as the process of conflict resolution, centred on the appeasement of competing interests. This is done through what Crick refers to as structured compromise and negotiation. Where this process fails and coercion or violence take over, normal political engagement and legitimate governance no longer are possible. Politics reverts to being unstructured, chaotic, and prone to risk events.

The value of this model of politics lies in its capacity to be applied across a broad spectrum of organised communal behaviour. Indeed, the majority of states in world politics fit this model. However, such a model of politics is not without its shortcomings. The focus on government and ruling institutions tends to assume some level of power distribution across the governed populace. The model also depends on a degree of authoritative, or institutional, structure within an environment to be validly applied. As Crick (1964) notes, in the absence or weakness of this structure, such as in failed or failing states, this model of politics stops functioning. Yet, many commonly identified political risks, such as civil unrest and violent conflict, arise out of exactly these situations, creating a potential paradox. Perhaps the most prominent example of this shortcoming is the lack of an authoritative 'world government' to address competing interests, develop and enforce international policy, and ensure that conflicts are settled before they descend into violence. In this respect, this interpretation also is narrow in its recognised means of engagement, with force, coercion, and other non-peaceful forms of interaction not considered political in nature. Finally, the model implies the need for defined boundaries of jurisdiction and the identification of actors by their recognition of, or allegiance to, such governed boundaries. This causes problems for political risks that may involve actors who are stateless, such as terrorists, or where political risks and their events occur across multiple territories and are not governance focused, such as ethnic or religious conflicts.

Both of these interpretations of politics are valuable to the theory of political risk that I develop here. Individually, they sit at opposite ends of the conceptual spectrum (Held 1991). Together, they form a comprehensive overview of the concept of politics. Indeed, without understanding each view of politics, we will not be able adequately to understand the concept of political risk. Furthermore, the structures of interaction on which the governance model fundamentally depends act as a bridge between these two conceptualisations. Where the structures fail, are weak or failing, or in development, a process of overlap or 'transition' between the two models of politics can be seen as the first model is gradually 'switched off' and the second model 'switched on', or vice versa. In this process, actors transition from the structured and cooperative forms of interaction that form the basis for governed civil society to less cooperative and, ultimately, unstructured forms of interaction, such as the use of force.

In essence, politics in the context of political risk can be seen as a *competition of interests*. Actors vie with each other for the resources they need to facilitate these interests. The nature of these competitions depends on the presence, level, and strength of cooperative governance and other structures in a given environment or arena of competition, and the capacity of those structures to facilitate and satisfy competing interests. This, in turn, influences the behaviour of, or the action taken by, actors in pursuit of their interests. The diversity of manifestations of political risk therefore can be seen to arise out of the expression of this behaviour by political competitors pursuing their own interests.

Power

The third pillar of political risk is power. This is one of the most complex and debated concepts in political science, owing primarily to its vagueness and the fact that power defines and shapes the nature of 'the political' (Hay 2002). Building on the conceptualisation of power as the capacity to realise interests and instil a degree of control is the notion that the power to do so lies in the possession and distribution of scarce resources required to facilitate needs and interests. In this interpretation, the balance and distribution of power in a community is linked to the division of resources. These are 'resources of power', which allow their holder to exert influence over other actors (Dahl 1961). This view of power forms the basis for pluralist approaches to politics, particularly in advanced Western societies. But it also finds considerable application in commercial settings. Here, political competition is seen to arise out of the push and pull of different interest groups that seek political influence as the means to secure a greater share of resources. Those groups that have greater access to resources are more able to exert control and influence over what Lukes (2005) calls the 'three dimensions of power'. These are: (1) decision-making, (2) agenda-setting, and (3) preference-shaping.

This overview identifies some of the key elements of power relevant to my conceptualisation of political risk: the capacity of one actor to dominate and control the will of another actor to realise the former's interests, the structures of social order, cooperation and governance, and the value and distribution of scarce resources. Like politics, power exists along a spectrum, from primitive and base forms, such as violence, force, and coercion, to sophisticated and advanced manifestations, like persuasion, influence, and authority. What form of power actors exert, how, and under what conditions, is shaped by the environment in which they find themselves competing and, most importantly, by the degree and forms of structure present. This, in turn, influences the nature of political risk that arises out of that environment.

The relationship between the structures of an environment and the form of power actors wield in pursuit of interests corresponds with the political model I described in preceding sections of this chapter. This is supported by Tsebelis (1990), who observed that:

> individual action is assumed to be an optimal adaptation to an institutional environment, and the interaction between individuals is assumed to be an optimal response to one another. Therefore, the prevailing institutions determine the behaviour of the actors, which in turn produces political or social outcomes.

Further to this, Dowding (1996) notes that 'the power of individuals varies with the nature of the circumstances in which they find themselves … (and) can be enhanced through cooperative actions when it is to (their) advantage to do so.' Where an environment lacks governance structures, the forms of power exerted by actors become increasingly and correspondingly basic and, at a more detailed level, are influenced by the degree of the structural vacuum. There also will be a higher risk profile. In contrast, where an environment is highly structured, with cooperative, functioning, and transparent civil governance, the power employed by actors tends to be more predictable. The basis of a given political risk event always will be underpinned by the form of power, the nature of which can be viewed either as basic or advanced. How these expressions of power affect a corporation's capacity to respond through mitigation, defence, or retaliation will form the basis for perceived severity of the corresponding political risk that is represented by that action of power. To make further sense of this, it is important to consider the corporation as an actor in more detail.

The Corporation as a Political Actor: Powers, Advantages, and Vulnerabilities

Corporations exist in several forms, ranging from sole proprietorships to large, equity-based multinationals. Most commonly, corporations are incorporated legal entities, a status that, once conferred, allows them to function as a 'citizen' of the country of incorporation and pursue their interests in a similar manner to natural persons. According to Bakan (2004), 'without the state, the corporation is nothing.' As an actor, the corporation's *raison d'être* is underpinned by the presence of functioning governance structures, and the corporation's status and recognition as an incorporated legal entity depends on them. Furthermore, these structures underpin its *empowerment* as an actor (an aspect I address in more detail below), and justify the corporation's right to pursue its objectives, enter into and pursue contracts of obligation, and claim ownership to property. In organisational economics, a significant body of literature has emerged on the theory of the corporation, which is devoted to the explanation and interpretation of a corporation's origins, purpose, behaviour, and relationship to the market. It is sufficient to note here the central principle underpinning this body of literature, namely, that the corporation's primary objective and purpose is profit maximisation through the optimal allocation of resources into the production of saleable goods and/or services.

On the issue of power and the corporation, Morgenthau (1973) proposed that corporations seek 'submission to (their) purposes' in political interaction. This can be interpreted as power in the form of control. The corporation's interests consequently are centred on control over its resources and operations in order to maximise its capacity to pursue objectives. Political risk therefore can be understood to affect this control by constraining or impeding the corporation, either through deprivation or loss of resources, or through the inability to maximise resources to meet objectives. This can happen directly, indirectly, or collaterally. However, the corporation is not a passive player in competitions of interest. Its incorporation as a legal entity empowers it with the capacity to engage with other actors in the pursuit of shared interests. Furthermore, just as the corporation's reason for existence is underpinned by structured governance, so too are its powers. These are tangible and intangible resources, influence, and knowledge. In addition, its powers are fixed. The corporation cannot engage in competitions of interest with forms of power that are deemed illegitimate under the rules of structured governance that underpin it. To do so would lead to legal sanction by state institutions.

Being an entity the very existence of which arises out of functioning structures of governance, it follows that a corporation's power as a political actor also stems from those same structures; conversely, this is its key weakness if the environment is unstructured. Contemporary corporations cannot take up arms and use violent or coercive means to pursue or defend their interests. This is why conventional political risk rating systems tend to assess war and conflict ravaged states as posing a high to severe level of risk. At their most elementary, a corporation's primary vulnerabilities as a political actor are those that involve sources of power from the basic end of the spectrum that they cannot control.

Returning again to sufficiently structured environments, a key strength corporations tend to have—particularly larger corporations—is the benefit of an abundance of *property*, which serves as a basic resource. Arguably, property resources can be seen to constitute the corporation's very being, forming the basis of its value, operational capacity, and, ultimately, its objectives. Here, I make a distinction between *tangible resources*, commonly referred to as land, labour, and capital, and *intangible resources*—knowledge, influence, and reputation, with the latter generally arising out of the former (O'Callaghan 2016).

Building on the idea of power as the capacity to realise interests and instil control is the notion that this power lies in the possession and distribution of scarce resources required to facilitate interests. In this interpretation, the balance and distribution of power in a populace is linked to the division of resources. These become 'resources of power', which allow their holder to exert influence over other actors. These resources encompass 'anything that can be used to sway the specific choices or strategies of another (actor)' (Dahl 1961). In the case of the corporation, its capacity for influence in an operating environment tends to be proportional to the relevant 'resources of influence' it controls: property, income, and knowledge. This also includes the level of its contribution through its business activity to that environment's economic output (Dahl 1961). This power particularly is applicable in determining the corporation's capacity to influence policy-making. The corporation's resources, organisational strength, and economic contribution confer on it such power that it can be exercised in an attempt favourably to affect the policy-making process, either by targeting political elites involved in 'decision-making' and 'agenda-setting', or by seeking to influence popular opinion through 'preference-shaping' in the form of media and other public campaigns (Dahl 1961).[1] A corporation's resource wealth also enables it to access specialist expertise, such as legal and communications professionals. As such, large corporations frequently can be regarded as political actors. Indeed, it is in the interests of corporations to be active in the political process due to the impact that a governing body's policy-making can have on their business activities. This perspective is supported by Liebeskind (1996), who says, 'knowledge is arguably the most important asset a corporation possesses.'

Knowledge underpins the corporation's other powers, providing the necessary inputs strategically to guide and direct the corporation in employing its resources to exert influence. I use the term here in the sense of the corporation's understanding of, or familiarity with, its operating environment. More specifically, I refer to the corporation's understanding of an operating environment's rules of the game—that is, the governance structure, and legal, regulatory, and socio-cultural norms that form the basis for how political interaction and power distribution occurs. Knowledge, in this sense, also is a type of power, and forms a primary aspect of relations between actors as the basis for how they develop strategies of action and response in competitions of interest. The need for corporations to develop an understanding of their operating environments can be seen in early definitions of

[1] The Mineral Council of Australia's (MCA's) campaign against the Australian Government's introduction of a Minerals Resource Rent Tax is a good example of this policy shift.

political risk, particularly by Drake and Prager (1977), who argue that political risks arise out of 'operating in countries where the environments are strange and not well understood.'

The corporation's capacity to engage in competition with an understanding of what forms of power are considered legitimate, and on what basis, is a key factor in maximising its control over its operations and minimising vulnerability to political risk. This particularly is the case when there is disparity between the corporation's home and foreign operating environments in terms of governance structures, regulations and laws, and other institutional, societal, and cultural norms. In such situations, the importance of this knowledge increases in proportion to the level of disparity between the corporation's home and foreign environments. In a foreign environment, the corporation is at a disadvantage operationally, especially in terms of knowledge and influence. It is for this reason that political risk conventionally has been viewed as a problem of 'foreign investment climates'. In various foreign jurisdictions, the corporation has a diminished capacity for understanding what is a legitimate action and what potentially falls outside of ethical standards. This can cause misunderstanding and lead to poor strategy execution. Consequently, it may result in friction and greater potential for political risk. Knowledge of an operating environment therefore is critical, due to the influence this understanding can have on whether the corporation perceives political risks (such as legal or regulatory change) to be legitimate or prejudicial. This can cause unnecessary tensions and reputational issues for both the corporation and the host jurisdiction, which may have ramifications for many years to come.

In summary, a corporation's powers arise from its existence as a product of structured governance. These powers primarily are sourced from its ownership of property and other intangible resources, such as knowledge, influence, and reputation. As governance structures weaken, so too do the strength and effectiveness of the corporation's powers. The corporation's ability to exert influence through legal or popular avenues, for example, depends on an investment climate with functioning laws and regulations, as well as a cohesive and cooperative social structure. Its capacity to understand its operating environment also hinges on this cohesion. Weak or failing governance structures tend to affect the distribution of resources, such as income and property, leading to unrest, discontent, and the use of increasingly basic forms or power, all of which ultimately precipitate risk and unpredictability in the jurisdiction seeking investment. This instability adversely affects the corporation's capacity to understand and forecast governance and populace behaviour, and to mobilise and utilise its resources to exert effective influence. Finally, the corporation's means of power are exclusively advanced and inherently fixed. As governance structures fail, and the disparity between the use of basic and advanced forms of power increases, so too does the corporation's vulnerability. This is due to the corporation's inability to compete against basic forms of power, and, subsequently, defend against political risks arising out of them. It is for this reason that political risks of a conventionally 'severe' categorisation can be seen to arise predominantly out of weak or failing governance structures. The more robust and effective a state's governance structures, the less likelihood there is of political risks arising from basic forms of power in actor interaction, and the greater the ability of the corporation to exercise its available forms of power to realise its interest of maximising control and profits.

Conclusion

The concept of political risk has suffered from a lack of conceptual clarity for many years. The purpose of this chapter has been to begin a conversation about how to develop a theoretical framework that can serve as a foundation both for the relevance of the concept and as a stimulus for new research in this field. In essence, only by breaking political risk into its foundational elements—property, politics, and power—is it possible to progress the concept of political risk.

By identifying how property, politics, and power can conceptually underpin political risk, I have presented a viable and broad reaching theoretical framework that can function as a conceptual structure relating politics to the firm. This framework demonstrates the conceptual relationship between politics and the corporation by showing that the latter's existence as an actor inherently is underpinned by the idea of structured governance, which provides the basis for its capacity and strength to engage politically, and the forms of power with which to do so.

Through this chapter, I demonstrated that the literature on political risk to date has been characterised by its reliance on *a posteriori* analysis of political risk phenomena and events as a means of interpreting and understanding political risk in operating environments. I established that this approach is flawed, due to the unpredictable and subjective nature of political risk phenomena. I suggested that conceptual analysis must move beyond such an approach to develop a wide-reaching understanding of political risk. I provide an alternative approach, proposing a conceptualisation that interprets the sources and nature of political risk, and the nature of its effect on corporations, by considering how such risks arise out of actor interaction in operating environments. I establish that the conceptual basis of political risk phenomena lies in them being a product of actions of ownership and use of power that are exercised in competitions of interest, and that the specific source of political risk therefore can be seen to be the form or action of power exercised by an actor.

I also used a unique approach to interpreting the conceptual nature of the effect of political risk on the corporation. By drawing on economic theory of the corporation and theories of power, I showed that corporations seek control over their operations and capacity to maximise the utility of their resources to meet objectives. In doing so, I established that political risk can be understood to affect a company's control of its operations, either through deprivation or loss of resources or through other constraints that impede its ability to maximise them and meet objectives. Finally, I showed how companies can be exposed to political risk by demonstrating that the corporation can be affected either directly, where it is engaged in competition as an actor, or indirectly, where it is exposed collaterally to competitions of interest between other actors in the operating environment.

While I have not focused my attention on the mining industry directly, I have contributed theoretical clarity to a concept that informs, and is central to, the industry. Mining companies encounter political risks as a matter of course. I have chosen to engage the mining industry at a fundamental theoretical level. I have sought to demonstrate that the current political risk approaches adopted by mining and other companies fail coherently to appreciate the foundational elements of the concept of political risk. But this is only a start. I have presented the beginnings of a more comprehensive framework to understand the concept of political risk. The next phase of work requires that we bring the three foundational elements of political risk down to earth. New ways of thinking about political risk will need to be developed that more closely focus on property, politics, and power.

References

Bakan, J. (2004). *The corporation: the pathological pursuit of profit and power*. New York: The Free Press.

Bauzon, K. (2000). Political forecasting and the Third World economies: a critical assessment. Paper presented at the Sixth Annual Meeting of the Global Awareness Society International, Montreal, 23-25 May 1997. http://journals.upd.edu.ph/index.php/kasarinlan/article/viewFile/1427/pdf_68. Accessed 23 August 2009.

Becker, W., McClenahan, W. (2003). *Market, the state and the Export-Import Bank of the United States: 1934-2000*. Cambridge: Cambridge University Press.

Boublikoff, A. (1935). Doing business with Russia. *The North American Review*, 239(3), 269-274.

Bouchet, M., Clark, E., Groslambert, B. (2003). *Country risk assessment: a guide to global investment strategy*. Cornwall: Wiley Finance Series.

Bremmer, I., Keats, P. (2009). *The fat tail: the power of political knowledge for strategic investing*. Oxford: Oxford University Press.

Brewer, T. (1981). Political risk assessment for foreign direct investment decisions: better methods for better results. *Columbia Journal of World Business*, Spring, pp.5-12.

Crick, B. (1964). *In defence of politics*. London: Penguin.

Dahl, R. (1961). *Who governs?: democracy and power in an American city*. New Haven: Yale University Press.

Drake, R., Prager, A. (1977). Floundering with foreign investment planning. *Columbia Journal of World Business*, 12(2), 66-77.

Dowding, K. (1996). *Power*. Buckingham: Open University Press.

Greene, M. (1965). Export Credit Insurance: its role in expanding world trade', *The Journal of Risk and Insurance*, Vol. 32(2), pp. 177-193.

Hay, C. (2002). *Political analysis: a critical introduction*. New York: Palgrave Macmillan.

Halperin, G. (1940). Mexico shifts her foreign policy. *Foreign Policy*, 19(1), 207-221.

Held, D. (1991). Editor's introduction. In D. Held (Ed.), *Political theory today*. Stanford: Stanford University Press.

Herz, J. (1941). Expropriation of foreign property. *The American Journal of International Law*, 35(2), 243-262.

Herz, J. (1940). Review: the true facts about the expropriation of the oil companies' properties in Mexico by government of Mexico. the facts and the law: Roscoe B. Gaither. The *American Journal of International Law*, 34(4), pp. 769-771.

Heywood, A. (1994). *Political ideas and concepts: an introduction*. New York: St Martin's Press.

Kobrin, S. (1979). Political risk: a review and reconsideration. *Journal of International Business Studies*, 10(1), 67-80.
Knight, F. (1927). *Risk and uncertainty in profit*. Chicago: University of Chicago Press.
Lasswell, H. (1936). *Politics: who gets what, when, how*. Cleveland: Meridian Books.
Leftwich, A. (2004). The political approach to human behaviour: people, resources and power. In A. Leftwich (Ed.), *What is politics?* Cambridge: Polity Press, pp. 100-119.
Liebeskind, J. (1996). Knowledge, strategy and the theory of the corporation. *Strategic Management Journal*, 17, 93-107.
Lukes, S. (2005). *Power: a radical view*. 2nd edn. London: Palgrave McMillan.
Meldrum, D. (2000). Country risk and foreign direct investment. *Business Economics*, 35(1), 33-40.
Morgenthau, H. (1973). *Politics among nations*. 5th edn. New York: Alfred A. Knopf.
North, D. (1995). The new institutional economics and Third World development. In J. Harriss, J. Hunter and C. Lewis (Eds.). *The new institutional economics and Third World development* (17-26). London: Routledge.
O'Callaghan, T. (2016). *Reputation risk and globalisation: exploring the idea of a self-regulating corporation*. London: Edward Elgar. Forthcoming.
Simon, J. (1985). Political risk forecasting. *Futures*, 17(2), 132-148.
Spagnoletti, B. and O'Callaghan, T. (2011). Going undercover: the paradox of political risk insurance. *Journal of Risk and Insurance*, 5(2), 1-23.
Truman, H. (1949). The President's message to the congress on 'Point Four' legislation. *The Annals of the American Academy of Political and Social Science*, 268(184), 184-187.
Tsebelis, G. (1990). *Nested games: rational choice in comparative politics*. Berkeley: University of California Press, Berkeley

Chapter 5
Social Risk and Business Risk in Mining Projects

Geordan Graetz and Daniel M. Franks

Abstract In this chapter, we argue that conceiving of social risk as the risk(s) *to* businesses conflates social risk with the better known concept of business risk and, in so doing, potentially will have negative consequences for both businesses and host communities. Contributing, then, to a contemporary theorisation of social risk, this chapter is organised in five sections. The present section provides the context of the work, while the following section reviews the literature on risk and discretely conceptualises social risk and business risk associated with mining projects. It also addresses the processes through which impacts and risks are identified. The third section examines the relationships between risks, rights, and impacts, as well as translation mechanisms and what we term 'mediators' of risk. The fourth section discusses the consequences of conflating social risk with business risk, while the fifth section concludes the chapter.

Abbreviation

CSR Corporate social responsibility

Introduction

The term, 'social risk', commonly is associated with the health and welfare indicators of individuals and communities at risk of certain illnesses and illicit drug use (Jenkins 1976; Botvin and Botvin 1992), or of needing social protection (Holzmann and Jorgensen 1999; Mckinnon 2004). It also is used by risk scholars (for example,

G. Graetz (✉)
Centre for International Risk, School of Communication, International Studies and Languages, The University of South Australia, Adelaide, SA 5001, Australia
e-mail: geordan.graetz@gmail.com

D.M. Franks
Stones for Development, United Nations Development Programme, Bruxelles, Bruxelles-Capitale, Belgium

T. O'Callaghan, G. Graetz (eds.), *Mining in the Asia-Pacific*, The Political Economy of the Asia Pacific, DOI 10.1007/978-3-319-61395-6_5

Slovic 1987; Macgill and Siu 2004) to distinguish between the physical/empirical risks of certain activities, developments, or technologies identified through scientific consensus or by so-called 'experts', and lay perceptions of risks associated with these activities. This understanding of social risk informs theory of the 'risk society', as well as theories on the social amplification of risk, in which '[i]ndividuals or groups collect and respond to information about risks, and act as "amplification stations" through behavioural responses or communication' with others (Renn et al. 1992).

In contrast, many corporations and other proponents of private and public sector developments, such as mining and infrastructure projects, interpret social risk as the risk(s) *to* their businesses/operations arising from their interactions with, and the actions of, host communities. This interpretation has been adopted in the academic literature by scholars including Joyce and Thomson (2000), Kytle and Ruggie (2005), and Bekefi et al. (2006). In other recent contributions to the literature, social risk is used to refer to the risk of negative impacts of such developments *on* individuals and groups (Lapalme 2003; Schafrik and Kazakidis 2011). In this chapter, we focus on the latter understanding of social risk, arguing that conceiving of social risk as the risk(s) *to* businesses conflates social risk with the better known concept of business risk and, in so doing, potentially will effect negative consequences for both businesses and host communities.

Contributing, then, to a contemporary theorisation of social risk, this chapter is organised in five sections. The present section provides the context of the work, while the following section reviews the literature on risk and discretely conceptualises social risk and business risk associated with mining projects. It also addresses the processes through which impacts and risks are identified. The third section examines the relationships between risks, rights, and impacts, as well as translation mechanisms and what we term 'mediators' of risk. The fourth section discusses the consequences of conflating social risk with business risk, while the fifth section concludes the chapter.

Risk in the Literature

There are numerous definitions of risk (for example, Kaplan and Garrick 1981; Graham and Weiner 1995; Rosa 1998; Sandman 1999; ISO 2002), but arguably the definition that best encapsulates the concept and the inherent consequences of risky activities, developments, or technologies is that proffered by Aven and Renn (2009), albeit with a minor addition: 'Risk refers to uncertainty about and severity of the consequences (or outcomes) of an activity with respect to something that humans (or their institutions) value.' Implicit in this definition is an understanding that risk is context dependent. Risk also should be understood to be the product of the interplay between three dimensions: (1) property; (2) power; and (3) politics, where property is viewed as the tangible and intangible assets and values of individuals, both on their own and collectively, and institutions (Feher in this volume).

The *perception* that a certain activity, development, or technology may result in negative (or positive) outcomes for one's (personal, communal, or corporate) property, values, or interests is central to the acceptability of an activity, development, or technology, and the level of associated risk that an individual, community, or corporation is willing to accept at any point in time (Slovic 1987). For example, individuals, landowners, and/or community groups opposed to mining may object to a proposal to develop a mine in a region due to fears that it will, among other things, drive down property prices, create noise and dust pollution, contaminate water ways, affect existing businesses and land uses, or precipitate unwanted health impacts. Conversely, the directors of a mining company may decide to abandon plans to develop the mine if they are unable to secure community support for the project, if they anticipate political risks with deleterious financial impacts on the business, or in order to avoid damage to the company's reputation due to public protests and social media campaigns, for example.

At the same time, risk entails the element of power, with power understood here as the capacity of an actor or actors to influence the course of certain events or actions and/or the behaviour of other actors. Staying with the example of the proposed mine, the company may be able to exert power over the community by asserting its legal right to develop the mine. It also may secure the support of municipal, state, or national government(s), the legal expertise and legislative, executive, and regulatory authority of which dwarfs the power and agency of landowners and opposition community groups. Moreover, the asymmetrical financial resources of the respective actors may mean that the mining company can develop marketing campaigns to challenge community members' perceptions of risk and promote the mine's social and economic benefits, or offer monetary incentives to the landowner(s) on whose property the mine is to be constructed in order to influence decision-making processes.

This example additionally demonstrates the inherently political nature of risk (Sapolsky 1990). Like risk, politics has various definitions (for example, Lasswell 1936; Oakeshott 1962). However, for our purposes, politics is defined as '[t]he struggles which result from the collisions between human purposes, and the expedients and practices which human beings have invented to cooperate and compete with one another in pursuing their purposes' (Dunn 2000). On the question of the proposed mine, the primary competing purposes are the company's desire to mine the resource in order that it can service its customer contracts and, in turn, make a profit as a fiduciary duty to its shareholders and, conversely, the community's desire to prevent the development and its perceived associated risks from becoming impacts. Municipal, state, and national governments also may have an interest in receiving resource rents in order to fund policy commitments. The expedients and practices referred to in Dunn's definition are the institutions and processes through which claims and counter-claims can be heard and mediated, principally, in this case, the judicial system and the deliberations of the different tiers and branches of government, notably, municipal governments, planning agencies, regulators, and the executive.

The role of perception and past experience in identifying and, subsequently, responding to risk is consequential. For lay members of society, as opposed to discipline experts, to accept the risks of certain activities, developments, or technologies, and to have certitude that the benefits of such activities will outweigh the costs (risks), community members must have trust in the experts, institutions, or organisations that provide them with information and/or the promise of benefits or opportunities (Guillaume and Charron 1999; Yim and Vaganov 2003). Conversely, distrust is a strong motivating factor (Guillaume and Charron 1999), and can lead concerned actors (for example, individuals, communities, and non-government organisations) to commence protests or take other political actions with potential attendant risks for the proponent of the development. The balance of power between the different actors and concerns about the legitimate use of power also can play a role in the formation of stakeholders' perceptions and the level of trust in which they hold experts and organisations.

In his work on the psychology of risk perception, Slovic (1987) asserts: 'people make quantitative judgments about the current and desired riskiness of diverse hazards and the desired level of regulation of each.' These judgements are influenced by an individual's qualitative assessment of (1) the 'characteristics of risk (e.g., voluntariness, dread, knowledge, controllability), (2) the benefits that it provides to society, (3) the number of deaths it causes in an average year, and (4) the number of deaths it can cause in a disastrous accident or disastrous year' (Slovic et al. 1984). Other factors influencing perceptions of risk associated with activities, developments, or technologies include worldviews/ideologies (Yim and Vaganov 2003); ethics and one's culture (Brenot et al. 1998; Guillaume and Charron 1999); the time period over which a risk will last; outrage (Sandman 1999); and the level of one's familiarity with activities, development, or technologies, for example (Parkins and Haluza-DeLay 2011).

The concept of risk is ubiquitous in the life, activity, and organisation of human society. We encounter risks while driving (Titchener and Wong 2010); as a consequence of tobacco use (Sugiishi and Takatsu 1993) or the location of a telecommunications tower (Collins 2010) or mine; due to falling trees (Ball and Watt 2013); and in taking a punt on a horse race (Golec and Tamarkin 1998). Perception, and individual or communal assessment of the risks associated with such activities, developments, or technologies, plays a role in the acceptance, avoidance, prevention, or mitigation of such risks. Thus, many of us continue to drive in the knowledge that we may have an accident, bringing harm to our person or property or to the lives or property of others. To reduce the risks to ourselves and other road users, we drive according to the speed limit, turn our lights on in conditions of low visibility, use our indicators, and take out insurance to compensate us in the event that we have a collision.

The inverse of risk is opportunity and reward, and the example of betting on a horse race demonstrates the potential rewards of *risk-taking* most clearly. In placing a wager on a horse race, the bettor is taking a risk not only that he or she will win and thereby recover the value of their wager, but also that they will make a profit. The factors of risk perception (voluntariness, controllability, benefits, worldview, ethics, culture, time, fear, outrage, familiarity, and past experience, etc.) feature heavily in the responses of individuals and communities to these examples and to the relative merits and 'acceptability' of each.

Cognisant of the omnipresent nature of risk and the role and cogency of perception in the identification and assessment of risk, in this chapter, we focus on the social risks to individuals and groups, and the business risks to companies, that arise in the context of mining projects. Such risks have the potential to derail projects, result in unwanted social impacts for individuals and communities, and occasion reputational damage and a diminution in financial capital for mining companies.

Social Risk

Corporate managers long have known that businesses can sustain risks to their profitability, market value, and reputation as a consequence of endogenous operational decisions and exogenous reactions to these decisions (Graetz and Franks 2013). By reducing these risks, firms can improve their revenue, profits, and market value (Amit and Wernerfelt 1990). However, in the mining industry, corporate leaders only recently have come to design and operate their projects cognisant of the understanding that mining activities also can precipitate risks for individuals and the communities in which they operate. Such risks, we contend, should be labelled *social risks*.

As noted in the introduction, the term, social risk, predominantly has been associated with social protection, social welfare, social amplification, and social perceptions of risk vis-à-vis the assessment and rating of risk by discipline experts. Social risk also has been conceptualised as the risk *to* companies/projects *from* individuals and host communities. This understanding informs leading mining giant, Rio Tinto's, *Social Risk Analysis Guidance Note* and contributions to the academic literature. According to Rio Tinto:

> The term 'social risk' covers a range of threats or opportunities for the business that may result from how the business impacts upon and interacts with communities and stakeholders. Social risks can both directly and indirectly cause lost or enhanced production, affect development timelines and impact capital expenditure, which in turn can affect NPV [net present value]. For example, failing to adequately manage cultural heritage impacts can lead to community dissent, leading to protests and hence production delay or suspension ... Social risks can also have non-economic consequences for Rio Tinto's 'social license to operate' with local communities and impact our relationships with government authorities and other stakeholders. For example, poor relationships with communities can lead to withdrawal of support for establishing or expanding operations. This can make permitting, agreements and other consents more difficult, more costly and potentially impossible to acquire (Rio Tinto 2011).

In the literature, Joyce and Thomson (2000) state: 'The viability of projects becomes threatened because they are considered socially unacceptable, a phenomenon that we describe as social risk.' In another contribution, Kytle and Ruggie (2005) assert:

> From a company perspective, social risk, like any other risk, arises when its own behavior or the action of others in its operating environment creates vulnerabilities. In the case of

social risk, stakeholders may identify those vulnerabilities and apply pressure on the corporation for behavioural changes.

Kytle and Ruggie (2005) additionally propose an equation for formulating social risk from 'the vantage point of a company': social risk = [threat (stakeholder + issue)] × [vulnerability]. Moreover, in a paper for the Harvard University Corporate Social Responsibility Initiative, Bekefi et al. (2006) conceive of social risk as a source of 'strategic' (read business) risk. They state:

> A key strategic risk that companies often miss or misdiagnose is 'social risk.' 'Social risk' is defined as challenges by stakeholders to companies' business practices due to real or perceived business impacts on a broad range of issues related to human welfare—for example, working conditions, environmental quality, health, or economic opportunity. The consequences may include brand and reputation damage, heightened regulatory pressure, legal action, consumer boycotts, and operational stoppages—jeopardizing short and long-term shareholder value (Bekefi et al. 2006).

In the context of resource developments, definitions that focus on social protection, welfare, amplification, and disparities in risk ratings arguably are inadequate when accounting for the threats *to* individuals, who perceive or expect the risk of unwanted social impacts from development interventions, and in designing and implementing systems for their prevention or management. Moreover, definitions that focus on the risks individuals pose to businesses such as those we have just seen are problematic, as they conflate the domain of social risk with the domain of business risk.[1]

Perhaps recognising the limitations of such understandings, other scholars have written of social risk as risk(s) *to* individuals and groups in the context of mining developments. Indeed, both Lapalme (2003) and Schafrik and Kazakidis (2011) conceptualise social risk in this way, but only to a degree. While Lapalme does not offer her own definition of social risk, she identifies social risks to individuals and groups arising from development interventions, and also acknowledges that the actions of individual and group actors can pose risks to businesses. She labels this social risk too (Lapalme 2003). This is confusing for a number of reasons, not least of which is the resultant lack of conceptual clarity between social risk *to* individuals/groups and social risk *from* individuals/groups as a type of business risk. Lapalme's discussion additionally suffers from a confusion of social risk with social impact, the differences between which are discussed below. Similar conceptual confusion is present in the work of Schafrik and Kazakidis (2011).

Not content with the aforementioned definitions and conceptualisations for the reasons stated above, we define social risks as the perceived or expected potential future threats to, and unwanted impacts on, individuals and groups of individuals

---

[1] Here, we draw a distinction between 'domains' of risk and 'types' of risk. The 'domain' of risk refers to the broad rubric of risk under which types of risks sit. The conventional domains are: environmental risk, business risk, social risk, and political risk. It is our contention that environmental risk should be conceived as risk *to* the environment; business risk as risk *to* the business and its operations; social risk as risk *to* individuals/groups of individuals; and political risk as risk *from* the state, principally, *to* corporate entities operating within the political society.

Table 5.1 Types of social risk and examples

Type	Examples
Socio-economic risk	Threat of future loss of income; risk of social inequality; inequitable distribution of goods, services and monies; dependence on resource rents/royalties for subsistence (in the case of mining developments)
Socio-environmental risk	Threat of environmental contamination and lost access to lands; potential future loss of land use; potential land use change
Cultural risk	Threats to cultural heritage; potential desecration of sacred sites; threats to traditional ways of life; loss of language; threat of assimilation; community fragmentation
Risk of human rights transgressions	Threats to property; denial of free association and expression; denial of right to consent and to control development(s)
Health and safety risk	Threats to human health, including stress, cancers, respiratory illness, and substance abuse; risk of injury or death
Livelihood risk	Threat to one's livelihood, including employment, access to goods and services, and housing; threats to traditional ways of life (also a cultural risk)

arising from the processes of social change precipitated by development interventions and the decisions of external actors, namely, businesses, industry organisations, financiers, executive governments, regulators, and non-government organisations. This definition draws a distinction between the recipient(s) of risk (individuals and groups) and the actor(s) responsible for giving rise to risk(s) (businesses, etc.), and draws attention to the related concepts of social change and social impact.

Types of social risk include socio-economic risk, socio-environmental risk, cultural risk, the risk of human rights transgressions, and risk to health and safety. Livelihood risk synthesis and overlays the other five types of risk (Table 5.1). The range and type of social risks that are perceived or expected may vary throughout the life-cycle of a development (Schafrik and Kazakidis 2011).

Social risk may be distinguished from two further concepts: social change and social impact. *Social change* is the evolution over time in, among other things, societal attitudes, beliefs, values and norms, culture, patterns of behaviour, organisation/structure, and demography. Vanclay (2002) observes that there are at least seven processes through which social change occurs (see Table 5.2).

In contrast, *social impact* 'refers to the impacts actually experienced by humans (at individual and higher aggregation levels) in either a corporeal (physical) or cognitive (perceptual) sense' *as a consequence of* the processes of social change (Vanclay 2002). Vanclay notes:

> Direct social impacts result from social change processes that result from a planned intervention. They may be the intention of specially designed activities to influence the social setting (intended impacts), or may unintentionally result from these activities.

Social impacts can be negative (unwanted) or positive (wanted) (Table 5.3), and pertain to various domains of social life (Burdge and Vanclay 1995). To be sure, the distinction between *social change* and *social impact* is important, because

Table 5.2 Social change processes and examples

Process	Examples
Demographic processes	In-migration; out-migration; presence of newcomers; presence of temporary workers; resettlement; displacement and dispossession
Economic processes	Conversion and diversification of economic activities; impoverishment; inflation; concentration of economic activity; globalisation
Geographical processes	Conversion and diversification of land use; urbanisation; gentrification; enhanced transportation and rural accessibility; physical splintering
Institutional and legal processes	Institutional globalisation and centralisation; decentralisation; privatisation
Emancipatory and empowerment processes	Democratisation; marginalisation and exclusion; capacity building
Socio-cultural processes	Social globalisation; segregation; social disintegration; cultural differentiation ('othering')
Other processes	Normalisation; new technologies and social phenomena

Adapted from Vanclay (2002)

Table 5.3 Examples of social impacts arising from private sector developments

Negative	Positive
Desecration of cultural heritage/ knowledge	Acquisition of education and skills leading to employment
Lost access to property	Increased personal/communal wealth/assets; provision of housing
Transgression of human rights	Enjoyment/realisation of rights (individual and collective)
Loss of human and social capital	Social development; human and social capacity building
Diminution in the value of environmental amenities	Preservation/conservation/enhancement of environmental amenities and cultural heritage
Crime	Increased policing, resulting in a reduction in criminal activity
Health impacts, e.g. stress; spread of sexually transmitted diseases; substance abuse; asthma (from dust)	Increased access to health care and facilities

social change is experienced differently by different people and, therefore, impact is a combination of change and the response to, or experience of, that change by different people.

In summary, *social risks* are the perceived or expected potential future threats to, and impacts on, individuals and groups of individuals arising from the processes of social change precipitated by development(s); *social change* refers to the processes through which culture, societal organisation, and norms evolve over time, including as a result of development activity; and *social impact* pertains to individuals' and groups of individuals' past and present experiences of activities, developments, or technologies.

Business Risk

In contrast to social risks, here, we conceive of business risks as the potential threats *to*, and unwanted impacts *on*, a company's operations, reputational capital, market share, and profitability, as a consequence of operational decisions and strategies, and the exogenous responses of other actors to those decisions and strategies. There are clear arguments for reducing these risks, with the principal benefits being higher shareholder value and profits, enhanced reputational capital, and reductions in operational stoppage time, due to lost time injuries, social disruption, or labour unrest (Amit and Wernerfelt 1990).

There are five types of business risk: social, technical, economic/financial, environmental, and political/legal (Table 5.4).

Importantly, social risks *can translate* into business risks. Indeed, a social risk may become a business risk if a business fails (or is perceived to fail) to prevent social risks from becoming experienced negative impacts. As shown in Table 5.4, business risks arising from unwanted social impacts may include the loss of community support, denial of access to the site of a (proposed) mining development, the risk of litigation costs and compensation claims arising from human rights violations, and diminution in reputational capital.

Developing reputational capital is a valuable function of a company's bargaining power. When a company has a good reputation (in terms of participating in good faith negotiations, maintaining best practice in social and environmental impact management, contributing to the local economy and demonstrating respect for stakeholders' rights, etc.), it likely will become a developer of choice 'with preferred access to prospective areas and projects' (Joyce and Thomson 2000). Conversely, companies with low or diminishing reputational capital likely will find that their operations or exploration and project development proposals are—or become—impeded by the actions of executive governments, regulators, non-government organisations, and individual and community-based actors.

Table 5.4 Types of business risk and examples

Types	Examples
Social	Social disruption/conflict (campaigns, demonstrations, strikes, legal action and other public pressure action) affecting the ability to operate; diminution in reputational capital
Technical	Threat of plant/equipment failure; loss arising from design and engineering failure
Economic/financial	Threat of reduced profits; threat of reduced access to capital; threat of damage to property resulting in financial losses; loss of market share; rise in business levies/taxation; threat of fines and/or litigation costs
Environmental	Consequences arising from environmental contamination/destruction, such as the threat of environmental fines and increased rehabilitation costs; threat to required environmental resources
Political/legal	Threat of loss of regulatory licence to operate; sovereign risk; nationalisation; change of legislation/regulation; rise in business levies/taxation; risk of litigation and associated expenses

Fombrun and Van Riel (1997) define the corporate reputation as 'a collective representation of a firm's past actions and results that describes the firm's ability to deliver valued outcomes to multiple stakeholders. It gauges a firm's relative standing … in both its competitive and institutional environments.' Companies compete for reputational status and '[w]ell-reputed firms have a competitive advantage within their industries' (Fombrun and Shanley 1990). Communities' perceptions of a company's social performance can influence perceptions of said company's reputation, which Jackson (2004) labels the 'business integrity thesis'. He says:

> The strongest case for the business integrity thesis is that ethical conduct in the marketplace actually makes good business sense. Regardless of how you stand in terms of these points of debate, one thing is clear: Insofar as the reputational stakes are much higher for today's companies, the market will dictate the final outcomes. Some firms will commit to higher standards and enjoy the reputational rewards, while others will decline and accordingly miss out on the opportunities (Jackson 2004).

Summarising the thesis, O'Callaghan (2007) states:

> [This] suggests that while investors and stakeholders want to see financially profitable enterprises, they must also adhere to values such as fair play, environmental and social responsibility and good conduct. Essentially, people are expecting corporations to be both profitable and ethical.

Thus, O'Callaghan (2007) argues that a good reputation is requisite to the success of a business and that reputational risks 'have the potential to undermine a corporation's ability to function as a commercial enterprise and impair its standing in the community.'

There is a correlation between communities' perceptions of social risks and corporate experience of risks to reputational capital as a function of business risk. Careful negotiation of the intersection of social risk and business risk, and respect for the rights and interests of stakeholders, then, is crucial to the success of mining projects and to the continued enjoyment of a company's reputational status.

In order to reduce their business risk profiles and as a matter of due diligence, companies conduct comprehensive appraisals of the legal and financial risks associated with their operations, either of their own volition or owing to legal obligation. Due diligence traditionally has been associated with corporate mergers and acquisitions (Graetz and Franks 2013; Perry and Herd 2004; Vivoda and Graetz in this volume), but the concept and language of due diligence increasingly is being embedded across the multiple operational facets of a business. For example, companies now may exercise due diligence with regard to their environmental liabilities (Freedman and Stagliano 2002), human rights performance (HRC 2011; Graetz and Franks 2013), organisational culture and human resources management (Horwitz et al. 2002), and social impacts (Esteves et al. 2012).

Under this contemporary understanding of due diligence, risk assessments are performed, comprising the identification of business risks (social, technical, economic/financial, environmental, and political/legal) to, and arising from, a company's operations or activities and, subsequently, the ascription of a quantitative and/or qualitative value of risk to a threat based on its severity and probability. Environmental, social, and human rights impact assessments are additional due diligence tools, which are used by mining companies to identify and understand the impacts of their operations on stakeholders and the ways in which risks and impacts are translated from one actor to another (Wathern 2004; Esteves et al. 2012; Graetz and Franks 2013). However, while impact assessment mostly is forward looking (ex-ante), hitherto, the tool insufficiently has incorporated the concept of risk in the social sense (Mahmoudi et al. 2013). Notwithstanding this point, risk and impact assessments aid decision-making (Koller 2005), with well-executed risk/impact assessments providing corporate managers/boards with a range of options about the direction of a particular aspect of the business or its operations from which to choose (Klassen and McLaughlin 1996).

As noted above, the inverse of risk is opportunity, and the inverse of unwanted impact is benefit. By performing risk and impact assessments, both business risks and social risks may be identified and, subsequently, risk management systems may be established to prevent risks from becoming unwanted impacts. This can have positive ramifications for the financial and reputational capital of a mining company, and has obvious benefits for the affected community and, potentially, the physical environment that is to host the development, potentially resulting in firms acquiring a reputation for excellence in corporate social performance. Higher returns for shareholders, a more responsive company to community priorities and concerns, the acquisition of social support, avoidance of litigation, a diminution in health and safety incidents, and biodiversity conservation are just some of the potential opportunities/benefits for stakeholders and the environment arising from the formalisation of risk management strategies in the mining industry.

In this view, business risk management is a valuable function of corporate social responsibility (CSR) (Lapalme 2003). Whereas, previously, the responsibility of the corporation was seen as the maximisation of profits for the benefit of shareholders in order to increase individual and societal wealth (Friedman 1970), according to Porter and Kramer (2006), 'CSR has emerged as an inescapable priority for business leaders.' The corporation now is seen as a participant in the society in which it operates and, on that account, is expected to abide by its rules and to 'engage in and contribute to society as a corporate citizen' (Warhurst 2001). Warhurst notes that:

> A strategy of [CSR] is defined here as the internalisation by the company of the social and environmental effects of its operations through proactive pollution prevention and social impact assessment so that harm is anticipated and avoided and benefits are optimised … The concept is about companies seizing opportunities and targeting capabilities that they have built up for competitive advantage to contribute to sustainable development goals in ways that go beyond traditional responsibilities to shareholders, employees and the law, and that internalise indirect socio-economic and bio-geophysical effects as well as direct impacts.

In so doing, the corporation may prevent social risks from becoming experienced negative impacts, in addition to reducing its business risk profile. However, many businesses' understanding of the benefits of investment in community relations capabilities and CSR programs still is lacking. Porter and Kramer (2006) argue:

> The fact is, the prevailing approaches to CSR are so fragmented and so disconnected from business and strategy as to obscure many of the greatest opportunities for companies to

> benefit society. If, instead, corporations were to analyze their prospects for social responsibility using the same frameworks [due diligence risk/impact assessment] that guide their core business choices, they would discover that CSR can be much more than a cost, a constraint, or a charitable deed – it can be a source of opportunity, innovation, and competitive advantage.

Porter and Kramer thus propose that companies engage in 'strategic CSR' to create a 'corporate social agenda', the aim of which is to identify new opportunities for innovation and collaboration. Such an agenda 'moves from mitigating harm to finding ways to reinforce corporate strategy by advancing social conditions' in tandem with community groups and other actors.

Theorising the Relationship Between Social Risk and Business Risk Associated with Mining Projects

Risk is a constant feature of the relationships between actors associated with mining projects; so too is the exercise and expression of rights (vis-à-vis other actors), and the experience of impacts. Focusing on the translation of risk from mining companies to individuals and groups and vice versa, in this section, we further theorise the relationship between social risk and business risk by examining the connections between rights, risks, and impacts, and the concepts of property, power, and politics. We also briefly discuss the role of mediators in the translation of risks.

Owing to the evolving expectations about the social responsibilities of the corporation, demonstration of respect for the rights of stakeholders has become an important obligation of mining companies and other businesses (HRC 2011). There are three generations of rights: individual civil and political rights; individual economic, social, and cultural rights; and group rights (Graetz and Franks 2013). While the philosophical foundations of human rights may be contested (Mendus 1995), individuals and groups have an expectation that their rights will be recognised, respected, and protected, and that there will be avenues for remedy if their rights are violated (HRC 2011).

An individual's or group's perception or experience of a rights violation at the hands of actors associated with the promotion or operation of a mining project represents a key social risk. Inversely, the expression of a grievance regarding a rights violation, whether that expression is informal (through traditional or social media, raising grievances with human rights advocacy organisations, protest, or established communication channels between the transgressor and the receiver) or formal (by taking legal action or petitioning political institutions), can translate into risks to/for the actor responsible for the violation.

A violation of a right constitutes a rights impact. Examples include the imposition of a mining project without the consent of affected peoples, resulting in the loss of property, forced resettlement, the failure to provide adequate compensation for lost access to property, and the reduction in the security of a person or persons due to adverse health, safety, socio-economic, or socio-environmental impacts.

Rights violations constitute a fundamental attack on an individual's or group's identity and place in the world. When that right is a group right—for example, the right of Indigenous Peoples collectively to exercise free, prior and informed consent regarding decisions about developments on traditional lands or to derive benefits in return for the provision of access—the violation of that right can have impacts on the group's relationships with external actors, the perception and experience of the development, the internal cohesion of the community, the exercise of the individual cultural rights of group members, and may have attendant risks for the responsible actor.

We saw above that property, power, and politics are fundamental elements of risk. They also are fundamental to, and the basis for the philosophical foundations of, the recognition and exercise of an individual's or group's rights. Mining projects frequently come into conflict with the property rights of individuals and the land rights of groups (in the case of Indigenous Peoples). Moreover, they are a channel for the exertion of power of one actor over another, and are situated within political, societal and institutional settings in which there are laws and regulations governing the activities of corporations.

Perceptions of social risks and fears about expected future impacts can drive individuals and groups to take political action(s) to prevent mining projects from proceeding. A common action is conflict (protest, boycott, or blockade), which also serves to translate risk from one actor (individuals and groups) to another (mining companies). For companies, the costs of conflict with host communities can be substantial, as shown in Franks et al. (2014). Particularly problematic are the costs of lost productivity—in terms of the value of production and time devoted by personnel to responding to conflict—due to delay, which, for the company directly and indirectly for other actors (for example, loss of taxation revenue for governments), can be acute (Franks et al. 2014). Another potential significant cost for businesses is the opportunity cost resulting from the denial of regulatory and social licences granting proponents the right to pursue projects (Franks et al. 2014).

At the same time, there also are risk 'mediators', which serve to regulate the translation of risks and impacts from one actor to another. Mediators include the human relationships between actors, land tenure, and the threat of legal punishment and social disapproval. If the relationship between the development proponent (mining company) and/or their agent—for example, a corporate community relations officer—and an individual or community actor is positive and characterised by trust and mutual respect, this relationship can serve to mediate (prevent) the translation of risk from the development proponent to the host community and vice versa. Such a relationship can allow for the early communication of grievances, requests, and concerns, etc., and/or the subsequent mitigation of experienced impacts in a way that is sensitive to the wishes of the host community. Such a relationship also can facilitate the communication of project proposals to a more open receiving audience and tailored message delivery based on the host community's needs and concerns; as well as assist in the identification of potential joint initiatives and opportunities for the creation of shared value (Porter and Kramer 2006). Conversely, if the relationship between the development proponent and its stakeholders is characterised by distrust, communication channels may not be established or open, and risks to actors may present or be exacerbated. Good communication therefore is crucial to the mediation of risk translation, with Sandman (1999) arguing that communication is not about 'puff pieces' promoting the worthiness of the development or the proponent, but rather about 'responding to worries and grievances'.

With regard to land tenure, strong legislated individual property rights and doctrines recognising the sovereignty of Indigenous Peoples over their traditional lands can serve to guard against unwanted mining projects and to mediate the translation of social risk(s) from the mining company to the host community. Whether or not the affected First Nation is able to exercise their right to free, prior and informed consent over land use and access has a bearing on the success of a project. However, in Australia, for example, freehold title, customary land rights, and mineral leases can coexist, and there may be few avenues for landowners to prevent unwanted mining developments on their lands. While landowners may be entitled to receive compensation and/or may enter into negotiations about benefit agreements, the monetary value of compensation may not reflect the true costs (risks and impacts) of the development, for example, future land use, lost access to land for agricultural or cultural purposes, reduced earnings, diminution in soil condition, legal fees, loss of visual amenity, dislocation, and stress and other health impacts.

It also is noteworthy that the threat of legal punishment and social disapproval can serve to dissuade mining companies from progressing project plans (Grasmick and Green 1980). Legal punishment and social disapproval can result in the diminution of reputational and financial capital. For this reason, companies may withdraw or amend a project proposal in order to maintain respectful relationships with other actors, avoid litigation and associated costs, and retain their status as a developer of choice.

Problems with Conflation

The central argument of this chapter is that, in the context of mining projects, there are costs to companies and host communities arising from the conflation of social risk with business risk. Yet, despite there being clear theoretical distinctions between the two domains, some companies continue to interpret social risk as the risk that individual and group actors pose to their business. As noted above, Rio Tinto is one firm that focuses on the business risk consequences of the company's potential negative social impacts in its guidance note on social risk; and, as we saw, this interpretation has a basis in the literature.

The problem with conceptualising social risk in such terms is that it does not allow for the acknowledgement that social risk is a domain of risk in its own right. Nor does it allow for the recognition of the unique types of social risk, potentially leading to ignorance of the threats—and, for that matter, opportunities—that mining projects pose to/present host communities. The conflation of social risk with business risk stems from the failure to distinguish between the risks to individual and group actors and the risks that such actors can precipitate for companies/projects. This distinction is important. Conflation can precipitate problems when it comes to the discrete identification of the different domains of risk and, subsequently, in the prevention and management of experienced impacts. More generally, failure to treat social risk and business risk discretely can have implications for a company's engagement with its host community, as well as its operations and reputational and financial capital.

Forecasting the connections (opportunities for translation) between social risk(s) and business risk(s) is difficult, though risk and impact assessments may assist with prediction. The problem with the existing suite of impact assessment and management tools, though, is that they are divorced from social risk analysis. Companies routinely perform risk analyses, but such analyses predominantly assess and identify potential risks to their business or operations. Moreover, companies may perform social impact assessments, but risk as a concept does not feature prominently in these assessments. Consequently, companies need to get better at incorporating the consideration of social risk concerns into their social and human rights impact assessment processes, as well as corporate assessments of risks that their operations are subject to or pose to other actors. Strengthened risk and impact assessment tools would allow for a better understanding of the links between social risk and business risk, as well as the circumstances in which one domain of risk is translated into another.

To be sure, not all social risks will become business risks. However, translation most likely will occur when the host community perceives threats to their property, values, or interests; where there are real or perceived asymmetrical power relations between the host community and the project proponent; and when the political institutions and other mediators of risk are inadequate to regulate both the translation of risks and impacts and the life-cycle of the project. There also may be political circumstances where social actors choose not to mobilise against projects, essentially preventing the translation of social risk into business risk. For example, under authoritarian political regimes in which social disruption of projects is not tolerated by a government or other agents of the state, threats of retribution and personal or communal harm may lead to situations of self-censorship and physical opportunities for social actors to translate social risks into business risks may be restricted, potentially exacerbating existing, or precipitating new, rights violations.

Notwithstanding our comments regarding the deficiencies of existing risk and impact assessment tools and processes, given that identification of risks and prediction regarding risk translation is difficult, companies would be best served by performing comprehensive due diligence (risk) assessments in order to aid prediction and provide guidance on where risk management efforts should be directed (HRC 2011). That being said, social risks arising from human rights violations arguably can be said with a higher degree of certainty to result in unforeseen project risks. Accordingly, priority should be given to identifying and preventing future threats to, and impacts on, stakeholders' rights.

Conclusion

This chapter adds necessary clarity to the literature on social risk and business risk in the context of mining projects. Hitherto, much of the literature on social risk has focused on social protection, social welfare, and the social perception of risk vis-à-vis the assessment of risk by discipline experts. However, in this chapter, we argue that social risks should be understood as the perceived potential future threats to, and unwanted impacts on, individuals and groups of individuals arising from the processes of social change precipitated by mining projects. Challenging the prevailing understanding among corporate leaders that social risk is a function of business risk, we draw a distinction between social risk and business risk, the latter of which we define as the potential threats to, and unwanted impacts on, a company's operations, reputational capital, market share, and profitability as a consequence of operational decisions and strategies, and the responses of other actors to these decisions and strategies.

In conflating social risk with business risk, we suggest that mining companies might perpetuate risks and costs for individuals, groups, and their own operations, as risks go undiagnosed and, thus, unabated. Our central contention, then, is that corporate leaders and the academic literature would be wise to treat social risk and business risk as discrete domains of risk in order to avoid conflation and, accordingly, to establish appropriate due diligence systems to assess and manage risks, thereby preventing risks from becoming impacts. Companies that lag in this area likely risk diminution in reputational capital and, consequently, market share and profitability.

Acknowledgements This chapter is an edited version of a paper published in the *Journal of Risk Research*. The original paper may be cited as: Geordan Graetz and Daniel M. Franks (2016): Conceptualising social risk and business risk associated with private sector development projects, *Journal of Risk Research*, 19(5).

References

Amit, R., Wernerfelt, B. (1990). Why do firms reduce business risk? *The Academy of Management Journal,* 33(3), 520-533.

Aven, T., Renn, O. (2009). On risk defined as an event where the outcome is uncertain. *Journal of Risk Research,* 12(1), 1-11.

Ball, D., Watt, J. (2013). The risk to the public of tree fall. *Journal of Risk Research,* 16(2), 261-269.

Bekefi, T., Jenkins, B., Kytle, B. (2006). Social risk as strategic risk. Corporate Social Responsibility Initiative, Working Paper No. 30. Cambridge, MA: John F. Kennedy School of Government, Harvard University. https://www.hks.harvard.edu/m-rcbg/CSRI/publications/workingpaper_30_bekefietal.pdf. Accessed 4 March 2013.

Botvin, G., Botvin, E. (1992). Adolescent tobacco, alcohol, and drug abuse: prevention strategies: empirical findings, and assessment issues. *Journal of Developmental and Behavioral Pediatrics,* 13(4), 290-301.

Brenot, J., Bonnefous, S. and Marris, C. (1998). Testing the culture theory in France. *Risk Analysis,* 18(6), 729-739.

Burdge, R., Vanclay, F. (1995). Social impact assessment. In F. Vanclay and D. Bronstein (Eds), Environmental and social impact assessment (pp. 31-66). Chichester: Wiley.

Collins, J. (2010). Mobile phone masts, social rationalities and risk: negotiating lay perspectives on technological hazards. *Journal of Risk Research,* 13(5), 621-637.

Dunn, J. (2000). *The cunning of unreason: making sense of politics.* New York: Basic Books.

Esteves A., Franks, D., Vanclay, F. (2012). Social impact assessment: the state of the art. *Impact assessment and project appraisal,* 30(1), 34-42.

Fombrun, C., van Riel, C. (1997). The reputational landscape. *Corporate Reputation Review,* 1(1), 5-13.

Fombrun, C., Shanley, M. (1990). What's in a name? Reputation building and corporate strategy. *Academy of Management Journal*, 33(2), 233-258.

Franks, D., Davis, R., Bebbington, A., Ali, S., Kemp, D., Scurrah, M. (2014). Conflict translates environmental and social risk into business costs. *PNAS*, http://www.pnas.org/content/early/2014/05/08/1405135111.abstract. Consulted 29 July 2014: 7576-7581. Accessed 7 March 2014.

Freedman, M., Stagliano, A. (2002). Environmental disclosure by companies involved in initial public offerings. *Accounting, Auditing & Accountability Journal*, 15(1), 94-105.

Friedman, M. (1970, September 13). The social responsibility of business is to increase its profits. *The New York Times Magazine.*

Golec, J., Tamarkin, M. (1998). Bettors love Skewness, not risk, at the horse track. *Journal of Political Economy*, 106(1), 205-225.

Graetz, G., Franks, D. (2013). Incorporating human rights into the corporate domain: due diligence, impact assessment and integrated risk management. *Impact Assessment and Project Appraisal*, 31(2), 97-106.

Graham, J., Weiner, J.B. (1995). *Risk versus risk: tradeoffs in protecting health and the environment.* Cambridge, Massachusetts: Harvard University Press.

Grasmick, H., Green, D. (1980). Legal punishment, social disapproval and internalization as inhibitors of illegal behavior. *The Journal of Criminal Law and Criminology,* 71(3), 325-335.

Guillaume, B., Charron, S. (1999). Exploring implicit dimensions underlying risk perceptions of waste from mining and milling of uranium ores in France. Fontenay-aux-Roses: Institute for Protection and Nuclear Safety. http://www.irsn.fr/FR/Larecherche/publications-documentation/Publications_documentation/BDD_publi/DSDRE/Documents/IRPA10-P10-176.pdf. Accessed 25 May 2015.

Holzmann, R., Jorgensen, S. (1999). Social protection as social risk management: conceptual underpinnings for social protection strategy paper. SP Discussion Paper 9904. Washington, DC: The World Bank. http://documents.worldbank.org/curated/en/1999/01/437984/social-protection-social-risk-management-conceptual-underpinnings-social-protection-sector-strategy-paper. Accessed 5 July 2015.

Horwitz, F., Anderssen, K., Bezuidenhout, A., et al. (2002). Due diligence neglected: managing human resources and organizational culture in mergers and acquisitions. *South African Journal of Business Management*, 33(1), 1-10.

Human Rights Council. (2011). Guiding principles on business and human rights: implementing the United Nations' protect, respect and remedy. *Framework. Report of the Special Representative of the Secretary-General on the issue of human rights and transnational corporations and other business enterprises, John Ruggie.* A/HRC/17/31. Guiding principles on business and human rights: implementing the United Nations' protect, respect and remedy. Accessed 24 December 2015.

International Organization for Standardization. (2002). Risk management vocabulary. ISO Guide 73: 2009 - Risk Management Vocabulary. Geneva: International Organization for Standardization. http://www.iso.org/iso/catalogue_detail?csnumber=44651. Accessed 24 December 2015.

Jackson, K. (2004). *Building reputational capital: strategies for integrity and fair play that improve the bottom line.* New York: Oxford University Press.

Jenkins, C. (1976). Recent evidence supporting psychologic and social risk factors for coronary disease: Part II. *The New England Journal of Medicine*, 294(19), 1033-1038.

Joyce, S., Thomson, I. (2000). Earning a social licence to operate: social acceptability and resources development in Latin America. *The Canadian Mining and Metallurgical Bulletin* 93(1037). http://oncommonground.ca/wp-content/downloads/license.htm. Accessed 20 May 2014.

Kaplan, S., Garrick, B.J. (1981). On the quantitative definition of risk. *Risk Analysis*, 1, 11-27.

Klassen, R., McLaughlin, C. (1996). The impact of environmental management on firm performance. *Management Science*, 42(8), 1199-1214.

Koller, G. (2005). *Risk assessment and decision making in business and industry: a practical guide*. 2nd Ed. Boca Raton, Florida: CRC Press.

Kytle, B., Ruggie, JG. (2005). Corporate social responsibility as risk management: a model for multinationals. Working Paper No. 10, Corporate Social Responsibility Initiative. Cambridge, Massachusetts: John F. Kennedy School of Government, Harvard University. https://www.hks.harvard.edu/m-rcbg/CSRI/publications/workingpaper_10_kytle_ruggie.pdf. Accessed 24 September 2015.

Lapalme, L. (2003). The social dimension of sustainable development and the mining industry: a background paper. Ottawa: Minister of Public Works and Government Services, Natural Resources Canada, Minerals and Metals Sector.

Lasswell, H. (1936). *Politics: who gets what, when, how*. New York: McGraw-Hill Book Company.

Macgill, S., Siu, Y. (2004). The nature of risk. *Journal of Risk Research*, 7(3), 315-352.

Mahmoudi, H., Renn, O., Vanclay, F., et al. (2013). A framework for combining social impact assessment and risk assessment. *Environmental Impact Assessment Review*, 43, 1-8.

Mckinnon, R. (2004). Social risk management and the World Bank: resetting the 'standards' for social security? *Journal of Risk Research*, 7(3), 297-314.

Mendus S. (1995). Human rights in political theory. *Political Studies*, 43(5), 10-24.

O'Callaghan, T. (2007). Disciplining multinational enterprises: the regulatory power of reputation risk. *Global Society*, 21(1), 95-117.

Oakeshott, M. (1962). *Rationalism in politics and other essays*. New York: Basic Books.

Parkins, J., Haluza-DeLay, R. (2011). Social and ethical considerations of nuclear power development. Staff Paper No. 11-01. Edmonton: Department of Rural Economy, Faculty of Agricultural, Life and Environmental Sciences, University of Alberta. http://ageconsearch.umn.edu/bitstream/103237/2/StaffPaper11-01.pdf. Accessed 31 July 2015.

Perry, J., Herd, T. (2004). Reducing M&A risk through improved due diligence. *Strategy and Leadership*, 32(2), 12-19.

Porter, M., Kramer, M. (2006). Strategy and society: the link between competitive advantage and corporate social responsibility. *Harvard Business Review,* 78-92.

Renn, O., Burns, W., Kasperson, J., et al. (1992). The social amplification of risk: theoretical foundations and empirical applications. *Journal of Social Issues*, 48(4), 137-160.

Rio Tinto. (2011). Social risk analysis guidance note. Rio Tinto.

Rosa, E. (1998). Meta-theoretical foundations for post-normal risk. *Journal of Risk Research*, 1(1), 15-44.

Sandman, P. (1999). Risk = hazard + outrage: coping with controversy about utility risks. *Engineering News-Record*, 4, A19-A23. http://www.psandman.com/articles/amsa.htm. Accessed 31 July 2015.

Sapolsky, H. (1990). The Politics of Risk. *Daedalus*, 119(4), 83-96.

Schafrik, S., Kazakidis, K. (2011). Due diligence in mine feasibility studies for the assessment of social risk. *International Journal of Mining, Reclamation and Environment*, 25(1), 86-101.

Slovic, P. (1987). Perception of risk. *Science*, 236(4799), 280-285.

Slovic, P., Fischhoff, B., Lichtenstein, S. (1984). Behavioral decision theory on risk and safety. *Acta Psychologia*, 56(1-3), 183-203.

Sugiishi, M., Takatsu, F. (1993). Cigarette smoking is a major risk factor for coronary spasm. *Circulation*, 87, 76-79.

Titchener, K., Wong, I. (2010). Driver distractions: characteristics underlying drivers' risk perceptions. *Journal of Risk Research*, 13(6), 771-780.

Vanclay, F. (2002). Conceptualising social impacts. *Environmental Impact Assessment Review,* 22(3), 183-211.

Warhurst, A. (2001). Corporate citizenship and corporate social investment: drivers of tri-sector partnerships. *Journal of Corporate Citizenship*, 1(1), 57-73.

Wathern, P. (2004). An introductory guide to EIA. In P. Wathern (Ed.), *Environmental impact assessment: theory and practice*. New York: Routledge.

Yim, M., Vaganov, P. (2003). Effects of education on nuclear risk perception and attitude: theory. *Progress in Nuclear Energy*, 42(2), 221-235.

Chapter 6
Security of Resource Supply and the Obsolescing Bargain

Tom Johnson

Abstract Policy responses of governments across the Asia-Pacific Region to the security of supply problem amount to a new form of resource nationalism. Resource nationalist policies transfer resource rents, assets, skills, and technology to the government at the expense of mining companies. In the coming years, governments will be under increasing pressure to introduce policies that nationalise mineral and energy supplies. This will occur as resource (material) intensity increases and reserves decline. My argument is that Vernon's obsolescing bargain model needs to be modified to take into account the security of supply issues facing host governments around the Asia-Pacific Region. This is a novel interpretation of the way resource nationalism is understood. Consequently, I modify the Vernon's model to reflect recent policy responses to the potential shortfall in resource supply in the Region. I begin by considering how these policies influence the bargaining power of companies and host governments in mining projects. Vernon's model is a useful tool for predicting and explaining future resource nationalist policies brought about by the growing gap between mineral and energy production and consumption in the Region. In response to this problem, I suggest that industry clusters may reduce the likelihood of resource nationalism affecting mining projects. This is something that has not, as yet, been trialled as a possible solution to resource nationalism in the region.

Abbreviations

ADB	Asian Development Bank
CE	Creeping expropriation
CNOOC	Chinese National Overseas Oil Company
CNPC	Chinese National Petroleum Corporation
DE	Direct expropriation
DMC	Domestic material consumption
EIA	Energy Information Administration

T. Johnson (✉)
Centre for International Risk, School of Communication, International Studies and Languages, The University of South Australia, Adelaide, Australia
e-mail: Tom.Johnson@sca.com.au

T. O'Callaghan, G. Graetz (eds.), *Mining in the Asia-Pacific*, The Political Economy of the Asia Pacific, DOI 10.1007/978-3-319-61395-6_6

FDI	Foreign direct investment
GDP	Gross domestic product
IEA	International Energy Agency
IOC	International Oil Company
LNG	Liquefied natural gas
MRRT	Minerals resource rent tax
NOC	National Oil Company
OBM	Obsolescing bargain model
OPEC	Organization of Petroleum Exporting Countries
RN	Resource nationalism
SOE	State-owned enterprise
SWF	Sovereign wealth fund
UNEP	United Nations Environment Programme

Introduction

A reliable supply of mineral and energy resources increasingly is pivotal to achieving economic growth and security in the Asia-Pacific Region. The Region's appetite for mineral and energy resources is growing as affluence increases and economies seek to develop more resource-intensive industries. However, growing dependence on these resources forces governments to prepare for a potential shortfall in supply. Governments can secure resources by introducing policies to shift access and control of domestic and foreign mineral and energy supplies from mining and energy companies to host governments. A range of policies recently introduced across the Region has achieved this very goal. These policies indicate a tendency toward resource nationalism (RN). Since 2011, resource nationalist policies have been introduced in Australia, China, Indonesia, Mongolia, Papua New Guinea, Sri Lanka, and Vietnam. In other parts of the world, the degree of RN appears to be growing as well.

Across the Region, Australia, China, and Indonesia are the largest suppliers of mineral and energy resources. Despite the size of the Region's mining industry, it consumes more mineral and energy resources than it is able to produce and, therefore, faces a security of supply problem. Dorian, Franssen, and Simbeck (2006) define security of supply as a 'reliable supply of energy resources at reasonable prices to support the domestic economy and industry.' The chapter extends this definition beyond energy resources, to include the minerals necessary to achieve ongoing economic growth and to bolster national security.

In this chapter, I argue that the policy responses of governments across the Region to the security of supply problem amounts to a new form of RN. As Wilson (2011) explains, RN is a mercantilist approach to resource management because mineral and energy resources are non-renewable.[1] RN policies transfer resource

[1] On mercantilism see Griffiths, O'Callaghan, and Roach (2014).

rents, assets, skills, and technology to the government at the expense of mining companies. In the coming years, governments will be under increasing pressure to introduce policies that nationalise mineral and energy supplies. This will occur as resource (material) intensity increases and reserves decline.

Material intensity is a measure of how many resources are used to produce the goods and services that make up a country's gross domestic product (GDP). My argument is that Vernon's (1971) obsolescing bargain model (OBM) needs to be modified to take into account the security of supply issues facing host governments around the world, particularly, in the Asia-Pacific Region. This is a novel interpretation of the way RN is understood. Vernon introduced the OBM to explain why bargaining power gradually shifted from mining companies to host governments over the lifespan of a resource project. I define bargaining power as the dependence of one party on the resources of another party in an economic and political relationship. From the commencement of a project, a host government gradually takes more wealth and control from a mining company, until the project is finally expropriated. By this stage, the original bargain has completely obsolesced.

In this chapter, I modify the OBM to reflect recent policy responses to the potential shortfall in resource supply in the Region. I begin by considering how these policies influence the bargaining power of resource companies and host governments in mining projects. The contemporary OBM is a useful tool for predicting and explaining future RN policies brought about by the growing gap between mineral and energy production and consumption in the Region.

Following this, I apply the concept of industry clusters to the Region's mining industry. I suggest industry clusters may reduce the likelihood of RN affecting mining projects, because they can increase the bargaining power of mining companies in relation to host governments. Clusters can provide a broader range of benefits to their members, such that government expropriation will have a negative effect on a country's economic policy by reducing the revenue streams from a project. The cluster concept can help to balance power between companies and host governments. This concept has not yet been applied to the mining industry in the Asia-Pacific Region.

Porter (2000) introduced the concept of industry clusters in the 1990s. At the time, he defined industry clusters as 'geographic concentrations of interconnected companies, specialised suppliers, service providers, firms in related industries, and associated institutions in a particular field that compete but also co-operate.' The companies in a cluster mutually support one another by sharing knowledge, technology, and skills. Mining industry clusters in the Region increase company bargaining power by facilitating the transfer of relevant goods and services to host governments, such that expropriation becomes too expensive.

The chapter is organised as follows. Section "RN and Its Consequences in International Political Economy" details the various forms of RN and explains why RN is a pressing challenge for the mining industry. Section "Security of Supply and the Obsolescing Bargain Model in the Region" examines the size and scope of the mining industry and the recent wave of RN in the Region. Section "The Contemporary Obsolescing Bargain Model in the Asia-Pacific Region" discusses how policy responses to security of supply issues influence the OBM. Finally, section "Strengthening the Bargaining Power of Mining Companies Using Industry Clusters in the Region" applies the concept of industry clusters to the mining industry, arguing that clusters present a way to restore the bargaining power of mining companies and lessen the impact of RN on their businesses. I contend that mining industry clusters may benefit all stakeholders, including infrastructure providers, local communities, host governments, companies interested in the development of new technologies, and relevant non-government organisations.

RN and Its Consequences in International Political Economy

RN shifts access, control, and ownership of domestic or foreign resource supplies to a government. In many cases, foreign mining companies may lose ownership of their projects, with or without compensation, if RN policies affect their projects (UNCTAD 2012). RN results from 'regulatory taking', expropriation, or nationalisation policies. Moran and West (2008) define regulatory taking as policies that change a regulatory framework to take away property rights from a private owner, without taking possession of the property itself. Kantor (2008) explains that regulatory taking may not amount to expropriation straight away, but over time deprives a proprietor from their ownership and control of a company. The four most common forms of regulatory taking in the mining industry are:

- mandated beneficiation, which forces mining companies to process raw materials within a host country
- gradually increasing government ownership of existing projects or introducing minimum government ownership laws for new projects
- restricting raw material exports
- windfall taxes and increased royalty rates, most often in periods of high commodity prices.

Less common forms of regulatory taking include minimum local labour requirements, price ceilings on raw materials, foreign ownership limitations, and restrictions on imported components in the mining company's supply chain.

Unlike regulatory taking, which only transfers partial control and ownership of a mining project to the government, expropriation transfers a mining project in its entirety from the mining company to the government. Lax (1983) argues that expropriation is 'the seizure or modification of the property rights of a group of companies as an exercise of the sovereign power of the state' (also Feher in this volume).

There are two types of expropriation: direct expropriation (DE) and creeping expropriation (CE). DE occurs when a host government seizes a mining project outright, usually by legislative decree or military force. Bunn and Mustafaoglu (1978) describe DE as 'sudden', as it can happen within a short timeframe. A good example of DE is the government of Venezuela's seizure of assets belonging Exxon Mobil and ConocoPhillips in 2007 (Arsenault 2012; Elliott 2007). CE occurs when a mining company walks away from its project once regulatory taking measures make the operation financially unviable. According to Jodice (1985), regulatory taking affecting the profitability, ownership, and subsequent behaviour of a company can lead to CE. The government of Ecuador's expropriation of the Fruta del Norte gold mining project in 2013 is an example. This occurred when Canadian company Kinross Gold walked away from the project, which it had purchased for $1.2 billion, after the government of Ecuador imposed a 70% windfall profits tax on mining projects (Kinross Gold Corporation 2013).

The United Nations Conference on Trade and Development (UNCTAD 2012) argues that nationalisation is the large-scale takeover of all private property in an entire industry. In other words, nationalisation is a collection of expropriations. As Wortley (1959) points out, 'nationalization differs in its scope and extent rather than its judicial nature from other types of expropriations.' The Bolivian military's occupation of domestic oil and gas fields in 2006 is an example of this. The military forced mining companies to hand over majority ownership of their projects to Yacimientos Petrolíferos Fiscales Bolivianos, Bolivia's state-owned oil and gas company (Prada 2006).

There are three types of RN: 'producer country' RN, 'consumer country' RN, and 'investment target country' RN. Ward (2009) defines producer country RN as host governments increasing their ownership and control of their domestic mining industry. There are economic and political incentives for a government to nationalise domestic resources. Friedman's (2006) first law of petro-politics argues that natural resource wealth and ownership leads to more authoritarian regimes and less political accountability. Similarly, Smith (2004) finds a correlation between oil wealth and the longevity of a regime in developing countries, which suggests resource wealth consolidates political power. Arezki and Gylfason (2013) argue that natural resource rents are associated with higher levels of corruption, while Luciani (2011) suggests governments nationalise mineral and energy resources to manipulate the rate of extraction. This occurs when governments conclude that mining companies either are overexploiting or under-exploiting the country's resources. The Government of Western Australia's introduction of policies requiring liquefied natural gas (LNG) producers to set aside 15% of LNG production for future domestic use is one example of this practice (Government of Western Australia 2011). Another is China's intention to tax coal production on a price basis, rather than by volume. This will precipitate a fivefold increase in tax paid by coal producers in China.

Ward (2009) defines consumer country RN as a government gaining exclusive access to foreign resource supplies. This is achieved when a country's foreign policy enables the government to dictate the trade of mineral and energy resources extracted in another country. One way of achieving this is through diplomacy, such as China's negotiation of oil and gas agreements with Iran totalling more than $120 billion since 2004. In return, China pledges to favour Iran in United Nations Security Council resolutions, including those intended to curtail Iran's nuclear proliferation policies (Simpson 2010). The decline of domestic mineral and energy reserves in the Region may lead to an increase in consumer country RN as governments look internationally for supplies. Collins and Erickson (2011) argue that consumer country RN already is occurring in the Region, as governments are using their naval presence to deter any potential interference with resource imports. This form of RN also can lead to diplomatic and security issues. China's implementation of new deep-water fields in disputed areas of the South China Sea and East China Sea has heightened tensions with Japan and other South East Asian countries.

Investment target country RN is achieved by using a sovereign wealth fund (SWF) to nationalise resource supplies. A SWF is an account that reinvests government revenue, often from resource rents, into programs that aim to achieve economic, political, and social objectives (Davis, Ossowski, and Fedelino 2003). The Sovereign Wealth Fund Institute (2014) calculates the combined asset value of global SWFs at $6.7 trillion. The financial resources of SWFs make them a powerful political tool for countries to achieve economic and political policy objectives, such as RN. For example, the world's biggest SWF, Norway's Government Pension Fund, is funded by Norway's nationalised oil fields. In the Region, SWFs have the potential to manipulate, or nationalise resource projects. Two events demonstrate this point. First, the Korea Investment Corporation's $100 million investment in an iron ore project in Brazil in 2011 (Kong 2011). Second, the Chinese Investment Corporation becoming the second largest shareholder in Russia's Uralkali, the world's largest potash producer (Weaver et al. 2013).

RN is a growing concern in the global mining industry. Ernst & Young's (2013) report, *Business risks facing mining and metals 2013–2014*, identified RN as the third biggest risk facing mining companies. Only a year earlier, RN was named the number one risk to the mining industry. Similarly, a Multilateral Investment Guarantee Agency (2013) survey found foreign investors considered political risk to be the second biggest barrier to investment in developing countries over the next 3 years.

RN has consequences for mining companies beyond the loss of profit and ownership. Expropriation of a project reduces the global market power of mining companies. Vernon (1977) argues that mining companies can keep export prices high or low by manipulating the supply and, therefore, prices of commodities in certain markets around the world. Rio Tinto and BHP Billiton have been oversupplying the iron ore market in order to drive down prices, for example. The loss of a project reduces the total supply of a particular resource controlled by the mining company, thereby reducing the market power of the company. Additionally, mining companies lose some of their managerial and technological advantages over host governments when RN affects their projects. When a fully functioning and technologically advanced mining project is expropriated, and trained workers remain with that project, a government can use its newfound operational capacity to compete in markets against mining companies. This forces mining companies to compete for contracts against state-sponsored competitors in the market, and, consequently, have fewer firm-specific advantages.

Host governments also face negative consequences from RN policies. First, RN policies discourage foreign direct investment (FDI) in the future. As Vernon (1977) points out, FDI is important for governments, because mining companies create economic activity for local contractors, suppliers, and distributors in other industries. These linkages are lost if there is no additional investment in the economy.

Second, RN is financially costly for a government. In a study of the cost of compensation payments and loss of operating efficiency caused by government ownership, Duncan (2006a) found that countries with a history of expropriation produce 6% less in their mining industry than countries with no history of expropriation. Third, RN transfers market risk to the host government. Curry (1976) warns that market forces, such as international commodity prices, which are out of the control of the government, may turn a project into a loss-maker. Finally, capturing a greater share of oil wealth by nationalising a country's oil industry reduces the democratic power of ordinary citizens. Using data from 133 countries between 1971 and 1997, Ross (2001) shows an increase in oil wealth corresponds to a decline in democratic practices, especially in lower-income countries. This is partly because oil-rich states tend to be authoritarian.

Security of Supply and the Obsolescing Bargain Model in the Region

The Asia-Pacific Region is a significant contributor to the global mining industry. Data obtained from the US Energy Information Administration (EIA) reveals that the Region produced nearly 30% of the world's primary energy in 2011. This is significantly higher than in North America, which produces 20% of the world's primary energy, and is double the primary energy production of the Middle East. In particular, the Region is the world's largest coal producer, and is a significant producer of petroleum and dry natural gas.

Reports by BP (2015) and the British Geological Survey (Brown et al. 2014) confirm the Region is the world's largest producer of coal, tin, and iron ore, and produces substantial quantities of other commodities. China is an important contributor to the Region's mining industry. China is the world's biggest producer of coal and the world's fourth largest producer of oil. Additionally, China dominates global production of many non-energy resources. The British Geological Survey's *World Mineral Production 2008–2012* shows China is the world's leading producer of 44 out of the 81 minerals and metals surveyed in the report (Brown et al. 2014). Australia and Indonesia are the two other major producers in the Region. According to the US EIA, in 2013, Australia was the world's second largest coal exporter and in 2015 it was the second largest LNG exporter. It is tipped to be the world's largest exporter by 2020 (APPEA 2015). Australia exports 70% of its hydrocarbon production and has the largest uranium reserves in the world. Australia also has large reserves of iron ore, nickel, bauxite, gold, opal, and zinc. Indonesia is one of the largest exporters of coal, bauxite, silver, nickel, and tin.

As is clear from the chapters of this book, the mining industry is a crucial component of many economies in the Region. The economies of Brunei, East Timor, and Mongolia depend almost entirely on commodity exports. In Australia, Papua New Guinea, New Caledonia, and the Solomon Islands, over half of export revenue is derived from commodity exports. However, some countries, including Japan and

Cambodia, have only negligible commodity exports, and rely more heavily on imports to sustain economic growth.

Despite the large size and scope of the mining industry, the gap is widening between mineral and energy production and consumption. Economic growth demands more resources, but reserves continue to deplete. Stern (2011) and Fondja Wandji (2013) find that resource consumption is directly and proportionately linked to economic growth. This means security of supply is a major challenge for governments looking ahead.

A steady and reliable supply of mineral and energy resources is even more important in economies of the Region, due to high levels of domestic material consumption (DMC). DMC is the total amount of materials directly used in the economy, minus exported materials. A United Nations Environment Programme (UNEP) report found that DMC in the Asia-Pacific grew 4.9% annually from 1975 to 2005 (UNEP 2011). This is significantly higher than the 0.5% annual growth recorded for the rest of the world over the same period. The Region now is the single largest raw material consumer in the world, with DMC increasing from below 25% in 1975 to 53% in 2005. The report attributes the increase to the Region's growing affluence. Additionally, the increasing rate of material intensity in the Region is another factor contributing to the widening gap between mineral and energy production and consumption. A second UNEP (2013) report finds that material intensity is increasing. That report shows that the shift of material intensive industries, such as manufacturing and transport, to countries in the Region is accelerating the depletion of resources. The Asian Development Bank (ADB 2011) report suggests this trend is likely to continue, predicting that, 'by the middle of this century, Asia could account for half of global output, trade and investment' (see Table 6.1).

A major reason for the increases in DMC and material intensity is China's growing hunger for mineral and energy resources to feed its industrialising economy. China's abundance of domestic mineral reserves satisfied the country's demand for resources in the 1980s and 1990s. However, China has become a net importer of mineral and energy resources since the 2000s (OECD 2012; Dungery, Fry-McKibben, and Linehan 2013). Although China produces 44% of the world's iron ore, it accounts for two-thirds of world iron ore demand (Brown et al. 2014). BP (2015) further notes that China (including Hong Kong) consumed 50.2% of global coal production in 2015, despite producing 47.7% of the world's total coal supply in 2015.

The gap between production and consumption is even wider in other minerals and energy sectors. BP (2015) found that China's net imports of oil reached 7 million barrels per day in 2013, surpassing the US as the world's largest net oil importer. China also increased natural gas production by 9.5% in 2013, which was the biggest global increase in natural gas production. However, China's natural gas consumption increased by 10.8% in the same year. There are few signs the gap will close in the future (KPMG 2011), and China acknowledges it faces a severe shortage of key mineral resources by 2020 (ChinaDaily USA 2012).

The widening gap between mineral and energy resource production and consumption across the Region is expected to increase in the future. The International

Table 6.1 Summary of raw material exports and their value

Country	Main raw materials export	Value of raw material exports ($B)	Total value of exports ($B)	Percentage of export revenue
Australia	Iron ore, coal briquettes, petroleum gas, crude petroleum	176	249	71%
Brunei	Crude petroleum, petroleum gas	12.4	12.9	96%
Cambodia	Gold, sand, wood	0.26	12	2%
China	Refined petroleum, petroleum gas, coal briquettes, gold, plywood, iron, rare earth elements	222	2120	10%
Indonesia	Coal briquettes, petroleum, gold	92	212	43%
Japan	Refined petroleum, iron, gold, silver	35.5	794	4%
Laos	Copper, iron ore tin	0.18	2.13	8%
Malaysia	Refined petroleum, petroleum gas, crude petroleum, tin	77	240	32%
Myanmar (Burma)	Petroleum gas, tin, iron ore	5.24	7.39	71%
New Zealand	Wood, crude petroleum, aluminium, coal briquettes	6.75	37.8	18%
Papua New Guinea	Gold, copper, platinum, silver, crude petroleum	6.84	9.37	73%
Philippines	Gold, copper, nickel, crude petroleum	8.16	72.2	11%
South Korea	Refined petroleum, hot-rolled iron, gold, petroleum coke	114	562	20%
Thailand	Refined petroleum, gold, iron	35	218	16%
Vietnam	Crude petroleum, coal briquettes	17	112	15%

Source: United Nations Commodity Trade Statistics Database (2012).

Energy Agency (IEA) predicts South East Asia's net oil imports will more than double by 2035 in order to fuel the Region's growing economies (IEA 2013). Additionally, the IEA warns that natural gas exports from the Region's major producers—Brunei, Burma, Indonesia, and Malaysia—will reduce significantly as domestic consumption outpaces production. While Australia has substantial reserves of uranium, coal, and natural gas, its reliance on oil imports is reaching critically high levels. Data from the Australian Bureau of Resources and Energy Economics reveals that 91% of the fuel and oil used in the country's transport industry is imported (Blackburn 2014). The lack of domestic oil production in Australia represents an energy security challenge. If supply were suspended for any reason, Australia only would have about three weeks oil in reserve (Stewart 2014).

Table 6.2 Asia-Pacific countries introducing RN policies, 2004–2014

Australia
China
India
Indonesia
Kazakhstan
Mongolia
Papua New Guinea
South Korea
Sri Lanka
Vietnam

In the Region, growing anticipation by governments of a potential shortfall in mineral and energy supplies has led to the introduction of RN policies. The wave of nationalisation is occurring across countries with varying income levels, stages of development, and political systems. Table 6.2 shows the spread of RN by highlighting the countries that have introduced RN policies in the period 2004–2014.

The mining industry of some countries in the Region, including China, Indonesia, New Zealand, the Philippines, and Sri Lanka, have state-owned enterprises (SOE) that are granted preferential access to resources.

Indonesia is an eager contributor to the latest wave of RN. In January 2014, Indonesia banned exports of nickel, tin, and bauxite in an attempt to force investment into domestic smelters and refineries (Evans-Pritchard 2014). From 2014, exports of unprocessed or semi-processed minerals attract a 20–25% tax, with the tax to increase to 60% by 2016 as the government sets out to increase local beneficiation (Sambijantoro 2014).[2] In 2012, Indonesia forced foreign-owned mining companies to shift majority ownership of projects to local investors after 10 years of operation (Thaher and Chatterjee 2012). In addition, Indonesia dissolved the country's oil and gas regulator, BPMigas. The regulator was deemed unconstitutional, because it was perceived to compromise the government's ability to use oil and gas reserves for the benefit of Indonesian citizens (Paramaditha and Jensen 2012).

China also has shown a commitment to RN. The US Chamber of Commerce (2012) noted that China's 12th Five Year Plan demanded the establishment of 'national champions' in the mining industry. This was to be achieved by demanding minimum government ownership in projects and the transfer of advanced technology to China. Outside of its geographic borders, Chinese foreign policy dictates a clear commitment to consumer country RN. This is achieved through the foreign expansion of Chinese SOEs, including Sinopec, China National Offshore Oil

[2] Beneficiation is the process of extracting the valuable minerals from the gangue, or waste material, which surrounds the mineral. Beneficiation techniques include crushing, grinding, magnetic separation and flotation. Mandated beneficiation increases economic activity because it includes more stages of production within the host country, but can create extra costs for MNCs if the country is poorly equipped for beneficiation. Mandated beneficiation also benefits the host country from the transfer of skills and knowledge (PricewaterhouseCoopers Indonesia 2013).

Corporation (CNOOC), and the China National Petroleum Corporation (CNPC). Burgess and Beilstein (2013) assert that China is attempting to nationalise resource supplies from Southern Africa and predict that Western mining companies will be blocked from that region.

Elsewhere in the Region, Papua New Guinea expropriated the country's largest company, Ok Tedi Mining, by increasing state control from 37% to 100% in 2013 (Fox 2013). In 2012, Mongolia capped foreign ownership of mining projects to a maximum of 49%, unless a government panel makes an exception (Edwards 2012). In Australia, the Gillard Labor government introduced in 2010 a Mineral Resources Rent Tax (MRRT) to tax 30% of iron ore and coal mining 'super profits' above $75 million (The Australian 2010). Although the subsequent conservative Abbott Coalition government repealed the MRRT in September 2014, the tax indicated Australia's willingness to engage in RN (Cormann 2014). In 2011, Vietnam introduced a 10% tax on gold exports to retain gold production and consumption within the country (Pham Muoi 2010). In the same year, Sri Lanka passed legislation enabling the government to expropriate 'underperforming' private companies. Thirty-seven companies immediately were expropriated (ABC 2011).

The Contemporary Obsolescing Bargain Model in the Asia-Pacific Region

In addition to the aforementioned examples of RN, governments across the Region specifically are introducing policies in response to the security of resource supply challenge. These policies amount to a newer form of RN, as they reduce the bargaining power of mining companies and, as a result, modify the OBM in favour of host governments.

The first policy response of governments in the Region to the security of supply challenge is the expansion of SOEs. This reduces the bargaining power of mining companies, because it increases the intensity of competition between companies for access to remaining reserves. Grosse (1989) argues that more competition in the mining industry weakens the bargaining power of mining companies, because they must compete to offer the best deal to a host government. Governments, on the other hand, have the power to provide access to mineral and energy resources to the highest bidder. As a result, mining companies are less able to negotiate favourable terms and conditions with a government. This issue is exacerbated when a government offers favourable investment conditions and support for its SOE. In countries with a strong SOE presence, such as China, the market is more difficult to enter, because governments tend to favour their own companies, and SOEs are able to take more risks than private companies when exploring for new reserves.

The global expansion of SOEs is most evident in the oil and gas industry. According to the World Bank, national oil companies (NOCs) control 90% of the world's oil and gas reserves and 75% of production (Tordo et al. 2011). The World Bank also estimates 60% of the world's undiscovered reserves are in countries

Table 6.3 National oil companies in the Asia-Pacific Region

Brunei	Petroleum Brunei
China	China National Petroleum Corporation
	China National Offshore Oil Corporation
	China National Refinery Corporation
	Kunlun Energy
	PetroChina
	Sinopec
East Timor	Timor Gap Empressa Publico
Hong Kong	Kunlun Energy (formally China National Petroleum Corporation Hong Kong)
Indonesia	Pertamina
Japan	Japan Oil, Gas and Metals National Corporation
Laos	Lao State Fuel Corporation
Malaysia	Petronas
Myanmar	Myanmar Oil and Gas Enterprise
Philippines	Philippine National Oil Company
Singapore	Singapore Petroleum Company
Sri Lanka	Ceylon Petroleum Corporation
South Korea	Korea General Petroleum United Corporation Korea National Oil Corporation
Thailand	PPT (Formally the Petroleum of Thailand)
Vietnam	PetroVietnam

where NOCs have preferential access to resources. The Region currently has a high concentration of NOCs, including in Brunei, China, Indonesia, Japan, Lao, Malaysia, Myanmar, the Philippines, Singapore, South Korea, Sri Lanka, Thailand, East Timor, and Vietnam. Only a handful of regional countries do not have a NOC. This list includes the Pacific Island states, Australia, Mongolia, and New Zealand (see Table 6.3).

NOCs likely will expand their reach outside the Region in the future, thereby creating more competition for international oil companies (IOCs) worldwide. The IEA (2013) expects net oil imports in South East Asia to double by 2035, meaning governments are more likely to expand the global reach of their NOCs to secure future supplies. The Region's NOCs already have started this process. According to Downs (2008), Chinese NOCs participate in oil and gas production contracts in 33 countries with the intention of securing future energy supplies. Chinese NOCs target oil and gas reserves across South America, Africa, the Middle East, Russia, and Canada. In the Region, Lim (2010) argues that China has pursued cordial political relationships with South East Asian oil and gas producers to ensure the steady flow of foreign energy supplies into the country. Fullbrook (2006) provides examples of Chinese NOCs investing heavily into Cambodia's oil and gas industry, while PetroChina has joined with Pertamina to operate oil and gas projects in Indonesia.

Elsewhere in the Region, the Vietnamese Government cited future energy security as the reason for PetroVietnam's expansion into foreign markets. Indonesia's Pertamina recently entered Burma to extract oil and gas reserves (Amin 2014).

Harris (2010) predicts traditional net oil exporters, Malaysia, Indonesia, and Vietnam, will become 'substantial net importers as their domestic demand increases' in the coming years. A 2009 Australian Government energy white paper anticipated that regional powers would become more proactive in their pursuit of energy security in years to come (Government of Australia 2009).

SOEs also have large presences in other mineral and energy sectors. Chinese SOEs, Chinalco and Baotou Steel, have established dominance over the world's rare earth elements sector (Wubbeke 2013; Ting in this volume). China Gold actively pursues gold supplies in South East Asia and Australia (Ng 2012). The China Metallurgical Group Corporation has a copper mining project in Afghanistan (World Bank 2013). Additionally, Vietnamese SOE, Vinacomin, plans to exploit coal, bauxite, and iron reserves in Laos and Cambodia (Vinacomin 2014). Non-oil and gas SOEs are found in China, Indonesia, New Zealand, Papua New Guinea, the Philippines, South Korea, Sri Lanka, and Vietnam.

The second policy response by governments to the security of supply challenge is the shifting of economic policy to favour long-term wealth creation over immediate resource rents. The reason for this is that resource supply is becoming a more important component of wealth creation than short-term rent maximisation. This is different to the argument made by Duncan (2006b) and others. High commodity prices are one of the principal causes of an obsolescing bargain. Governments seek to capitalise on resource rents when commodity prices are high, and may use the windfall revenue to fund programs to increase their popularity.

Rising levels of DMC and material intensity in regional economies indicate that countries would derive greater value from securing reliable supplies of mineral and energy resources than from maximising short-term resource rents. Guriev, Kolotilin, and Sonin (2007) argue that expropriation occurs during commodity price booms because 'the immediate prize is too valuable relative to future revenues.' However, the value of a steady resource supply for future economic growth flips this equation. The value of future revenues, brought about by securing supplies, is more valuable than immediate windfall profits. This creates a more imminent obsolescing bargain for mining companies because the incentive for RN is to guarantee future supply. Unlike an obsolescing bargain caused by high commodity prices, the supply of non-renewable resources diminishes unless new reserves are discovered. This is in contrast to fluctuating commodity prices, which offer a reprieve to mining companies when commodity prices are low. Low prices mean governments must renew attractive investment terms to bring FDI back into the sector. However, because mineral and energy supplies are depleting, there is no defined turning point for mining companies in the contemporary OBM.

Economic policy favouring access to resource supplies for long-term wealth creation weakens the bargaining power of mining companies. This especially is the case in mining that is critically important to an economy. For example, IOCs face a greater likelihood of RN, because of the importance of oil in the Region's manufacturing and transport sectors. The Region has less than 4% of the world's proven oil reserves, and therefore relies heavily on imports from the Middle East (Wu et al. 2008). The Region's rapid depletion of oil reserves is symbolised by Indonesia's withdrawal from OPEC after turning from a net oil exporter to an oil importer (BBC 2008). Despite the Region's declining reserves, the ADB predicts regional demand for oil will increase by 1.9% annually until 2035 (Tan 2013). This encourages governments to introduce policies aimed at securing future oil supplies, which can amount to RN and a quicker obsolescing bargain for IOCs.

Equally, mining companies in critical mineral sectors more likely will be targeted and affected by RN. According to Eggert et al. (2007), a critical mineral is one that is 'essential in use … and subject to supply restriction.' Mining companies in these industries face a quicker obsolescing bargain after the initial transfer of capital and skills to the host country. One of the main advantages offered by mining companies to governments after this point is the efficient operation of the project, along with tax and royalty revenue. However, with security of supply becoming a priority for governments, mining companies offer fewer on-going advantages, as they cannot increase the quantity of resources needed by governments besides discovering new reserves. As a result, there is strong incentive for governments to expropriate or nationalise the operations of these mining companies. The bargain will obsolesce irrespective of commodity prices, because a government aims to increase its quantity of resources, rather than revenue from the sale of those resources.

The third policy response of the Region's governments to the security of supply challenge is the development of national security policies. O'Brien and Williams (2010) define national security as the preservation of a state's sovereignty from external threats. Mineral and energy resources increasingly are important to achieving national security objectives. Deutch, Schlesinger, and Victor (2006) argue that a greater dependence on mineral and energy imports reduces a country's national security. This is because of the impact of import dependence on foreign policy. Governments that are dependent on resource imports must appease producer countries to ensure that they receive an adequate supply of mineral and energy resources, even if those policies contradict the purchasing country's national security objectives. Klare (2002) argues that 'the relentless expansion in worldwide demand, the emergence of significant resource shortages, and the proliferation of ownership contests' will cause tension and conflict between states in the future. An example of this is the territorial dispute between Indonesia and Malaysia over the oil-rich Ligitan and Sipadan islands. The two countries deployed fighter jets and warships over the islands before the International Court of Justice (2002) awarded the islands to Malaysia in 2002.

The correlation between resource supply and national security policies weakens the bargaining power of mining companies. This is because control over the production and distribution of mineral and energy resources, achieved through RN, enhances a government's capacity to achieve national security objectives. Essentially, the benefits brought about by mining companies cannot easily match the importance of national security. The problem is exacerbated by the desire of mining companies to export resources to as many customers as possible in order to maximise their profits. Governments, on the other hand, aim to secure exclusive access to mineral and energy resources in order to use them to achieve national security—including economic—objectives. This creates a quicker obsolescing bargain for mining projects, because host governments are motivated to expropriate resource projects for control and ownership purposes.

Strengthening the Bargaining Power of Mining Companies Using Industry Clusters in the Region

The policy responses of governments to the security of supply issue weaken the bargaining power of mining companies. This modifies the OBM to favour host governments, and negatively affect mining companies. As a result, mining projects are more likely to be affected by RN, especially as reserves deplete. Mining companies therefore must look to new ways to increase their bargaining power in order to reduce the likelihood that they will be affected by RN.

The concept of industry clusters may assist with mitigating the risk posed by RN. Porter (1998, 2000) first introduced the concept to demonstrate the benefits of a group of interconnected companies, suppliers, and institutions competing, but also cooperating, in a small geographic space. As the companies in the cluster develop new skills, knowledge, and technology to remain competitive, the cluster develops new industries, thereby creating new products. Industry clusters also foster efficiency by enabling individuals to network and share knowledge and expertise, and removing barriers to communication between suppliers and consumers.

Industry clusters across the Region may increase mining companies' bargaining power if they can provide on-going transfer of resources, knowledge, and technology to host governments. Industry clusters successfully have been developed in mining industries in South America, Europe, and North America. For example, the Chilean Mining Cluster Program has mitigated RN by facilitating the on-going transfer of resources from mining companies to the host government. In states within the Asia-Pacific Region, industry clusters could use the impending depletion of mineral and energy resources (as a consequence of the resources being mined) as an opportunity to assist host governments to prepare for a sustainable post-mining economy. This could be achieved by developing non-mining industries concurrently with a mining project's operation. Further research on this subject is required.

Conclusion

The impending security of supply issue facing governments in the Asia-Pacific Region has implications for the future of the Region's mining industry. The policies introduced by governments, intending to guarantee a supply of mineral and energy resources to achieve economic and political objectives, have redefined the bargaining power of mining companies and host governments. Mining companies face a tougher obsolescing bargain, because their capacity to offer resources to host governments diminishes as security of supply becomes a greater concern. The spread of SOEs, along with the prioritisation of long-term economic growth and national security objectives, reduces the bargaining power of mining companies.

Mining companies must develop new ways to transfer valuable resources to host governments if they are to avoid RN. I have suggested that the concept of industry clusters may provide a means to address the priorities both of mining companies and host governments to create mutually beneficial relationships. Industry clusters take advantage of the security of supply nexus by offering a long-term solution to a growing challenge facing governments in the Region, while also acting as a mechanism for mitigating RN.

The exploitation of natural resources long has led to conflict and cooperation between mining companies and governments. However, there will likely more conflict when governments identify security of supply as a key priority. To turn a conflictual relationship into a cooperative one, mining companies must understand why policies amount to RN, and the potential mechanisms for preventing RN from affecting their operations. The impetus now is on mining companies to take steps toward strengthening their bargaining power vis-à-vis host governments. Should mining companies fail to strengthen their bargaining power, the security of supply issue may cause RN to become a more common feature of the mining industry.

References

APPEA. (2015). LNG Export Markets. http://www.appea.com.au/oil-gas-explained/benefits/benefits-of-lng/export-revenue/. Accessed 8 June 2016.

Amin, K. (2014, 21 April). State-owned enterprises aim to be Asian tigers. *Jakarta Post*. http://www.thejakartapost.com/news/2014/04/21/state-owned-enterprises-aim-be-asian-tigers.html. Accessed 21 April 2014.

Arezki, R. Gylfason, T. (2013). Resource rents, democracy, corruption and conflict: evidence from Sub-Saharan Africa. *Journal of African Economies*, 22(4), 552-569.

Arsenault, C. (2012, 6 January). Exxon 'loses' Venezuela nationalisation case. *Al Jazeera*. http://www.aljazeera.com/indepth/features/2012/01/201215194512924679.html. Accessed 21 April 2014.

Asian Development Bank. (2011). *Asia 2050: realising the Asian Century.* Manila: The Philippines. http://www.adb.org/publications/asia-2050-realizing-asian-century. Accessed 5 July 2015.

Australian Broadcasting Corporation. (2011). Sri Lanka passes nationalisation law. http://abcasia-pacificnews.com/stories/201111/3360812.htm?desktop. Accessed 20 February 2015.

Blackburn, J. (2014). Australia's liquid fuel security part 2. *National Road and Motorists' Association*. http://www.mynrma.com.au/images/About-PDF/Fuel-Security-Report-Pt2.pdf. Accessed 24 February 2015.

BP. (2015). BP statistical review of world energy. London. http://www.bp.com/en/global/corporate/energy-economics/statistical-review-of-world-energy/2015-in-review.html. Accessed 24 February 2015.

British Broadcasting Corporation. (2008). Indonesia to withdraw from OPEC. http://news.bbc.co.uk/2/hi/business/7423008.stm. Accessed 20 February 2015.

Brown, T., Idoine, N., Raycraft, E., et al. (2014). World mineral production 2008-12. *British Geological Survey*. Keyworth: Nottingham. http://nora.nerc.ac.uk/507092/. Accessed 3 November 2015.

Bunn, D., Mustafaoglu, M., (1978). Forecasting political risk. *Management Science*, 24(15), 1557-1567.

Burgess, S., Beilstein, J. (2013). This means war? China's scramble for minerals and RN in Southern Africa. *Contemporary Security Policy*, 34(1), 120-143.

ChinaDaily USA. (2012, 23 April). China mineral stocks running low. http://usa.chinadaily.com.cn/business/2012-04/23/content_15113037.htm. Accessed 26 August 2015.

Collins, G., Erickson, A. (2011). Energy nationalism goes to sea in Asia, NBR Special Report No. 31. The National Bureau of Asian Research. Seattle: West Washington. http://www.nbr.org/publications/element.aspx?id=538. Accessed 29 August 2015.

Cormann, M. (2014) *Repeal of the Minerals Resource Rent Tax*. Canberra.

Curry, R. (1976). Global market forces and the nationalisation of foreign-based export companies. *Journal of Modern African Studies*, 14(1), 137-143.

Davis, J., Ossowski, R., Fedelino, A. (2003), Fiscal policy formulation and implementation in oil-producing countries. International Monetary Fund, Washington, D.C. http://www.imf.org/external/pubs/nft/2003/fispol/index.htm. Accessed 24 June 2015.

Deutch, J. Schlesinger, J. Victor, D. (2006). National security consequences of U.S. oil dependency. Council on Foreign Relations, New York. www.cfr.org/content/publications/attachments/EnergyTFR.pdf. Accessed 6 June 2015.

Dorian, J., Franssen, H., Simbeck, D. (2006). Global challenges in energy. *Energy Policy*, 34(15), 1984-1991.

Downs, E. (2008). China's NOCs: lessons learned from adventures abroad. *Fundamentals of the global oil and gas industry*, pp. 27-31.

Duncan, R. (2006a). Costs and consequences of the expropriation of FDI by host governments. Australian Agricultural and Resource Economics Society 2006 conference. http://ageconsearch.umn.edu/bitstream/139524/2/2006_duncan.pdf. Accessed 5 May 2015.

Duncan, R. (2006b). Price or politics? an investigation of the causes of expropriation. *Australian Journal of Agricultural and Resource Economics*, 50(1), 85-101.

Dungey, M., Fry-McKibbin, R., Linehan, V. (2013). Chinese resource demand and the natural resource supplier. *Applied Economics*, 46(2), 167-178.

Edwards, T. (2012, May 18). Mongolia passes watered-down foreign investment law. *Reuters*. http://uk.reuters.com/article/2012/05/18/mongolia-mining-idUKL4E8GI3HV20120518. Accessed 20 August 2015.

Eggert, R., Carpenter, A., Freiman, S., et al. (2007). Minerals, critical minerals, and the U.S. economy. National Research Council of the National Academies. Washington, D.C. http://dels.nas.edu/resources/static-assets/materials-based-on-reports/reports-in-brief/critical_minerals_final.pdf. Accessed 4 October 2015.

Elliott, L. (2007, 2 May). Venezuela seizes foreign oil fields. *The Guardian*, http://www.theguardian.com/business/2007/may/02/oilandpetrol.venezuela. Accessed 24 July 2015.

Ernst & Young. (2013). Business risks facing mining and metals 2013-2014, London. http://www.ey.com/Publication/vwLUAssets/Business_risks_facing_mining_and_metals_2013percentE2percent80percent932014_ER0069/$FILE/Business_risks_facing_mining_and_metals_2013percentE2percent80percent932014_ER0069.pdf. Accessed 3 August 2015.

Evans-Pritchard, A. (2014, 13 January). RN alive and well as Indonesia bans key metal exports. *The Telegraph*. http://www.telegraph.co.uk/finance/commodities/10569653/Resource-nationalism-alive-and-well-as-Indonesia-bans-key-metal-exports.html. Accessed 24 July 2015.

Fondja Wandji, Y. (2013). Energy consumption and economic growth: evidence from Cameroon. *Energy Policy*, 61, 1295-1304.

Fox, L. (2013, 18 September). PNG government takes control of Ok Tedi mine, repeals laws protecting BHP from legal action over pollution. *Australian Broadcasting Corporation*. http://www.abc.net.au/news/2013-09-19/png-government-takes-control-of-png-ok-tedi-mine/4967004. Accessed 6 April 2015.

Friedman, T. (2006). The first law of petropolitics. *Foreign Policy*, 154, 28-36.

Fullbrook, D. (2006, 3 October). China's growing influence in Cambodia. *Asia Times Online*. http://www.atimes.com/atimes/SouthEast_Asia/HJ06Ae01.html. Accessed 3 August, 2015.

Government of Australia. (2009). Defending Australia in the Asia Pacific century: force 2030. Canberra: Commonwealth of Australia. http://www.aph.gov.au/About_Parliament/Parliamentary_Departments/Parliamentary_Library/pubs/rp/rp1516/DefendAust/2009. Accessed 5 May 2015.

Government of Western Australia. (2011). Policies - Department of State Development. http://www.dsd.wa.gov.au/8482.aspx - 8837. Accessed 24 August 2015.

Griffiths, M., O'Callaghan, T., Roach, S. (2014). *International relations: the key concepts*. 3rd edn. London: Routledge.

Grosse, R. (1989). *Multinationals in Latin America*. London: Routledge.

Guriev, S., Kolotilin, A., Sonin, K. (2007). Determinants of expropriation in the oil sector: a theory and evidence from panel data. Working Paper No. 115, Centre for Economic and Financial Research, Moscow. http://papers.ssrn.com/sol3/papers.cfm?abstract_id=1141660&rec=1&srcabs=427682&alg=5&pos=8. Accessed 6 July 2015.

Harris, S. (2010). Global and regional orders and the changing geopolitics of energy. *Australian Journal of International Affairs*, 64(2), 166-185.

International Court of Justice. (2002). Sovereignty over Pulau Ligitan and Pulau Sipadan (Indonesia/Malaysia), The Hague, 17 December. http://www.icj-cij.org/docket/files/102/7714.pdf. Accessed 5 july 2014.

International Energy Agency. (2013). South East Asia Energy Outlook. Paris. http://www.iea.org/publications/freepublications/publication/weo-2013-special-report-southeast-asia-energy-outlook.html. Accessed 4 April 2014.

Jodice, D. (1985). Political risk assessment: an annotated bibliography. Westport, Conneticutt: Greenwood Press.

Kantor, M. (2008). Are you in good hands with your insurance company? In T. Moran, G. West, K. Martin (Eds.), *International political risk management: needs of the present, challenges for the future* (pp. 137-170). International Bank for Reconstruction and Development/World Bank, Washington, D.C.

Kinross Gold Corporation. (2013, 10 June). Kinross announces it will cease development of Fruta del Norte. Toronto, Ontario. Kinross announces it will cease development of Fruta del Norte. http://www.kinross.com/news-and-investors/news-releases/press-release-details/2013/Kinross-announces-it-will-cease-development-of-Fruta-del-Norte/default.aspx. Accessed 9 July 2015.

Klare, M. (2002). *Resource wars: the new landscape of global conflict*. New York: Henry Holt and Company.

Kong, K. (2011, 13 May). South Korea to invest in Brazil iron ore project. *Wall Street Journal*. http://online.wsj.com/news/articles/SB10001424052748703864204576320502167310150. Accessed 15 May.

KPMG. (2011). China's 12th Five-Year Plan: overview. http://www.kpmg.com/CN/en/IssuesAndInsights/ArticlesPublications/Publicationseries/5-years-plan/Documents/China-12th-Five-Year-Plan-Overview-201104.pdf. Accessed 29 November 2015.

Lax, H. (1983). *Political risk in the international oil and gas industry*. Boston, Massachusetts: International Human Resources Development Corporation.

Lim, T. (2010). The emergence of the Chinese oil industry in South East Asia: China-ASEAN sub-Regionalism - Pan Pearl River Delta (PPRD) Regionalism and cooperation in oil. *Oil and gas in China: the new energy superpower's relations with its region* (pp. 39-56). Singapore: World Scientific Publishing.

Luciani, G. (2011). Global oil supplies: the impact of RN and political instability. Working Document No. 350. Centre for European Policy Studies, Brussels. https://www.ceps.eu/publications/global-oil-supplies-impact-resource-nationalism-and-political-instability. Accessed 31 March 2015.

Moran, T., West, G. (2008). Overview. In T. Moran, G. West, K. Martin (Eds), *International political risk management: needs of the present, challenges for the future* (pp. 107-113). International Bank for Reconstruction and Development/World Bank, Washington, D.C.

Multilateral Investment Guarantee Agency. (2013). World investment and political risk, The International Bank for Reconstruction and Development/The World Bank, Washington, DC. https://www.miga.org/documents/WIPR13.pdf. Accessed 5 November 2015.

Ng, E. (2012, 9 April). China Gold panning for overseas options. *South China Morning Post*. http://www.scmp.com/article/997758/china-gold-panning-overseas-options. Accessed 21 March 2015.

O'Brien, R., Williams, M. (2010). *Global political economy: evolution and dynamics*. 3rd edn. London: Palgrave Macmillan.

Organisation for Economic Co-operation and Development. (2012). Steelmaking raw materials: market and policy developments. Paris. http://www.oecd.org/sti/ind/50494260.pdf. Accessed 4 April 2014.

Paramaditha, A., Jensen, F. (2012, 13 November). Indonesia court scraps energy regulator, shocks industry. *Reuters*. http://www.reuters.com/article/2012/11/13/indonesia-energy-bpmigas-idUSL3E8MD21320121113. Accessed 24 July 2015.

Pham Muoi, N. (2010, 2 December). Vietnam to impose 10per cent tax on gold exports. *Wall Street Journal*. http://online.wsj.com/news/articles/SB10001424052748703377504575650273767336374. Accessed August 2 2015.

Porter, M. (2000). Location, competition, and economic development: local clusters in a global economy. *Economic Development Quarterly*, 14(1), 15-34.

Porter, M. (1998). *The competitive advantage of nations*. 2nd edn. London: Macmillan.

Prada, P. (2006, 2 May). Bolivia nationalises the oil and gas sector. *New York Times*. http://www.nytimes.com/2006/05/02/world/americas/02bolivia.html. Accessed 13 May 2015.

PricewaterhouseCoopers Indonesia. (2013). *mineIndonesia 2013*. Jakarta.

Ross, M. (2001). Does oil hinder democracy? *World Politics*, 53(3), 325-361.

Sambijantoro, S. (2014, 30 January). Indonesia defies Freeport on export tax. *Jakarta Post*. http://www.thejakartapost.com/news/2014/01/30/indonesia-defies-freeport-export-tax.html. Accessed 25 October 2015.

Simpson, G. (2010). Russian and Chinese support for Tehran. *Middle East Quarterly*,17(2), 63-72.

Smith, B. (2004). Oil wealth and regime survival in the developing world, 1960-1999. *American Journal of Political Science*, 48(2), 232-246.

Sovereign Wealth Fund Institute. (2014). Sovereign wealth fund rankings. http://www.swfinstitute.org/fund-rankings/. Accessed 25 August 2015.

Stern, D. (2011). The role of energy in economic growth. *Annals of the New York Academy of Sciences*, 1219(1), 26-51.

Stewart, C. (2014, 24 February). Defence 'to rely on' fuel imports'. *The Australian*. http://www.theaustralian.com.au/national-affairs/defence/defence-to-rely-on-fuel-imports/story-e6frg8yo-1226835509983. Accessed 30 October 2015.

Tan, F. (2013, 14 October). Oil imports by 2035 to rise close to Mideast 2011 crude output - ADB. *Reuters*. http://www.reuters.com/article/2013/10/14/asia-pacific-energy-idUSL4N0I03D620131014. Accessed 9 September 2015.

Thaher, R., Chatterjee, N., (2012, 7 March). Indonesia rule change rattles foreign mine owners. *Reuters*. http://www.reuters.com/article/2012/03/07/us-indonesia-mining-idUSTRE8260FP20120307. Accessed 24 July 2015.

The Australian. (2010, 2 July). Full statement and details of new mining tax. http://www.theaustralian.com.au/archive/politics/full-statement-and-detail-of-new-mining-tax/story-e6frgczf-1225887000521. Accessed 26 August 2015.

Tordo, S., Tracy, B., Arfaa, N. (2011). National oil companies and value creation, Working Paper No. 218, The International Bank for Reconstruction and Development/The World Bank, Washington, D.C. http://siteresources.worldbank.org/INTOGMC/Resources/336099-1300396479288/noc_volume_I.pdf. Accessed 4 March 2015.

U.S. Chamber of Commerce. (2012). China's approval process for inbound foreign direct investment: impact on market access, national treatment and transparency. Washington, D.C. https://www.uschamber.com/chinapercentE2percent80percent99s-approval-process-inbound-foreign-direct-investment-impact-market-access-national-treatment. Accessed 24 July 2015.

United Nations Commodity Trade Statistics Database. (2012). *OEC: the observatory of economic complexity. Accessed* 19 August 2014, http://atlas.media.mit.edu/.

United Nations Conference on Trade and Development. (2012). United Nations Commodity Trade Statistics Database. OEC: the observatory of economic complexity. http://atlas.media.mit.edu/. Accessed 14 Sempember 2015.

United Nations Environment Programme. (2011). Resource efficiency: economics and outlook for Asia and the Pacific. Bangkok. The United Nations. http://www.unep.org/roap/Portals/96/REEO_AP_Key.pdf. Accessed 29 February 2015.

United Nations Environment Programme. (2013). Recent trends in material flows and resource productivity in Asia and the Pacific. Bangkok: The United Nations. http://www.unep.org/pdf/RecentTrendsAP(FinalFeb2013).pdf. Accessed 24 December 2015.

Vernon, R. (1977). *Storm over the multinationals: the real issues*. Cambridge, Massachusetts: Harvard University Press,

Vernon, R. (1971). *Sovereignty at bay: the multinational spread of U.S. enterprises*. New York: Basic Books.

Vinacomin. (2014) Business areas. http://www.vinacomin.vn/en/about/Business-areas.html. Accessed 26 July 2015.

Ward, H. (2009). Resource Nationalism and sustainable development: a primer and key issues. International Institute for Environment and Development, London. http://pubs.iied.org/G02507.html. Accessed 5 Novenmebr 2014.

Weaver, C., Terazono, E., Noble, J. (2013, 24 September). CIC Chinese fund obtains 12per cent equity stake in Uralkali. *Financial Times*, http://www.ft.com/intl/cms/s/0/6eb1eed6-24e8-11e3-bcf7-00144feab7de.html - axzz3HORDp9PS. Accessed 22 October 2015.

Wilson, J. (2011). RN or resource liberalism? explaining Australia's approach to Chinese investment in its minerals sector. *Australian Journal of International Affairs*, 65(3), 283-304.

World Bank. (2013). Q&A: Aynak and mining in Afghanistan. http://www.worldbank.org/en/news/feature/2013/04/02/qa-aynak-mining-afghanistan. Accessed 24 January 2015.

Wortley, B. (1959). *Expropriation in public international law*. Cambridge: Cambridge University Press.

Wu, K., Fesharaki, F., Westley, S., Prawiraatmadja, W. (2008). *Oil in Asia and the Pacific: production, consumption, imports, and policy options*. East-West Centre, Honolulu. http://www.eastwestcenter.org/publications/oil-asia-and-pacific-production-consumption-imports-and-policy-options. Accessed 5 November 2014.

Wübbeke, J. (2013). Rare earth elements in China: policies and narratives of reinventing an industry. *Resources Policy*, 38(3), 384-394.

Chapter 7
Mining and Human Rights

Eboni Tiller

Abstract This chapter identifies a gap in the literature between the theoretical and philosophical motivations of corporate community development (CCD) and their practical applications. In recent years corporate community relations (CCR) has undergone a substantial change, moving from simply considering the fundamental ethical requirements of corporations, towards seeking to establish a role for corporations in contributing to community development. This change has led to a polarising debate, with proponents supporting the potential for corporations genuinely to contribute to a community and opponents perceiving the move as little more than a public relations exercise. This chapter suggests one catalyst for this debate may be the lack of a solid theoretical and philosophical foundation underpinning and guiding the development actions of corporations. While the rhetoric of human rights is often utilised as motivation for CCD the adequacy of this as a development framework is questioned. Consequently, this chapter explores the limitations of human rights as a development framework and proposes the capability approach as a more appropriate theoretical and philosophical framework to underpin the development aspirations of mining companies.

Abbreviations

CCD	Corporate community development
CCR	Corporate community relations
CSR	Corporate social responsibility
ICMM	International Council on Mining and Metals
ISO	International Standardization Organization
MNC	Multinational corporation
NGO	Non-government organisation
OHCHR	Office of the United Nations High Commissioner for Human Rights
TNC	Transnational corporation

E. Tiller (✉)
Centre for International Risk, School of Communication, International Studies, and Languages, University of South Australia, Adelaide, Australia
e-mail: ebonitiller1@tpg.com.au

T. O'Callaghan, G. Graetz (eds.), *Mining in the Asia-Pacific*, The Political Economy of the Asia Pacific, DOI 10.1007/978-3-319-61395-6_7

UDHR	Universal Declaration of Human Rights
UN	United Nations
UNCTAD	United Nations Conference on Trade and Development

Introduction

In recent years, there has been a substantial change in the focus of corporate community relations (CCR).[1] The scope of CCR has expanded to, not only include the ethical requirements for corporations, but also to consider the corporations role in contributing to development within the community they are operating. This change is demonstrated by calls from international organisations, non-government organisations (NGOs), and governments, as well as the growing prominence of corporate community development activities initiated by corporations (CCD) (see BHP Billiton 2014; Newmont 2009; Rio Tinto 2013).[2] The debate on whether corporations have implemented these development actions out of a genuine desire to 'do good' or merely as a public relations exercise still remains contentious, in large part this may have arisen due to the lack of a solid theoretical and philosophical foundation. There has been growing discussion on how development actions are best delivered; for example, the various development frameworks, such as the 'partnerships for development' and the 'cluster-approach' within the mining industry (See Johnson and O'Callaghan and Spagnoletti in this volume). However, consideration of what constitutes development and how the role of development extends to corporations has largely been neglected.

It is not surprising that the underlying theory has been neglected. The demands from the international and local community have been so persistent that corporations have been required either to accept and implement the courses of action endorsed by these actors, or risk reputational, and social damage and, consequently, incur economic costs. This is not to lay blame on any actor. It is simply to recognise that in a world of rapid change, there is now a greater emphasis on achieving results, especially when people's well-being is concerned. However, there is still significant importance in creating an underpinning theory that adequately captures the role of

[1] In this chapter, CCR is used in the broadest sense to describe the relations between corporations and the community. This encompasses the various understandings of corporate social responsibility (CSR), as well as alternatives to CSR, such as sustainable development, community engagement and development partnerships, among others. Whereas CSR is often used as a broad sense equivalent to CCR, it also has a more tightly defined meaning, as well as being associated with international standard ISO 26000 (International Organization for Standardization 2010). Therefore the term CCR is used to avoid confusion between the formal understanding of CSR and the more general concept.

[2] In this chapter, CCD is used to describe the social contributions a corporation makes to the community in which it operates. Whereas CCR represents the broad, overarching idea of corporate community interactions, CCD specifically reflects the changing focus of CCR to a greater emphasis on community development activities.

development. Without establishing or identifying the theoretical and philosophical motivations for contributing to development there is little to guide the objectives and actions of corporations.

The value of a theoretical framework is that it articulates the purpose for engaging in development and who is intended to benefit. This is important as it guides the actions of those creating and implementing policies. Therefore, in order to determine how corporations can best contribute to development and enable this to take place, an effective normative framework must first be established.

While there is a reasonably established body of literature on development theories, this literature is almost always centred on the role of the state. However the corporate role in development differs from that of the state. A CCD agenda must integrate a business and development agenda. The corporation cannot and should not be expected to take on a larger development role than the state. Therefore, a development theory requires a solid understanding of what constitutes development and how this is understood within the corporate context.

Typically, corporations do not engage in the debate about the theoretical and philosophical underpinnings of the programs they are being asked to implement. Rather, much of this debate occurs in the area in which advocates such as the United Nations (UN) and other NGOs operate. The language of human rights is powerful due to its ubiquity. However, over time, the human rights rhetoric appears to have been appropriated, either implicitly or explicitly, to attempt to establish a normative development framework. Yet whether this is actually an adequate and/or appropriate framework is questionable.

The purpose of this chapter is to explore the limitations of using the human rights frameworks to assess CCD and to discuss an alternative way of theorising this issue. While much of this chapter has general applicability to any industry, here the specific focus is on mining industry. This field is of particular interest, as the largely invasive nature of the industry has received specific attention since the inception of CSR. Therefore, analysing CCD within the context of the mining industry provides a good frame of reference. In view of this, the first section of this chapter is a background into the industry and development.

The second section explores the human rights framework, starting with a discussion on the origins of human rights before delving into the issues of applying this framework within the mining industry. This section includes consideration of the human rights rhetoric and challenges whether it should continue to be used in relation to business. Ultimately it is suggested that the nature of the human rights rhetoric is difficult for businesses to comprehend and implement, and may be detrimental or counter-productive.

The third section explores the capability approach as a more comprehensible rhetoric and framework for CCD actions. It begins with a brief examination of the capability approach before distinguishing between the capability approach and the human rights framework. This includes a discussion of the merits of the capability framework as a more useful alternative for corporations to engage with. This section concludes by identifying the key elements mining companies will need to consider when implementing the capability approach.

The Mining Industry and CCD

In the last few decades the mining industry has gone through significant changes. Rapid globalisation and the rise in mineral and energy commodity prices has driven the expansion of the mining industry (Davis and Franks 2014). This rapid growth has resulted in the mining industry becoming increasingly influential in many of the countries and communities where they operate. The issue of the corporation's obligations beyond its shareholders has arisen as a consequence of this increasing power and influence. While the role of corporations traditionally is primarily based on the Friedman view of profit maximisation, this is no longer an adequate business model. Instead, a focus on the ethical area of business and how it relates to the community has become increasingly prevalent.

The invasive nature of mineral and energy extraction has the capacity to impact the surrounding community, especially where there are environmental and social sensitivities. As a result, the mining industry has been under increasing pressure to take CCR seriously. Over time the role of CCR has evolved. Since at least the 1970s corporations have been implored to consider and mitigate the impact of their actions on the freedoms of individuals and the community in which they operate. However, more recently attention has turned to the role of the corporation and its contribution and participation in development within a particular community.

Despite calls for the mining industry to protect human rights, or at least not infringe on individuals human rights, in 2014 the mining industry comprised the highest percentage of total human rights abuse allegations (29%) (Kamminga 2015). Yet, at the same time as continuing reports of human rights abuse, globally within the industry there has also been a proliferation of industry-led codes of conduct based on the principles of CSR in its various forms (Paton 2000; Sethi 2003; Sullivan 2003). Thus, it would appear that within the industry there is, at the very least, a desire to appear to be doing good, and probably a genuine desire to do good. However, there is an apparent disconnect between the desire to do good and the continued allegations of human rights abuse.

The disconnect between these two matters calls into question whether the mining industry and CCD are simply incompatible or whether there is an issue in how the debate is framed. Currently there is significant polarisation between mining and development (Blowfield 2004; Kemp 2010; Sayer 2005). Some, especially from international organisations and within the mining industry, view CCD as having the potential to benefit corporations, as well as the wider community. While, for others, CCD is nothing more than a public relations exercise for corporations to protect themselves from reputational damage.

The UN, World Bank and International Council on Mining and Metals (ICMM) have all demonstrated support for corporate development activities with a 2006 joint report identifying that mining investments offered unique opportunities for economic growth, poverty reduction and participation in the global economy (ICMM 2006). Conversely, others have expressed concern about the prioritisation of economic growth, rather than a more concerted focus on human development

(Newell 2005; Utting 2005). Furthermore, critics have questioned whether claims about corporate contributions to human development are merely 'tokenistic' gestures rather than genuinely reflecting reality (Manteaw 2007).

While, there is an ongoing debate on the value of CSR and CCD it is not within the scope of this chapter to further consider this in depth. Rather the purpose of this chapter is to address the fundamental theoretical issues with CCD and to determine if there is a theoretical basis that can amalgamate business activities and a development agenda. Integrating these two ideas is not easy, but it is important, and ultimately necessary, if corporations are pragmatically to incorporate CCD into their business philosophy and framework.

Human Rights and CCD

Numerous theories have been used over the years as the theoretical foundation to expand the scope of CCR into a more ethical paradigm. While ethical frameworks such as normative stakeholder theory (Carroll and Buchholtz 2002; Weiss 2003), sustainable development (Van Marrewijk and Werre 2003; World Commission on Environment and Development 1987) and the common goods approach (Mahon and McGowan 1991; Velasquez 1992) have been used as the theoretical underpinning to motivate corporations to focus on the ethical element of CCR, it is the language of human rights that has dominated this space. There are a number of reasons why this may be the case. It may be the influence of the UN, other international organisations or NGOs, or the fact that the language of human rights has become to persuasive. Either way the past few decades have seen innumerable calls for corporations to respect human rights in the communities in which they operate. It may also be because of expected reputational benefits.

In the mid-1990s the scope of human rights expanded to include corporations. Traditionally, the protection of human rights was generally considered the responsibility of the state, and human rights abuse typically occurred at the hands of the state. The expansion of the role of corporations to include human rights is not necessarily a recent assertion. Since the inception of capitalism there have been calls for corporations to consider the effects of their activities, however most of these claims came from left-wing advocates. Today, these calls result not from those with redistributive ideologies but from international institutions and non-government organisations, many of who have no political affiliation to maintain their credibility (Ratner 2001).

Broadly, human rights are the inalienable rights inherently granted to individuals simply because they are human, irrespective of gender, race, colour, religion, political or social status. Human rights are universal and are imbedded in the Universal Declaration of Human Rights (UDHR) and two additional covenants; the International Covenant on Civil and Political Rights, and the International Covenant on Economic, Social and Cultural Rights (UN General Assembly 1948, 1966a, 1966b).

According the British Government, human rights 'consist of civil and political rights, freedom of opinion and expression, freedom of thought, conscience and religion, and the right to take part in the government of one's own country—and economic, social, and cultural rights—including the right to employment and to a decent standard of living, health and education. Human rights include the most fundamental right of all—the right to life'. Human rights are universally recognised with all countries having ratified at least one of the core human rights treaties stipulated by the United Nations (UN), and 80% of states having ratified four or more of the treaties (OHCHR 2015). The universal ratification of at least one of the human rights treaties creates legal obligations and specifies the role of the state in ensuring human rights are unassailable.

Given the state centric origin of the human rights rhetoric there are challenges and limitations in extending their scope to corporations.[3] Some of the issues that arise when considering human rights in the private sector are the result of the legalistic character of human rights and some are the result of conceptual difficulties with the human rights rhetoric.

Legally, there is a governance divide between corporate human rights abuse and international human rights law. John Ruggie suggests that the UN's Secretary-General's Special Representative for Business and Human Rights, 'given the lack of international legal personality of corporate actors, they cannot be directly bound by international law as such, and that, apart from certain narrowly drawn responsibilities in the field of international criminal law, corporations have no existing international obligations in the field of human rights, as most codes are voluntary in nature and are addressed to States' (Ruggie 2006). However, there are 'soft-law' measures that insure that corporations are not absolved of their duty to respect human rights. Thus while there is a governance gap between human rights law and the role of corporations there are some legal measures to ensure corporations act in an ethical manner. There is still an issue of whether the human rights framework is one that should be guiding corporate development actions.

When the social role of corporations was limited to simply not infringing on individuals freedoms it was easier to overcome the gap between human rights activism and theory. However as the social role of corporations has expanded to include development activities, the inadequacies of human rights as a development framework has become more obvious.

In 2008 Ruggie produced a report for the UN titled *Protect, Respect and Remedy: A Framework for Business and Human Rights* (Ruggie 2008). The purpose of this report was to outline the corporation's role in relation to human rights. This was followed by another report, which formed the basis of the *Guiding Principles on Human Rights and Business*, informally known as the Ruggie Framework (United Nations 2011). However, the Ruggie Framework did not come to fruition easily. Since the 1970s the UN has established various commissions and groups to identify and generate an agreed code of conduct for transnational and multinational corporations (TNCs and MNCs) (Deva 2012). The first attempt was presented in 1990 after a decade of

[3] The state-centric nature of human rights is embodied in the preamble and early chapters of the United Nations Charter.

deliberation, but was ultimately not adopted, due to numerous disagreements between developing and developed countries (Muchlinski 2007).

The next attempt to adopt some form of human rights policy for corporations was proposed by former UN General Secretary Kofi Annan in 1999 in the form of a 'Global Compact'. The Global Compact consists of ten principles in the areas of human rights, labour, environment and anti-corruption (United Nations Global Compact 2013). The Global Compact was a voluntary, opt-in approach to guide good corporate behaviour. In 2003 another UN working group submitted to the *Sub-Commission on the Promotion and Protection of Human Rights* for approval the final draft of the *Norms on the Responsibilities of Transnational Corporations and Other Business Enterprises with Regards to Human Rights* (hereafter referred to as the Norms). While, the Norms were accepted by the Sub-commission, significant resistance from both corporations and states led to the Commission on Human Rights concluding that the Norms held no legal standing (Deva 2012). It is in light of these previous attempts to construct an international standard on corporations and human rights that the Ruggie Framework was established.

Importantly, within the Ruggie Framework it was noted that corporations and states had different roles in relation to human rights. This was one of the areas of contention when the Norms were first presented to the public. They imposed a greater responsibility for human rights protection on corporations than on states. Therefore it was significant that the Ruggie Framework was based on the understanding that corporations and the state had different but complementary roles. Within the Ruggie Framework, the corporation must respect human rights. They must consider all rights their operations may impact and ensure they do not abuse or violate any of them (Ruggie 2008). In doing this, Ruggie stresses the importance of not reducing human rights to an arbitrary list, as doing so would significantly reduce the accountability of corporations. However, Handelsman (2002) suggests that this may be the only context in which corporations have the capacity to view human rights.

According to Handelsman there are a number of supposed origins of human rights, including the divine endowment of rights as represented in the American Declaration of Independence, the humanist view that rights are based on the dignity of human nature, the utilitarian view of rights and the arbitrary decisions of legislative bodies arbitrating rights. Corporations only have the capacity to view human rights in the final context. They lack the theological and humanitarian wisdom and the foresight to understand human rights in other contexts. Thus frequently human rights are reduced to an arbitrary list of rights. However there are a number of issues that arise when the concept of human rights is reduced to a list. When reduced to a list the risk of excluding relevant and important human rights becomes high, and there is a much greater danger of the list becoming a dictatorial Northern construct imposed on the South, with little consideration of Indigenous social or cultural values.

In 2011 the Ruggie Framework was unanimously approved by the United Nations Human Rights Council, making it the first corporate human rights framework to be endorsed by the UN and, as a result, making it the foremost framework for human rights integration with business. However the scope of this framework is limited to viewing the social role of corporations in respecting human rights. There is no mention of the role of corporations in contributing to development activities. As the earlier UN attempts suggest, there is great difficulty in establishing a universal, internationally recognised, set of principles to guide corporate community actions.

The universally accepted moral and ethical understanding of the role of corporations in relation to human rights only extends to corporations as human rights respecters. This raises questions about the limitations of the conceptual understanding of human rights as a development framework. The human rights framework is based on the assumption that every person is entitled to a certain basic standard of human dignity. Development is viewed as the fulfilment of rights that are depriving individuals of this dignity. However, the human rights approach does not establish what a 'right' necessarily means. The language of rights suggests that having a 'right to' something means being treated in a certain way, or achieving a certain level of well-being, or creating certain opportunities and capacities to live a life of choice (Nussbaum 1997)? However, without a definitive understanding of what constitutes a 'right' it is difficult to move beyond the rhetoric into a more practical application of rights-based policies.

Human rights are morally appealing, yet there is some concern about their usefulness in influencing and developing policy (Sen 2005). Rights are easy to state allowing them the ability to be used to defend almost any position. However, often the simplicity of the rights language does not reflect the complexity of the situation (Handelsman 2002). There is an obvious conflict between the facile language of rights and the complex nature of business. The overly rhetorical nature of the human rights framework makes it very hard to implement successfully, and measure the effects of human rights based policies. This is not to say there is no role for human rights, instead it is to suggest that attempting to articulate a practical role for corporations in community development, based on the fundamental ideas of a human rights framework, is fraught with difficulties.

Human rights are a set of universal ideals for individuals that can be held up as an ethical-political demand for justice. Therefore, while the major role for human rights may be in providing an ethical ideal for corporations to be held to, they do not provide the comprehensive theoretical underpinnings necessary to establish a solid corporate community development initiative. Given this, the purpose of the rest of this chapter is to determine if there is a more useful narrative that allows corporations to not only recognise the impact of their actions on the freedoms of individuals and the community, but also provides them with the opportunity to actively contribute to that community. The following section explores the capability approach as a means of providing a more useful working framework for CCD.

The Capability Approach and CCD

As CCR moves towards CCD there is a moral imperative to establish an understanding of what constitutes development in the corporate sphere. As established earlier, the human rights framework is limited in its ability effectively to establish a community development role for corporations. Yet there is still the necessity to determine an effective theory to underpin the actions of CCD. The development literature has moved away from understanding development as purely wealth-creation and material prosperity, towards recognising the fundamental role of development as enabling individuals to live a life of choice. In the following section it is proposed that it may be more productive to leave the ambiguous, abstract language of human rights to the ethical domain and, instead, focus on the language and ideas of the capability approach.

The capability approach was first developed by Amartya Sen in the 1980s (Sen 1985). The fundamental basis of this approach is the creation of freedoms or capabilities to enable individuals to live a life in which they have choice. While, the capability approach has been used on numerous occasions to form the theoretical foundation for development, education, poverty alleviation and other social issues, there has been little exploration done on applying the capability approach to the business sector. Yet, given the redefined role of corporations within communities, and the increasing focus on development activities, there is the space to employ the capability approach in the corporate and mining context. Similar to state-led development, the formation of a solid normative theory is important, as without it there is little to guide business practices, and it is hard to determine the 'right' course of action.

In broad terms, the capability approach comprises a number of elements primarily concerned with enabling the capabilities of individuals to live a life that they value and make the choices that fulfil them. According to Sen, when examining well-being, the most important factor to consider is what individuals are able *to be* or *to do*; which Sen labels 'functionings'. Functionings are activities or '*doings*' or '*states of being*' that can be assumed (Sen 2000). For example, 'states of being' may include being healthy, being well nourished, being famished, being literate or being able to socialise. Whereas 'doings' may include getting a job, taking a holiday, voting or running for public office.

While functionings describe what individuals are able *to be* or *to do*, capabilities describe the *freedoms of choice* or *opportunities* available to achieve these functionings. Therefore, an individual's capabilities represents their capacity to choose between various functionings to create a life they value. However, the capability set outlined by Sen is not merely a description of the functionings currently available to a person, but a set of functionings from which a person *should* have the ability to choose (Martins 2006). Therefore well-being should be measured by the capabilities of an individual, not their functionings, and should identify opportunities, not merely their achievements (Frediani 2010; Sen 1992). It is only when functionings and capabilities are examined together that a true understanding of the context is formed.

Sen has consistently refused to prescribe a predetermined list of core capabilities on the basis that doing so would undermine the fundamental understanding of his capability approach (Sen 2004). That is, that the capabilities need to be tailored to the specific context and depend upon personal value judgments (Clark 2005). Therefore Sen views a predetermined list of capabilities, compiled without public contribution or debate or social reasoning, to be fundamentally contrary to his approach. Instead, he counters, it should be the people directly affected that should participate and decide what capabilities are most important (Sen 2000). While this reasoning is philosophically sound, in reality it is difficult to apply and translate into policy. Sen does not make any distinction between capabilities that should be fundamental to every human life and those that are trivial. Without a starting point, operationalising or implementing the core ideas of his approach into policy becomes difficult.

In this regard Nussbaum disagrees with Sen's position. Table 7.1 highlights ten central capabilities that Nussbaum considers to be the minimum necessary requirements that every state should offer each citizen (Alkire 2002).

She maintains these capabilities are every person's moral entitlements; deeming them 'central capabilities' necessary for a human life to be 'not so impoverished that it is not worth the dignity of a human being' (Nussbaum 2000). Her list is comprised in such a manner that she deems all components to be separate; that is, one element cannot be satisfied merely by providing a larger amount of another (Anand et al. 2005). The list is intentionally general, allowing the capabilities to be interpreted and implemented into policies that are relevant and appropriate to each state. However, throughout her writings Nussbaum has reiterated that her list is not intended to be definitive, and suggests the significance of the capabilities should be discussed through public debate, in this regard she largely agrees with Sen.

Given the elements of the capability approach, how does this translate into an effective theoretical framework for CCD? The capability approach provides an answer to the question 'what constitutes development'. Within the capability approach the purpose of development is to improve the lives of individuals within a community. While this does not differ from the development objective of the human rights framework, what does differ is what comprises this improvement. Within the human rights framework, poverty is viewed as the deprivation of human rights. Therefore development is seen as the fulfilment of human rights. However, as noted earlier, there are issues with viewing development in this way. Alternatively, the capability approach views the improvement of peoples' lives as the expansion of individuals' capabilities. It is in expanding these capabilities that people have the capacity to live a life in which they have choice. This difference is more than semantics. Within the perspective of human rights the scope of development extends to fulfilling an individual's human right, there is no mention of the role of development beyond the attainment of these rights. However, the scope of development within the capability approach extends to include any capabilities that improve the lives of individuals. These capabilities are not confined to an established list but rather determined by the desires of the community. This is an important element that allows the capability approach to be operationalised more easily than the human rights framework.

The capability approach does not come with a predetermined list like that of rights. While Nussbaum's list of central capabilities is a starting point, this is just what she considers to be the minimum requirement for someone to live a life of human dignity. Ultimately both Nussbaum and Sen advocate for any list of capabilities to be tailored to the specific context in which they occur and result from open, public discussion. This is an important element, especially in business, where

Table 7.1 Nussbaum's list of central capabilities

1	*Life* The ability to live a life of normal length so as to not die prematurely or before one's life becomes so reduced as to not be worth living. *Bodily health* The ability to have good health, including reproductive health, sufficient nourishment, and adequate shelter.
2	*Bodily integrity* Being secure from violent assault, including sexual assault and domestic violence; having choice in reproduction and opportunities for sexual satisfaction.
3	*Senses, imagination, and thought* The ability to use the senses, to imagine, think, and reason and to achieve these in a competent way cultivated by an adequate education. To use imagination to experience and produce works of one's own choice with guarantees of freedom of expression.
4	*Emotions* Being able to have attachments to things and people outside ourselves and not having one's emotional development hindered by fear or anxiety.
5	*Practical reason* Being able to form a conception of good and to engage in critical reflection about the planning of one's life.
6	*Affiliation* (a) Being able to live with others, engage in social interaction, and recognize and show concern for other human beings. (b) Having the social bases of self-respect and non-humiliation; being able to be treated as a dignified human being with equal worth.
7	*Other species* The ability to live with concern for and in relation to animals, plants, and the world of nature.
8	*Play* Being able to laugh, to play, and to enjoy recreational activities.
9	*Control over one's environment.* (*a*) *Political*—Being able to participate in political choices that govern one's life; having the right of political participation, and protections of free speech and association. (*b*) *Material*—Being able to hold property and having property rights in equal basis with others; having the right to seek employment on equal basis with others; having the freedom from unwarranted search and seizure.

Source: (Nussbaum 2011)

corporations have the ability to impact very specific areas of an individual's freedom; which may differ from context to context. The open public discussion should consider the various perspectives of individuals within their specific community. It is in establishing the community's priorities that the development activities of corporations can be focused. According to Sen 'it is the people directly involved who must have the opportunity to participate in deciding what should be chosen, not local elites … or cultural experts' (Sen 2000). Hence, rather than development being motivated by an abstract idea that it is an individual's entitlement, with little understanding of a communities specific wants and needs, development is driven by the specific desires of the community.

Conclusion

A fundamental shift has occurred in recent years in the role of corporations in society. While CCR has undergone numerous transformations, today there is increasing focus on how corporations should contribute to development within the community in which they operate.

There have been numerous points of difference around this issue and it is these that have dominated the debate. One area of debate is whether CCD projects are motivated by a genuine desire to contribute to society or are a public relations exercise. Another area of dispute has been about how best to deliver these development activities. Yet among these debates there has been relatively little discussion about the fundamental theoretical and philosophical foundations around what actually constitutes development and how this role is extended to corporations. This domain has been largely neglected in the literature, however there is great importance in establishing an adequate theory that effectively defines development and the scope of corporations.

To date, much of the theoretical discussion has been based on the implicit assumption that a human rights framework provides an adequate underpinning for CCD development. Human rights are morally and ethically appealing. However, they have limitations that make their utilisation as a normative development theory difficult. While the ubiquitous nature of the human rights rhetoric often means it can be evoked to defend almost any position, it is this same universality that limits its ability to form an adequate development framework. In order to appeal universally, the language is often ambiguous and abstract. The facile language of human rights often fails to reflect the complexity of a situation. Further, when development is extended to include corporations, there is the need to amalgamate a corporate agenda with a development agenda, which often does not reflect the same 'state-led' development agenda. The human rights framework does not provide the space for this to occur. Consequently, it is more appropriate that the human rights rhetoric remains in the ethical realm and act as a moral guide for all corporations, and especially mining companies.

If the role of human rights is to provide the ethical and moral guidance to corporations there remains a space for the creation of a development theory that defines development in a way that considers, not only what individuals need, but also enables corporations to become engaged in communities. The capability approach views development as the expansion of capabilities. The expansion of capabilities allows individuals to live a life they value and choose. This understanding differs from the understanding of development within the human rights perspective. It expands the scope of development to not only include actions that lead to the fulfilment of human rights but the potential to expand the scope of development to capture those capabilities that the community truly values.

The establishment of a solid theoretical and philosophical foundation is vitally important. Without instituting a normative theory that sufficiently informs development actions there is little to guide the objectives of CCD. This area has been significantly neglected in the past. However, both the corporation and the community benefit from the establishment of a robust adequate theoretical framework. Corporations are not aided by the use of ambiguous, abstract ideas, and unfortunately this occurs, when the human rights framework attempts to become a foundation for development. Similarly, communities are not necessarily benefited by corporation's attempts to deliver development actions under the guise of human rights. The designation of the capability approach as a development framework to underpin CCD creates the opportunity for future research to explore the impact of this theory on the development actions of corporations.

References

Alkire, S. (2002). Dimensions of human development. *World Development,* 30(2), 181-205.

Anand, P, Hunter, G., Smith, R. (2005). Capabilities and well-being: evidence based on the Sen-Nussbaum approach to welfare. *Social Indicators Research,* 74(1), 9-55.

BHP Billiton. (2014). Value through performance: sustainability report 2014. http://www.bhpbilliton.com/home/society/reports/Documents/2014/BHPBillitonSustainabilityReport2014_interactive.pdf. Accessed 24 September 2014.

Blowfield, M. (2004). CSR and development: is business appropriating global justice? *Development,* 47(3), 61-68.

Carroll, A., Buchholtz, A. (2002). *Business and society: ethics and stakeholder management.* 5th edn. Cincinnati: South-Western Publishing.

Clark, D. (2005). The capability approach: its development, critiques and recent advances. *Global Poverty Research Group,* 1-18. http://www.gprg.org/pubs/workingpapers/pdfs/gprg-wps-032.pdf. Accessed 12 February 2012.

Davis, R., Franks, D. (2014). *Costs of company-community conflict in the extractive sector.* Cambridge, Massachusetts: Harvard Kennedy School.

Deva, S. (2012). Guiding principles on business and human rights: implications for companies. *European Company Law,* 9(2), 101-109.

Frediani, A. (2010). Sen's capability approach as a framework to the practice of development. *Development in Practice,* 20(2), 173-187.

Handelsman, S. (2002). Human rights in the minerals industry. mining, minerals and sustainable development. https://commdev.org/files/1152_file_Human__20Rights__20in__20the__20Minerals__20Industry.pdf. Accessed 9 June 2013.

International Council on Mining and Metals. (2006). Resource endowment initiative. http://www.icmm.com/page/84152/our-work/projects/articles/resource-endowment-initiative. Accessed 5 August 2013.

International Organization for Standardization, (2010). Social responsibility. International Organization for Standardization. http://www.iso.org/iso/home/standards/iso26000.htm. Accessed 5 August 2013.

Kamminga, M. (2015). Company responses to human rights reports: an empirical analysis. SSRN. http://papers.ssrn.com/sol3/papers.cfm?abstract_id=2559255. Accessed 5 August 2013.

Kemp, D. (2010). Mining and community development: problems and possibilities of local-level practice. *Community Development Journal,* 45(2), 198-218.

Mahon, J., McGowan, R. (1991). Searching for the common good: a process-oriented approach. *Business Horizons,* 34(4), 79-87.

Manteaw, B. (2007). From tokenism to social justice: rethinking the bottom line for sustainable community development. *Community Development Journal,* 43(4), 297-311.

Martins, N. (2006). Capabilities as causal powers. *Cambridge Journal of Economics,* 30(1), 671-685.

Muchlinski, P 2007, *Multinational enterprises and the law.* 2nd edn, Oxford: Oxford University Press.

Newell, P. (2005). Citizenship, accountability and community: the limits of the CSR agenda. *International Affairs,* 81(3), 541-557.

Newmont. (2009). Community relations review: global summary report. http://www.newmont.com/files/doc_downloads/crr/CRR-Global-Summary-Report-and-Appendices-English.pdf. Accessed 16 July 2013.

Nussbaum, M. (2011). *Creating capabilities.* Cambridge, Massachusetts: Harvard University Press.

Nussbaum, M. (2000). *Women and human development: The Capabilities Approach.* Cambridge: Cambridge University Press.

Nussbaum, M. (1997). Capabilities and human rights. *Fordham Law Review,* 66(2), 273-300.

United Nations. Office of the United Nations High Commissioner for Human Rights (2015). *What are human rights?* http://www.ohchr.org/EN/Issues/Pages/WhatareHumanRights.aspx. Accessed 10 January.

Paton, B. (2000). Voluntary environmental initiatives and sustainable industry. *Business Strategy and the Environment,* 9(5), 328-338.

Ratner, S. (2001). Corporations and human rights: a theory of legal responsibility. *The Yale Law Journal,* 111(1), 443-545.

Rio Tinto. (2013). 2013 Sustainable development: supporting our licence to operate. http://www.riotinto.com/sustainabledevelopment2013/_pdf/rio_tinto_2013_sustainable_development.pdf. Accessed 6 April 2014.

Ruggie, J. (2006). Interim report of the special representative of the secretary-general on the issue of human rights and transnational corporations and other business enterprises. United Nations. http://www1.umn.edu/humanrts/business/RuggieReport2006.html. Accessed 6 April 2014.

Ruggie, J. (2008). Protect, respect and remedy: a framework for business and human rights. United Nations. http://business-humanrights.org/en/un-secretary-generals-special-representative-on-business-human-rights/un-protect-respect-and-remedy-framework-and-guiding-principles. Accessed 6 April 2014.

Sayer, J. (2005). Guest editor's introduction: do more good, do less harm: development and the private sector. *Development in Practice,* 15(3/4), 251-268.

Sen, A. (2005). Human rights and capabilities. *Journal of Human Development,* 6(2), 151-166.

Sen, A. (2004), Capabilities, lists and public reason: continuing the conversation. *Feminist Economics*, 10(3), 77-80.

Sen, A. (2000). *Development as freedom.* United States of America: Anchor Books.

Sen, A. (1985). *Commodities and capabilities*. Amsterdam: North-Holland.

Sen, A. (1992). *Inequality re-examined.* Oxford: Oxford University Press.

Sethi, S. (2003). *Setting global standards: guidelines for creating codes of conduct in multinational corporations*. New Jersey: John Wiley & Sons.

Sullivan, R (Ed.), (2003). *Business and human rights: dilemmas and solutions.* Sheffield, England: Greenleaf Publishing.

UN General Assembly. (1948). Universal declaration of human rights. United Nations. http://www.un.org/en/documents/udhr/. Accessed 14 February 2015.

UN General Assembly. (1966a). International covenant on civil and political rights. United Nations. https://treaties.un.org/Pages/ViewDetails.aspx?src=TREATY&mtdsg_no=IV-4&chapter=4&lang=en. Accessed 14 February 2015.

UN General Assembly. (1966b). International covenant on economic, social and cultural rights. United Nations. https://treaties.un.org/Pages/ViewDetails.aspx?src=TREATY&mtdsg_no=IV-3&chapter=4&lang=en. Accessed 14 February 2015.

United Nations. (2011). Guiding principles on business and human rights: implementing the united nations "protect, respect and remedy" framework. United Nations. https://www.unglobalcompact.org/library/2. Accessed 14 February 2015.

United Nations Global Compact. (2013). Overview of the UN Global Compact. United Nations. https://www.unglobalcompact.org/aboutthegc. Accessed 20 February 2015.

Utting, P. (2005). Corporate responsibility and the movement of business. *Development in Practice,* (15)3, 375-388.

Van Marrewijk, M. and Werre, M. (2003). Multiple levels of corporate sustainability. *Journal of Business Ethics,* 44(2/3), 107-120.

Velasquez, M. (1992). International business, morality and the common good. *Business Ethics Quarterly,* 2(1), 27-40.

Weiss, J. (2003). *Business ethics: a stakeholder and issues management approach.* 3rd edn. Ohio: Thomson - South-Western.

World Commission on Environment and Development. (1987). *Our common future.* Oxford: Oxford University Press.

Chapter 8
Mergers and Acquisitions: The Politics of Megadeals in the Mining Industry

Vlado Vivoda and Geordan Graetz

Abstract Activity in the mining industry in the Asia-Pacific Region has surged during the last 25 years, as growing populations, higher personal incomes, and increased demand for goods, services, and infrastructure—particularly in China and India—have led to large-scale building programs and rising energy consumption. This has intensified the use of minerals, metals, and oil and gas. While this longitudinal trend has been interrupted by several global and regional financial shocks, including the 1997 Asian Financial Crisis and the 2008 Global Financial Crisis (GFC), both of which saw weakened demand for commodities, mining companies predominantly have experienced a sustained period of production and revenue expansion, and growth in their market capitalisation. Competition for resource deposits, the emergence of a plethora of juniors eager to capitalise on the burgeoning demand, and consolidation at the mid-cap and major level have been observable features of the mining industry market. With this consolidation in mind, here, we examine recent merger and acquisition (M&A) activity in the mining and energy sectors across the Asia-Pacific Region.

Abbreviations

CNOOC	Chinese National Offshore Oil Company
CNPC	Chinese National Petroleum Corporation
FDI	Foreign Direct Investment
GFC	Global Financial Crisis
IPE	International Political Economy
MNC	Multinational Corporation

V. Vivoda (✉)
Centre for Social Responsibility in Mining, Sustainable Minerals Institute, The University of Queensland, St Lucia, QLD 4072, Australia
e-mail: v.vivoda@uq.edu.au

G. Graetz
Centre for International Risk, School of Communication, International Studies and Languages, The University of South Australia, Adelaide, SA 5001, Australia
e-mail: geordan.graetz@gmail.com

T. O'Callaghan, G. Graetz (eds.), *Mining in the Asia-Pacific*, The Political Economy of the Asia Pacific, DOI 10.1007/978-3-319-61395-6_8

M&A	Mergers and Acquisitions
NOC	National Oil Company
Q2	Second Quarter
RN	Resource Nationalism
US	United States

Introduction

Activity in the mining industry in the Asia-Pacific Region has surged during the last 25 years, as growing populations, higher personal incomes, and increased demand for goods, services, and infrastructure, particularly in China and India, have led to large-scale building programs and rising energy consumption. This has intensified use of minerals, metals, and oil and gas. While this longitudinal trend has been interrupted by several global and regional financial shocks, including the 1997 Asian Financial Crisis and the 2008 Global Financial Crisis (GFC), both of which saw weakened demand for commodities, mining companies predominantly have experienced a sustained period of production and revenue expansion, and growth in their market capitalisation. Competition for resource deposits, the emergence of a plethora of juniors eager to capitalise on the burgeoning demand, and consolidation at the mid-cap and major level have been observable features of the mining industry market.

With this consolidation in mind, we examine recent merger and acquisition (M&A) activity in the mining and energy sectors across the Asia-Pacific Region. The chapter proceeds in five sections. The present section outlines the aims and scope of the chapter. In the second section, we survey broader market activity and likely future trends. In the third and fourth sections, we turn our focus to a subset of M&A activity cross-border M&A *megadeals*, which are defined as attempted or completed cross-border M&A with a value over US$10 billion. The final section concludes the chapter.

There have been six attempted cross-border megadeals in the region since 2005, three of which have been successful. We review the theoretical literature on M&A in order to understand why and how such deals manifest, paying particular attention to the political-economic grounds for their success or failure. Discussion of megadeals is important for several reasons. First, such deals allow for the examination of broader geo-political issues, such as the shifting balance of power within the Asia-Pacific Region. Second, by analysing the reasons for their success or failure, we also can draw attention to recent learning within the mining and energy industries regarding the navigation of regulatory and political impediments to transactions. Third, deals with values over US$10 billion are of regional economic significance.

Our review of the literature on M&A draws on the disciplines of international political economy (IPE), political science, international relations, and international business studies. This literature offers three plausible reasons for the success (or failure) of megadeals. The first is *market structure*; that is, the extent of the impact of an attempted M&A megadeal on industry (or market) concentration levels in the target country and the effect it had on its successful completion; second, *power*: corporate interests influence political actors and political structures in target countries to such an extent that they affect the successful completion of an attempted M&A megadeal; third, *ideas*: political/ideational issues in the target countries, related to an increase in nationalism during a resource boom period (Johnson in this volume), have an impact on the successful completion of an attempted megadeal.

By analysing attempted megadeals in the mining industry, we also speak to the debate about the extent of state involvement in the functioning of the markets for commodities in the Asia-Pacific Region during the past decade (Underhill 2000). In particular, we contribute to the existing body of work within the discipline of IPE on the extent of state protectionism in the mining industry, which suggests that there has been a resurgence of resource nationalism (RN) (Joffe et al. 2009; Vivoda 2009; Johnson in this volume). We add to this literature by showing why, and under what conditions, target states engage in RN and investment protection. We probe whether national governments seek to reconcile domestic populist pressure, win elections, and/or enhance the national interest, as well as whether material or ideational prerogatives dominate government decision-making.

Trends in M&A Activity in the Mining and Energy Sectors

Global Activity

Global M&A activity in the mining industry has been concentrated in the Asia-Pacific Region. However, recent conservative market conditions, disciplined capital management by firms (retention of capital, or return of monies to shareholders in the form of dividends), and volatile commodity prices have seen the suppression of activity. Nevertheless, mining companies have out-performed the financial market, and have led the post-GFC recovery. While activity across the Asia-Pacific Region in the mining industry reached all-time lows in 2013, there now are signs that M&A market activity is gaining momentum.

There were 702 global M&A deals in the mining industry in 2013, with a combined value of US$87.3 billion. This excludes the US$37.4 billion merger between Swiss mining giants Glencore and Xstrata. In second quarter (Q2) 2014, there were 112 deals in the industry, with a value of US$9.5 billion. This was down 21% on the previous quarter, and 41% on Q2 2013. However, total deal value was 33% higher than first quarter 2014.

If we look to Australia as an example of recent market trends, we see that there were 178 deals involving Australian mining companies in 2013, with a total value of US$5.5 billion. The number and value of deals were down 24% and 65% on 2012 figures, respectively. This was the lowest volume since 2003 and the lowest value since 2004. M&A activity declined again in 2014, reaching historical lows. Despite this, Australia continues to be viewed as a favourable investment destination

(KPMG 2014), and there is expected to be increased activity, particularly in the iron ore sector (Riley-Smith and Mak 2014).

There has been substantial M&A activity in Canada's gold and copper sectors, with some Canadian firms undertaking sizeable acquisitions in Australia, including the '$543-million offer for Australia's Papillon Resources by Vancouver-based mid-tier B2Gold' (Lazenby 2014). There were 12 gold deals in Canada in Q2 2014, with a combined value of US$4.6 billion. The largest of these deals (US$3.6 billion) was the acquisition of Osiko Mining by Yamanda Gold Inc. and Agnico Eagle Mines Ltd. (Lazenby 2014; Kean 2014). In the copper sector, the largest proposed deal (AUD$1.1 billion) was advanced by China's state-owned Guandong Rising Assets Management for Australia's PanAust. The deal did not proceed, however. Merger discussions between leading global gold producers Newmont Mining and Barrick Gold also broke down in April 2014 (The Canadian Press 2014).

In the oil and gas sectors, there were 1600 deals in 2012, with a value of US$400 billion (Hammerson and LaMaster 2014). New entrants to the market, including expansionist national oil companies (NOCs), financial investors, and oil traders, drove this activity (Hammerson and LaMaster 2014). In 2013, there were 1400 deals totalling US$337 billion (EY 2014a). There were four deals with values over US$10 billion in 2012, and three in 2013. The largest deal (US$270 billion) was a supply contract between China National Petroleum Corporation (CNPC) and Russia's Rosneft. In the first half of 2014, there were 101 deals in the sectors with a value of US$60.5 billion (Oil & Gas Journal 2014). This was the strongest level of activity in five years, and occurred at upstream, midstream, and downstream levels (Oil & Gas Journal 2014). Approximately 50% of the activity centred on 'shale plays' in the United States (US), with Texas being a hub of activity (Oil & Gas Journal 2014). Unconventional gas deals were among the highest value transactions (EY 2014a). By November 2014, deals in the oil and gas sectors for the year had totalled US$369 billion. There were 29 deals with values over US$10 billion, including Halliburton's US$38 billion acquisition of Baker Hughes. Activity in the sectors was concentrated in Australia, Canada, and the US; however, deal volume was down in all three countries (EY 2014a). The US and Canada together accounted for 61% of deals in the sectors. Total deal value in Australia fell from $US16.4 billion in 2012 to $1.4 billion in 2013.

China has been behind much of the M&A activity in the Region's mining activity. For example, China accounted for 66% of Australian in-bound mining sector M&A in 2013, and Chinese investors are predicted to 'continue an aggressive acquisition spree' (Grant Thornton 2014). Outbound investment from China became more feasible after Beijing 'abolished quotas on purchase of foreign exchange for overseas investment' in 2005 (The Beijing Axis 2014). However, despite the loosening of restrictions on outbound foreign direct investment (FDI), Chinese firms arguably have not had much success. Indeed, Wang Jiahua, vice chair of the China Mining Association, estimates that approximately 80 per cent of Chinese outbound investments in mining companies have failed (The Beijing Axis 2014). While we discuss several plausible reasons as to why China has faced impediments to its attempted acquisitions below, it is noteworthy at this juncture that China's growing demand for mineral and energy resources will necessitate the adoption of new approaches to FDI in order to realise its investment goals. One such strategy is to reduce the level of FDI by Chinese state-owned enterprises in favour of private sector-led investment (The Beijing Axis 2014). It also is noteworthy that many of China's overseas investments now proceed through Hong Kong, perhaps as a means to circumvent opposition to Chinese FDI in target countries.

Future Trends

Based on industry projections, we can point to a number of trends that likely will characterise the M&A market in the mining industry in the near to medium-term.

1. While the market has not yet recognised and rewarded the capital discipline of the extractive companies (Barings 2014), the period of conservative fiscal management is coming to an end. The rise of hedge funds, other private equity firms, and superannuation funds presents extractive companies with alternative options for capital-raising (KPMG 2014). In the first half of 2014, 21% of mining deal volumes were made by financial investors. These funds were estimated to have amassed US$20 billion to invest in the mining industry in 2015 alone (Kean 2014). However, capital will continue to be scarce for juniors.
2. According to global accounting firm, Grant Thornton (2014), one third of junior mining companies need to raise capital. Given that equity markets have been closed to juniors, due to the market's conservative fiscal stance, increased M&A activity is likely. While there will be further consolidation in the market at all levels, from juniors, through mid-caps, to majors, most activity will be concentrated at the junior level (Barings 2014). At the major level, it is noteworthy that GlencoreXstrata approached Rio Tinto about a potential merger in July 2014. There has been no further activity on this front. Another potential target for acquisition or merger is AngloAmerican. Notwithstanding the attractiveness of Anglo, Barings (2014) forecasts that another large-cap merger is a low probability in the near-term.
3. Junior and mid-cap firms have been undervalued by the market (KPMG 2014). As a consequence, a rerating of mining shares is likely (Barings 2014).
4. The bulk of activity will continue to be concentrated in the Asia-Pacific Region, specifically, in the Asian hemisphere. Gold firms will be principal targets (Mining Insight 2014).
5. Mergers and acquisitions increasingly will be seen by companies as a means to deliver sustainable earnings growth (Barings 2014). Accordingly, there will be a period of divestment of non-performing assets by the majors. While there already is a large volume of assets for sale, given market conditions, not all assets will be sold. The Deloitte Center for Energy Solutions (2014) suggests

that companies will divest of non-performing assets in order to raise capital to invest in their core projects.

6. Grant Thornton surveyed 250 global mining executives, with the consensus being that M&A activity in the mining industry will increase from 2015 (Batten 2014). However, activity is not predicted to return to previous highs (Mining Insight 2014).
7. In the oil and gas sectors, increased M&A activity is forecast to drive the next phase of production, which is due to start in 2018 (Herbert Smith Freehills 2014). However, according to the Deloitte Center for Energy Solutions (2014), oil sands developments will continue to struggle for capital.
8. Other challenges are predicted to come from shareholder activists and judicial scrutiny. As Riley-Smith and Mak (2014) observe:

   When a target is in play, its share register often changes significantly in a short period of time, as event-driven hedge funds buy target shares … Recently with emergence and higher profile of 'activist shareholders', the attention given to shareholders in an M&A context has changed from enlisting major shareholders to assist the bidder in putting pressure on the target, to dealing with the shareholder's own agenda. 'Activist' shareholders may have their own agenda which does not necessarily align with either the bidder's or target's objectives.

Explanations for the Success or Failure of Mergers and Acquisitions from the Literature

We move beyond the embrace of a particular orthodoxy (for example, classical or Keynesian, material or ideational) to provide meaningful contributions not only to theoretical knowledge, but also to public policy and corporate decision-making regarding the governance of, and M&A activity in, the mining industry (Sørensen 2008). To this end, we engage both material and ideational literatures. Such an approach allows for the integration of the distribution of capabilities (wealth and power) among the various stakeholders and institutions associated with M&A activity with ideational factors (history, culture, and identity).

Previous studies on M&A activity in the mining industry either have focused on single cases (Drysdale and Findlay 2009; Yao et al. 2010) or on the purely economic aspects of deals. These studies have found that target countries treat foreign M&A attempts differently than domestic M&A by discriminating against foreign investors (Evenett 2002; Aktas et al. 2007; Conybeare and Kim 2010). We contribute to this body of knowledge by examining the politics of megadeals in the mining industry in the Asia-Pacific Region. In order to ascertain why the six attempted cross-border M&A megadeals surveyed below either were successful or unsuccessful, we examine three plausible explanations that emerge from a review of the literature.

The Market Structure Explanation

Tensions arise from different levels of concentration of market power. Oligopolistic or monopolistic market structures historically have been present in the mineral and energy industries. Against a competitive market structure, oligopolistic or monopolistic structures can cause significant inflationary pressure for countries, and undermine macroeconomic balance. Such concerns are central to economic policy debates, which have been marked by shifts in influence between classical economists (Friedman 1962) and their institutional and Keynesian rivals (Galbraith 1967; Krugman 1995). Without corrective public policy responses, monopolistic or oligopolistic market structures in the mining industry can undermine the macroeconomic balance, as well as economic growth, as illustrated by the example of the failed acquisition of Potash Corp. by BHP Billiton (discussed below). Accordingly, any inflow of capital, while an advantage in the short term, may be a source of vulnerability over the longer term (Widmaier 2003). This potentially can have negative consequences for growth. To the extent that extractive companies can attract investment inflows and partner in joint ventures, they also can increase the potential vulnerability of their domiciled economies to shifts in investor sentiment and to foreign government influence in their affairs. Consequently, the first possible explanation is that *megadeals that would have had a smaller impact on industry concentration levels would have been less politicised in the target country and would have been more likely to be successful.*

The Firm Influence Explanation

The concerns for market and political power, as well as foreign policy independence, cannot be approached from any one disciplinary or theoretical paradigm, but requires an eclectic political-economy approach to integrate concerns regarding interests and ideas. With respect to *interests*, consider first that conflicts between governments and multinational corporations (MNCs) usually centre on the issue of the division of rents (Vivoda 2009). During a period of high commodity prices, when large rents can be earned, the phenomenon of RN surfaces, increasing the ability of the state to identify and extract a larger share of rents (Wälde 2008; Johnson in this volume). This corresponds to Wilson's (1986) model of the politics of the world oil market—the petro-political cycle. In this view, politics at the peak of the market differ substantially from politics in a trough. In rising markets, sellers, such as commodity-exporting governments, gain leverage; in falling markets, buyers, such as MNCs or commodity-importing governments, gain leverage (Wilson 1986).

Consequently, the second possible explanation is that powerful corporate interests influence political actors and political structures to such an extent that they are able to affect the successful completion of a particular megadeal. Such an instrumentalist understanding of the role of the state within global capitalism assumes that the state simply is an instrument in the hands of corporate elites, with governments captured to ensure necessary conditions for profit maximisation (Stokes 2007). In this view, the structural power of business influences the political process.

To be sure, there are various means through which MNCs can—and do—influence government policy. Channels of corporate influence in the formulation of government policy include: participation in, or direct access to, key policy-making units; recruitment of policy-makers into the business organisation, or vice versa; joint ventures with government; government contracts; consultation with policy-making units; lobbying; the provision of bribes to government officials; and covert collusion with intelligence agencies (Brookstone 1976). Using these and other channels of influence, large mineral and energy companies may, at their own behest, serve as initiators of policy agenda items, and/or have crucial influence over policy and partisan debate regarding the outcome of an attempted merger or acquisition, as the example of the CNOOC-Unocal deal demonstrates (discussed below).

The Resource Nationalism Explanation

It also is important to highlight the ways in which policy requires *politics*, as market power finds expression in partisan debate. From this viewpoint, economic prerogatives and the market power of firms do not simply 'speak for themselves' in determining the success or failure of a particular megadeal; other factors also have an influence. The coalitional distribution of power among extractive companies, manufacturers, and other constituents (for example, labour unions) and the acquirer's and target's domiciled countries and their respective relationships (Putnam 1988) are several examples of this. Economic power must find expression in the formation of partisan alignments/coalitions among interest groups and institutional allies in government, as well as in broader debates in the media and society regarding ideas about how the economy functions and how a particular industry should be regulated.

Shifts in market power can occur in an ideational context, as views regarding the proper scope for regulation provide a broader influence on state–market interactions (Tannenwald 2005). For example, Ruggie's (1982) analysis of the ideas shaping the Bretton Woods economic order—the institutions comprising the predominant global system for the management of commercial and financial relations—highlighted the Keynesian ideas that legitimised more assertive institutional efforts to stabilise commodity prices, including Keynes' own call for the establishment of an international organisation to maintain 'buffer stocks' for commodities to reduce price volatility. In this context, our analysis suggests that ideational shifts may act as an important influence on M&A megadeals in the mining industry, as they legitimise efforts to promote price stabilisation or resource ownership.

Consequently, the third possible explanation is that *attempted megadeals have collapsed due to political/ideational issues, related to an increase in nationalism during a resource boom period*. In the context of rising markets, in which sellers—commodity-exporting governments—gain leverage, it is suggested that countries treat foreign M&A attempts differently than domestic M&A, using stringent M&A laws to discriminate against foreign investors (Evenett 2002; Aktas et al. 2007; Conybeare and Kim 2010). Indeed, governments protect targeted domestic firms, especially 'national champions', from M&A activity, even if only for populist ideological or electoral reasons. In such circumstances, governments may even 'protect' firms that wish to be acquired (Conybeare and Kim 2010).

M&A Megadeals in the Mining Industry in the Asia-Pacific Region

With these explanations in mind, let us now turn to examine the empirical reasons for the success or failure of the six cross-border M&A megadeals that have been proposed in the mining industry in the Asia-Pacific Region during the course of the last 10 years (Table 8.1). From the evidence presented below, it is apparent that a combination of the plausible explanations outlined above account for the success or failure of these deals.

On 4 April 2005, the directors of Unocal, the twelfth largest oil company in the US, accepted a $16.5 billion acquisition offer by Chevron, the second largest US oil company. The offer was one quarter in cash and three quarters in Chevron stock. However, on 22 June 2005, the Chinese National Offshore Oil Corporation (CNOOC), a Chinese state-owned company, made a counteroffer of $18.5 billion in cash, financed in part by low interest rate loans from its state-owned parent company. CNOOC's offer was driven by the company's need to increase control of overseas oil and gas fields in order to double oil production between 2005 and 2010 to meet China's growing oil demand. The acquisition of Unocal would have made that

Table 8.1 M&A megadeals in the mining industry (Asia-Pacific Region) 2005–2014

Date	Acquirer	Acquirer country	Target	Target country	Industry	Value (US $bn)	Success
06/2005	CNOOC	China	Unocal	US	Oil and gas	18.5	No
11/2007	Rio Tinto	UK/ Australia	Alcan	Canada	Mining	38.1	Yes
02/2008	Alcoa and Chinalco	US/China	Rio Tinto	UK/Australia	Mining	14.3	Yes
07/2009	Chinalco	China	Rio Tinto	UK/Australia	Mining	19.5	No
11/2010	BHP Billiton	UK/ Australia	Potash Corp.	Canada	Mining	39.0	No
12/2012	CNOOC	China	Nexen	Canada	Oil and gas	15.1	Yes

objective possible in just one transaction. However, on 30 June 2005, a non-binding US Congressional House Resolution (H.R. 344), which recommended a presidential review of the CNOOC deal, was passed by a vote of 398 to 15. In a letter to President George W. Bush, House Energy and Commerce Committee Chairman Joe Barton wrote, 'we urge you to protect American national security by ensuring that vital U.S. energy assets are never sold to the Chinese government' (quoted in Mullins and Berman 2005). In mid-July 2005, Chevron increased its bid to $17.7 billion, turning up the heat on CNOOC to respond with a higher bid of its own. Although higher, CNOOC's offer faced unprecedented political opposition in Washington, creating a level of uncertainty, which made it impossible to compete with Chevron. CNOOC withdrew its bid on 2 August 2005, allowing Chevron to proceed with the takeover, which was completed on 10 August and at a cheaper price than was offered by CNOOC.

Three factors influenced the outcome. First, Unocal had symbolic value in America, and domestic US opposition to China's purchase of the company in part was based in the overall context of Sino-American relations. Specifically, by halting CNOOC's acquisition of Unocal, Washington was attempting to contain Beijing's emerging political, economic, and military power (Schorgten 2006). Many of Unocal's assets and operations were located in strategically important regions, and some US politicians feared that China would remove oil from the markets. Dorn (2005) referred to the surge of anti-China legislation and excessive politicisation of the deal as 'creeping protectionism'. Moreover, Kim (2007) described US protectionism against Chinese ownership as 'economic patriotism', and as a new breed of mercantilist fervour or a reincarnation of protectionism from the past. Second, adroit lobbying by Chevron arguably influenced the outcome. Third, the CNOOC offer was not strictly commercial, as the bid would not have been able to proceed without the financial backing of the Chinese government.

In contrast to its failed Unocal bid, in February 2013, CNOOC successfully completed a US$15.1 billion takeover of Nexen, a Calgary-headquartered energy producer, following a review by the Canadian government. The deal represented the largest single foreign acquisition by a Chinese company. With the memory of its failed 2005 bid for Unocal still fresh, CNOOC put lessons learned into practice, networking strongly with regulators in Canada and the US, thereby smoothing the way for its cash deal to acquire Nexen. As part of the deal, CNOOC made a series of pledges to the Canadians. It promised to keep its North American headquarters in Calgary, to increase its long-term investments in oil sands developments, and to list on the Toronto Stock Exchange. CNOOC also pledged that it would not reduce the company's labour force, and would pay a 61% premium above the value of the company stock price the day before the offer was made.

The political backlash in 2005 against Chinese efforts to acquire Unocal more broadly has encouraged Chinese firms to tread more carefully as they target foreign assets. In the Unocal transaction, CNOOC was competing with Chevron, a US company, while the bid for Nexen was an uncontested, negotiated deal, which had the full support of the company's board. Also, CNOOC reassured management and the Canadian government that Nexen would remain a Canadian company. Further contributing to the political and regulatory palatability of the deal, the operational assets of Nexen primarily were located in the Gulf of Mexico and the North Sea, not on Canadian territory, thereby reducing any immediate strategic national interest concerns. Moreover, Nexen held over 50% of its assets in the United Kingdom, effectively allowing CNOOC to bypass Canadian regulatory hurdles, as the assets of interest predominantly were not on Canadian soil.

While CNOOC heeded lessons from the Unocal debacle, it also closely studied BHP Billiton's failed $39 billion bid to acquire fertiliser maker Potash in late 2010. BHP Billiton's August 2010 hostile takeover of Potash failed due to rejection by the Canadian Government. The bid widely was expected to win approval until the Saskatchewan Premier, Brad Wall, launched a public campaign to oppose the deal, demanding that Ottawa 'stand up for Canada' to prevent the loss of a 'Canadian icon' and a 'strategic resource'. On 3 November 2010, Industry Minister Tony Clement announced that he would block the BHP Billiton bid, as he determined that the acquisition would not deliver a net benefit for Canada. It was only the second rejection of a foreign takeover in Canada in 25 years. While BHP Billiton promised good outcomes in terms of employment and Canadian participation at the company level, the government deemed that the loss of control of the company negatively would affect Canada's pricing power in the global potash sector and that the loss of provincial tax revenue would be construed as incompatible with provincial policy objectives (Dawson 2012).

Realising that a hostile bid of this size nearly would be impossible to achieve after BHP Billiton's failed approach for Potash, CNOOC approached Nexen directly and received the full support of the company's board. CNOOC hired Hill and Knowlton to lobby stakeholders in Canada and the US, and Bell Pottinger in the United Kingdom. Early conversations with policy-makers that were facilitated by the lobbyists led CNOOC to propose key concessions (mentioned above). The successful acquisition of Nexen suggests that Chinese NOCs learned from their early mistakes and that they adjusted their strategies accordingly (Moreira 2013). It also became apparent to the Chinese that deals for critical resources would be easier if they were done with companies in countries whose interests more naturally aligned with China's.

Apart from the aforementioned differences with the Unocal and Potash bids, CNOOC's acquisition of Nexen has a more fundamental distinction. The Nexen transaction involved a company from a consuming country (China) acquiring a company from a supplier country (Canada). By way of contrast, Unocal involved a company from a consuming country (China), acquiring a company in another consuming country, in this case the US. Similarly, the Potash transaction involved a company from a supplier country (Australia) attempting to acquire a company from another supplier country (Canada). In the Nexen deal, the interests of the two domicile countries were aligned, while it was more difficult to see an alignment of interests in either the Unocal or Potash deals. In Unocal, the domicile countries were—and still are—in competition for oil, while in Potash, the countries involved were/are competing for market share (Perkowski 2012).

The leaders of large consuming states—in particular, the US and China—understandably are concerned about their respective countries' ability to ensure access to the resources needed to maintain domestic economic growth. Indeed, it is why America's dependence (until recently) on imported oil was a constant theme of that country's political campaigns. It also is why China, as the largest energy consumer in the world, has been securing extractive resources across the globe. However, supplier countries—like Australia and Canada, which have large reserves of natural resources, but relatively small populations and economies—mostly are concerned with finding long-term, stable markets for their products. Developing and selling more of their natural resources is their path to prosperity.

Canada's principal problem is that the US market absorbs approximately 95% of the country's oil exports (BP 2014). Having one large and powerful customer can be an asset for any company/country. However, it also can be a vulnerability, as Canada experienced first-hand early in 2012 when the Obama administration rejected the construction of the Keystone XL pipeline, which would have transferred oil from Alberta to the Texas Gulf Coast. In February 2012, shortly after President Obama thwarted construction of Keystone, then Canadian Prime Minister Stephen Harper visited China to discuss oil sales and other economic issues. Harper's visit highlighted Canada's determination to diversify its energy sales. In light of the failure to secure the Keystone pipeline, Canada now will build a pipeline to the country's Pacific Coast, where tankers will be filled to supply customers in China and elsewhere in Asia.

Economic ties between Canada and China have increased in recent years. Indeed, Chinese state-owned companies invested more than $16 billion in Canadian energy projects between 2010 and 2012, and China's state-controlled Sinopec has a stake in the proposed pipeline to the Pacific Coast, which substantially would augment Chinese investment in Alberta's oil sands. It is against this background that CNOOC pursued its acquisition of Nexen. Arguably, then, the Nexen deal was able to proceed due to broader geo-political realities, as well as the newly formed Sino-Canadian 'strategic partnership' (Paltiel 2012).

While China's oil and gas companies have been active internationally for a relatively long time, Chinese mining companies burst onto the global M&A scene in the mid-2000s following a series of acquisitions. Since its entry into the market, China has made a number of investments in Australia's mining and energy industries, for example, which have been particularly sensitive due to adverse community and political sentiment regarding Chinese FDI in the country (Drysdale and Findlay 2009; Huang and Austin 2011). China largely has been able to manage these sensitivities by investing in low-value Australian companies/assets. The standout exception was the $19.5 billion bid in 2009 by the Aluminium Corporation of China (Chinalco) for 18% of Rio Tinto, which failed amid a sea of acrimony. The Australian Foreign Investment Review Board did not apply its 'national interest' test to reject the bid, but instead sat on its hands until the Rio Tinto board rejected the offer. The same process played out in the US in 2005, when the Committee on Foreign Investment watched CNOOC's offer for Unocal fall apart before it was required to make a determination on the deal.

Similar to its interest in Canada, resource security has been a prime impetus for Chinese investment in Australia, with the country seeking to ensure that Australia's natural resources flow to China in predictable quantities at low prices (Garrett 2012). Beginning in 2005, a burgeoning wave of Chinese investments precipitated a new mining boom in the Australian iron ore and coal sectors, which only recently has slowed. While normally a welcome development, the state-owned and strategic nature of the Chinese investors raised concerns in Australia about how such investments should be regulated. As a result, in February 2008, the Australian Government declared an intention to more closely screen FDI from state-owned sources, which both supporters and detractors alike claimed was evidence of increased RN in Australia's approach to its trade and investment relationships with China (Wilson 2012).

In late 2008 and early 2009, Rio Tinto found itself under significant funding pressure as a consequence of its July 2007 acquisition of Canadian aluminium producer, Alcan. Rio Tinto purchased Alcan for $38.1 billion in cash, topping a hostile offer from Alcoa by roughly one-third (Alcoa withdrew its offer, rather than opening a bidding war with Rio Tinto). Flush with cash due to the booming global appetite for mineral and energy resources, Rio Tinto acquired Alcan to position itself as a global leader in three metals—iron ore, copper, and aluminium—with the aim being to profit from demand emanating from China's rapidly growing economy. Rio Tinto offered $101 a share for each share of Alcan, a 65.5% premium on Alcan's market value prior to the Alcoa bid. It also was a premium of 32.8% on Alcoa's cash-and-stock offer. The board of Alcan, which did not allow Alcoa to perform due diligence assessments of the company, unanimously backed the offer from the Anglo-Australian mining giant. With the acquisition, Rio Tinto became the world's largest aluminium and bauxite producer, and the fourth largest producer of alumina. Under the terms of the deal, Dick Evans, Alcan's chief executive was to lead the aluminium group from Montreal and report to Tom Albanese, Rio Tinto's then chief executive. To help secure the deal, Rio Tinto agreed with the Government of Quebec to maintain Alcoa's head office in the province in return for water and power rights that the government previously had granted to Alcan (Berman and Glader 2007).

Rio Tinto's acquisition of Alcan has been described as 'the worst mining deal ever' (McCrae 2013). In an interview with the *Wall Street Journal*, Dick Evans stated that Rio's decision to proceed with the purchase of Alcan was the 'worst decision ever, the largest metals and mining transaction in the history of the world at the high point in the commodity cycle' (quoted in Miller 2013). Not only did Rio Tinto acquire Alcan at the peak of the aluminium cycle and after entering into a bidding war with Alcoa and another major, Brazil's Vale, but China also entered the aluminium industry in late 2007, flooding the market with cheap metal. With the GFC just around the corner, in hindsight, it would have been more prudent for Rio Tinto to maintain its cash reserves. The Alcan deal not only hampered Rio's ability to make subsequent acquisitions, but it also placed significant strain on the company's finances, as it became focussed on managing debt.

In February 2008, in an attempt to prevent the completion of BHP Billiton's proposed merger with Rio Tinto, a deal that would have brought together the world's second and third largest iron ore producers, Chinalco joined forces with US aluminium producer Alcoa to acquire a nine per cent stake in Rio Tinto for $14.3 billion (representing a 21% premium on Rio's share price at the time), becoming the company's largest shareholder. The move forced BHP Billiton to raise its all-share offer for Rio when it went hostile. In the end, BHP Billiton abandoned the proposal due to the economic downturn and Rio's $40 billion debt, and never had to resolve how it would deal with competition from Chinalco (Yao et al. 2010).[1]

In its search for finance, in February 2009, Rio Tinto announced a suite of transactions with Chinalco. However, Rio Tinto subsequently withdrew from negotiations with Chinalco to pursue refinancing through an equity issue announced in June 2009, opting to form an iron ore joint venture with BHP Billiton (PwC 2010). This deal subsequently was blocked after anti-competition complaints from regulators and customers, including Chinese firms. Reaction to Chinalco's February 2009 bid received similar reactions in Australia to those in the US in response to CNOOC's bid for Unocal in 2005. For example, one Australian Senator ran television advertisements opposing the deal, arguing that another sovereign state should not be able to own a domestic sovereign asset.

In response, China's State Council alleged that BHP Billiton had inflamed Australians' fear of 'Chinese colour' in order to undermine Chinalco's attempt to increase its share in Rio Tinto (Nyland et al. 2011). In retaliation, in July 2009, the Shanghai Public Security Bureau arrested four members of Rio Tinto's iron ore price negotiating team, three of whom later were jailed, on charges of bribing officials at Chinese steel mills to provide inside information on the Chinese position during price talks. The arrests, which came to be referred to as the 'Stern Hu Affair', proved contentious, drawing official criticism from the Australian Government on the basis that there was a lack of transparency and regarding the authenticity of evidence (Wilson 2012).

The preceding survey of attempted megadeals in the mining industry over the past decade reveals that their outcomes are interconnected. China is a common thread that runs through all six deals. Even if its state-owned companies did not participate directly in the Rio Tinto-Alcan and BHP Billiton-Potash deals, China indirectly was involved either in its capacity as the largest consumer of specific commodities, or as a keen observer, ready to interfere if the likely outcome were to be unfavourable to its interests. The six deals also show that there is a multiplicity of factors at play in determining the eventual outcome of M&A megadeals. In the cases of CNOOC-Unocal and Chinalco-Rio Tinto, the cashed-up Chinese offers were insufficient to overcome domestic political opposition and protectionist sentiment. In the cases of Rio Tinto-Alcan and CNOOC-Nexen, important concessions to host states were crucial to ensuring that the deals were successful. Another important factor is the complementarity of interests between the home and host economies. If the two economies are competing for market share, as in the cases of Australia and Canada in the BHP Billiton-Potash deal and China and the US in the CNOOC-Unocal deal, it is likely that the deal will fail, due to the target country's

[1] It is notable that China considered using this tactic when BHP Billiton bid for Potash. The rationale for China to derail the deal was clear, with the world's most populous country wanting to secure reasonably priced supplies of the key crop nutrient to feed its people with a shrinking supply of arable land. If Prime Minister Harper had not blocked the deal, China (through Sinochem) was prepared to acquire a stake in Potash in order to thwart the deal.

fears of excessive market control by the buyer if the two countries are energy/resource exporters, or due to opposition on energy/resource security grounds if both states are importers. Alternatively, if the (geo-political) interests of the home and host states are in alignment, as was the case with China and Canada on the CNOOC-Nexen deal, this may pave the road for a successful outcome. Finally, the above survey demonstrates the key role of institutional learning when embarking on a megadeal. This is evident in the substantive transformation in the approach adopted by CNOOC in its successful takeover of Nexen, compared to the approach that it took during its failed bid for Unocal.

Conclusion

In the foregoing discussion, we examined the contributing factors in the success or failure of six proposed cross-border megadeals in the mining industry since 2005. We used a political-economic lens to analyse the deals and, based on our survey of the literature, arrived at three plausible explanations as to why certain deals proceeded, while others did not. We found that in the first instance the structure of the market itself had an effect on the outcome of the megadeals—that is, deals that would have had a smaller impact on industry concentration levels arguably would have been less politicised in the target country and, in turn, more likely would have been successful. The second explanation centred on the level of power and influence that the company/ies had over political interests and processes governing M&A activity. Rising RN in target countries, particularly opposition to in-bound Chinese investment, was the third reason for deal failure. Conversely, we found that when the interests of the target and domiciled states and markets were aligned, the deals were successful.

Merger and acquisition activity in the Asia-Pacific Region's mining industry market has faced a number of challenges in recent years. However, there are signs of recovery, as our survey of recent market trends and likely future developments show. Regardless of the size of the deal, corporate leaders and potential financiers should be aware of these trends and challenges, and establish appropriate due diligence frameworks in order to avoid deal failure, and to maximise deal-making opportunities.

References

Aktas, N., de Bodt, E., Roll, R. (2007). Is European M&A regulation protectionist? *Economic Journal*, 117(522), 1096-1121.

Barings. (2014). Mining M&A could drive sector share price recovery. *Investment Insights*. London: Baring Asset Management Limited.

Batten, K. (2014). Mining M&A set to double in 2015. *MiningNews.net*. http://www.miningnews.net/storyview.asp?storyid=826938727. Accessed 16 Nov 2014.

Berman, D., Glader, P. (2007, 12 July). Rio Tinto nears deal to acquire Alcan of Canada; bid of about $37 billion would thwart Alcoa. *Wall Street Journal*. http://www.readabstracts.com/Business-international/Rio-Tinto-nears-deal-to-acquire-Alcan-of-Canada-bid-of-about-37-billion-would-thwart-Alcoa-consolida.html. Accessed 3 January 2015.

BP. (2014). *BP statistical review of world energy*. http://www.bp.com/en/global/corporate/energy-economics/statistical-review-of-world-energy.html. Accessed 7 December 2014.

Brookstone, J. (1976). *The multinational businessman and foreign policy*. New York: Praeger Publishers.

Conybeare, J., Kim, D. (2010). Barbarians at the gate: state control of global mergers and acquisitions. *The World Economy*, 33(9), 1175-1199.

Dawson, L. (2012). Potash and blackberries: should Canada treat all foreign direct investment the same? *Commentary*. A Macdonald-Laurier Institute Publication, June. http://www.macdonaldlaurier.ca/files/pdf/Should-Canada-Treat-All-FDI-The-Same-Commentary-June-2012.pdf. Accessed 3 March 2015.

Deloitte Center for Energy Solutions. (2014). Deloitte oil & gas mergers and acquisitions report–midyear 2014. Houston and Washington, DC: Deloitte Center for Energy Solutions. http://www2.deloitte.com/us/en/pages/energy-and-resources/articles/oil-and-gas-mergers-and-acquisitions-report-midyear-2014.html. Accessed 3 March 2015.

Dorn, J. (2005). How to improve U.S.-China relations in the wake of CNOOC. *Korean Journal of Defense Analysis*, 17(3), 63-85.

Drysdale, P., Findlay, C. (2009). Chinese foreign direct investment in Australia: policy issues for the resource sector. *China Economic Journal*, 2(2), 133-158.

Evenett, S. (2002). How much have merger review laws reduced cross-border mergers and acquisitions? In W. Rowley (Ed). *International merger control: prescriptions for convergence*. London: International Bar Association.

Ernst & Young. (2014a). Decade low for mining deals, gradual return to M&A expected in 2014. http://www.ey.com/AU/en/Newsroom/News-releases/Decade-low-for-mining-deals-gradual-return-to-MA-expected-in-2014. Accessed 16 Nov 2014.

Friedman, M. (1962). *Capitalism and freedom*. Chicago: University of Chicago Press.

Galbraith, J. (1967). *The new industrial state*. Boston: Houghton-Mifflin.

Garrett, G. (2012, 18 August). China investment elephant is well and truly in the room. *The Australian*. http://www.bing.com/search?q=China+investment+elephant+is+well+and+truly+in+the+room&form=IE10TR&src=IE10TR&pc=HPDTDFJS. Accessed 18 July 2015.

Grant Thornton. (2014). *Gathering momentum: the resurgence of M&A*. .Global Mining Report 2014. http://www.grantthornton.com.my/en/insights/industries/gathering-momentum/. Accessed 25 July 2015.

Hammerson, M., LaMaster, J. (2014). *Oil and gas M&A: a practical handbook*. London: Globe Business Publishing Ltd.

Herbert Smith Freehills. (2014). Trends for oil & gas in 2014. http://www.herbertsmithfreehills.com/insights/legal-briefings/trends-for-oil-and-gas-in-2014. Accessed 18 Nov 2014.

Huang, X., Austin, I. (2011). *Chinese investment in Australia: unique insights from the mining industry*. London: Palgrave-Macmillan.

Joffé, G., Stevens, P., Lux, G., et al. (2009). Expropriation of oil and gas investments: historical, legal and economic perspectives in a new age of resource nationalism. *Journal of World Energy Law & Business*, 2(1), 3-23.

Kean, A. (2014). Private capital funds to boost mining M&A in early 2015 with up to US$20B war chest. *Metals & Mining Daily*, 2(163), 1 and 19.

Kim, J. (2007). Fears of foreign ownership: the old face of economic nationalism. *SAIS Review of International Affairs*, 27(2), 167-177.

KPMG. (2014). *Mergers and acquisitions yearbook 2014 edition*. https://issuu.com/kpmgswitzerland/docs/pub-20140115-ma-yearbook-2014-en. Accessed 16 November 2016.

Krugman, P. (1995). Dutch tulips and emerging markets. *Foreign Affairs,* 74(4), 28-45.

Lazenby, H. (2014). Global mining M&A rebounds, but Canada takes the cake. *Mining Weekly*. http://www.miningweekly.com/article/global-mining-ma-rebounds-but-canada-takes-the-cake-2014-09-17. Accessed 16 Nov 2014.

McCrae, M. (2013, 15 February). The worst mining deal ever: Rio Tinto buying Alcan for $38.1B. *Mining.com*. http://www.mining.com/the-worst-mining-deal-ever-rio-tinto-buying-alcan-for-38-1b-37600/. Accessed 16 November 2014.

Miller, D. (2013). Politics. In D. Miller (Ed.), *Blackwell Encyclopaedia of Political Thought*. Oxford: Blackwell.

Mining Insight. (2014). Industry Trend Analysis-Mining M&A: Asia & Gold Shine-JUNE 2014. http://www.mining-insight.com/industry-trend-analysis-mining-ma-asia-gold-shine-june-2014. Accessed 16 November 2014.

Moreira, S. (2013). Learning from failure: China's overseas oil investments. *Journal of Current Chinese Affairs*, 42(1), 131-165.

Mullins, B., Berman, D. (2005, 29 June). Republicans urge White House to review CNOOC's Unocal bid. *The Wall Street Journal*. http://www.wsj.com/articles/SB112000629092472376. Accessed 1 October 2015.

Nyland, C., Forbes-Mewett, H., Thomson, S. (2011). Sinophobia as corporate tactic and the response of host communities. *Journal of Contemporary Asia*, 41(4), 610-631.

Oil & Gas Journal. (2014). PwC: 2Q oil and gas M&A activity strongest in 5 years. http://www.ogj.com/articles/2014/07/pwc-2q-oil-and-gas-m-a-activity-strongest-in-5-years.html. Accessed 18 Nov 2014.

Paltiel, J. (2012). Canada's China re-set: strategic realignment or tactical repositioning? re-energizing the strategic partnership. *Canadian Foreign Policy Journal*, 18(3), 267-272.

Perkowski, J. (2012, 25 July). Why Nexen is different: the politics of supply. *Forbes*. http://www.forbes.com/sites/jackperkowski/2012/07/25/why-nexen-is-different-the-politics-of-supply/#6688e53337d6. Accessed 21 July 2015.

Putnam, R. (1988). Diplomacy and domestic politics: the logic of two-level games. *International Organization*, 42(3), 427-460.

PwC. (2010). Mining deals 2009 annual review: mergers and acquisitions activity in the mining industry. http://www.pwc.com/cz/en/poradenstvi-pri-transakcich/assets/mining-deals-2009.pdf. Accessed 4 February 2015.

Riley-Smith, J., Mak, S. (2014). M&A Review: 2014 Mid-Year Update. Corrs Chambers Westgarth. http://www.corrs.com.au/thinking/insights/m-and-a-review-2014-mid-year-update/. Accessed 31 Oct 2014.

Ruggie, J.G. (1982). International regimes, transactions and change: embedded liberalism in the post-war economic order. *International Organization*, 36(2), 379-415.

Schortgen, F. (2006). Protectionist capitalists vs. capitalist communists: CNOOC's failed Unocal bid in perspective. *Asia Pacific: perspectives*, 6(2), 2-10.

Sørensen, G. (2008). The case for combining material forces and ideas in the study of IR. *European Journal of International Relations*, 14(1), 5-32.

Stokes, D. (2007). Blood for oil? global capital, counter-insurgency and the dual logic of American energy security. *Review of International Studies*, 33(2), 245-264.

Tannenwald, N. (2005). Ideas and explanation: advancing the theoretical agenda. *Journal of Cold War Studies*, 7(2), 13-42.

The Beijing Axis. (2014). China's continued quest for natural resources. *The China Analyst*. Beijing. http://www.thebeijingaxis.com/Newsletter/April_2014/PDF/The_China_Analyst_-_April_2014.pdf. Accessed 10 January 2015.

The Canadian Press. (2014). Barrick Gold and Newmont Mining blame each other for failed merger talks. http://www.globalpost.com/dispatch/news/the-canadian-press/140428/barrick-gold-says-newmont-mining-corp-has-terminated-merger-. Accessed 2 December 2014.

Underhill, G. (2000). State, market, and global political economy: genealogy of an (inter-?) discipline. *International Affairs*, 76(4), 805-824.

Vivoda, V. (2009). Resource nationalism, bargaining and international oil companies: challenges and change in the new millennium. *New Political Economy*, 14(4), 517-534.

Wälde, T. (2008). Renegotiating acquired rights in the oil and gas industries: industry and political cycles meet the rule of law. *Journal of World Energy Law & Business*, 1(1), 55-97.

Widmaier, W. (2003). Constructing monetary crisis: monetary understandings and state interests in cooperation. *Review of International Studies*, 29(1), 61-77.

Wilson, E. (1986). The petro-political cycle in world oil markets. In R. Enders and J. Kim (Eds), *Energy resource development: politics and policies* (pp. 1-20). Westport, Connecticut: Greenwood Press.

Wilson, J. (2012). Chinese resource security policies and the restructuring of the Asia-Pacific iron ore market. *Resources Policy*, 37(3), 331-339.

Yao, S., Sutherland, D., and Chen J. (2010). China's outward FDI and resource-seeking strategy: a case study on Chinalco and Rio Tinto. *Asia-Pacific Journal of Accounting & Economics*, 17(3), 313-326.

Part II
The Major Resources

Chapter 9
Iron Ore

Richard Leaver

Abstract As a country that has historically lived off international sales of primary commodities, Australians need to understand better than most the cyclical fashion in which commodity markets move and the risks entailed in that recurring pattern. To that end, this chapter investigates the way that political risks have been, and are still being, handled in the three nations from the Asia-Pacific region that have successively led the world in steel production through the last century—the United States, Japan and China. The implications for Australia, which has now emerged as the world's largest exporter of iron ore, associated with each of these three regions are sketched out, with greater detail on the 'super-cycle' derived from the phenomenal expansion of Chinese steel production in recent decades.

Abbreviations

EAAU	East Asian Analytical Unit
GATT	General Agreement on Tariffs and Trade
GFC	Global Financial Crisis
NSW	New South Wales
OPEC	Organization of Petroleum Exporting Countries
SA	South Australia
UK	United Kingdom
US	United States

R. Leaver (✉)
School of History and International Relations, Flinders University,
Bedford Park, SA 5042, Australia
e-mail: richard.leaver@flinders.edu.au

T. O'Callaghan, G. Graetz (eds.), *Mining in the Asia-Pacific*, The Political Economy of the Asia Pacific, DOI 10.1007/978-3-319-61395-6_9

Introduction

More than any other product, steel would emerge as the symbol of the second industrial revolution in the twentieth century. A reduced carbon alloy of cast iron, its superior properties, most notably, the unique combination of hardness and tensile strength—had been known for centuries, but the ability to produce it had always been a skilled art, better practised in the East than the West (Smith and Forbes 1957). Seemingly unamenable to the earliest intimations of mass production, all this changed in the second half of the nineteenth century, when a sequence of technical and metallurgical advances progressively opened previously shut doors. These advances—the introduction of coking coal into blast furnaces producing cast iron; the invention of the Bessemer Converter to de-carbonise this precursor of steel; its quick-order replacement by the larger scale open hearth furnace (Smil 2005)—cumulatively yielded previously unimaginable quantities of this highly-valued metal, and at low cost. Thus, in little more than a generation, steelmaking became the leading index of the sophistication of national manufacturing and, equally, a proxy for the war potential of those states able to contest dominance within the sector.

That contest was vigorous and, at least until 1914, peaceful enough. The fundamental idea behind the Bessemer converter emerged more or less simultaneously in the United Kingdom (UK) and the United States (US), and all the key technological advances were quick to diffuse across international borders. But raw material endowments were, by contrast, geographically fixed and, suddenly, much more significant. For if steel were to be mass produced, then it followed that there had to be mass consumption of its two constituent raw materials: coking coal and iron ore; both of which were, in their highest grades, relatively scarce and extremely bulky. Scarcity meant that none of steel's would-be giants could initially feel totally secure about their national endowments. Britain, for instance, had sufficient coal, but always looking abroad for better grades of iron ore, while Germany and France at best enjoyed different halves of the pie. Furthermore, the high bulk of these precursors inevitably implied that long-distance imports would be expensive to move, especially over land, perhaps to the point where the cost competitiveness of the final product would be seriously compromised in the new and more commercial environment. Hence the low value to weight of steel's bulky inputs set in motion a logic of localisation for the emerging steel industry that stood in opposition to the logic of globalisation generally associated with the second industrial revolution. Indeed, so strong was the pull of localisation that the sovereign control of high-grade raw materials, especially proximal deposits, often seemed like an objective worth fighting for (Berglund 1919).

So it was that the revolution that ushered in the mass production of steel simultaneously created strategic and commercial risks around the supply of its essential raw materials. This chapter investigates the ways that those risks have been, and still are being, handled in the three states of the Asia-Pacific Region that successively have led the world in steel production through the last century: the United States (US), Japan, and China. The implications for Australia are discussed.

The US Surge

Although US inventers had sound legal claims over the conversion process, the rollout of steelmaking on that continent was initially held in check by litigation between rivals. In the UK, by contrast, potential rivals ended up cooperating—an outcome that initially spirited UK steel output into the lead position. One consequence was that, as continental railway building became a growth pole for American industrialisation, imported iron and steel initially provided the lion's share of the rails (Misa 1995). However, by the end of the nineteenth century, the US was eclipsing the UK as the leading steel producer and, in the process, beginning to reconfigure the iron rule of localisation. Initially, American steel had appeared to conform to that rule. The industry quickly settled in northern Pennsylvania where it could exploit the massive high-grade coal reserves of the Appalachians, which, in turn, were conveniently close to the major domestic markets of the rapidly urbanising and industrialising northeast. Iron ore also was available locally, albeit neither in large quantities, nor particularly high grades, but without a superior alternative, it was put to use until something better, hopefully, came along.

As Rogers (2009) meticulously charts, 'something better' did arrive with the late-century discovery of the high-grade iron ore deposits of the Mesabi Region in northern Minnesota. These ores nonetheless posed a significant transportation challenge, since they were more than a thousand miles overland from 'Steel Valley' in Pennsylvania. Distance alone threatened to price these ores out of the market, or to force the industry to relocate. However, possibilities for the low cost transport simultaneously were emerging on the Great Lakes system where, as Foord's (1898) real-time calculations showed, the steel-hulled bulk freighter was allowing the US steel industry to liberate itself from the tyranny of proximity. This combination of cheap transport with high-grade ore quickly became the taproot of America's emerging competitive edge in steel and steel-based products.

However, in two other respects the rise of the US reconfirmed the logic of localisation, albeit on an enhanced scale. First, in comparison to all the leading European industrial states, the sheer geographical size of the American state gave the nation's steel industry a high degree of self-sufficiency in raw materials, which, in turn, suppressed the level of international trade. Hence, in spite of impressively high rates of domestic steel production, a high level of national self-sufficiency persisted well into the post-World War Two period. Second, the US steel industry was one of the pioneers of the internalisation of supply chains through vertical integration. The main driver of this process in US industry was typically the control of transport, the principal factor in the rise and rise of John D. Rockefeller. And so it proved to be in steel where Rockefeller's threat to control Great Lakes transport prodded Andrew Carnegie into a preemptive move that ended up creating US Steel, the American industry's heavyweight (Mancke 1972). There was a pronounced surge in the pace of vertical integration across the industry and the capture of economic rents down the value chain by means of the internalisation of commodity markets. This, in turn, generated a fearsome American lead in global steel production, and a quantum leap in US manufactured exports that featured the intensive use of natural resources (Irwin 2003). Here, then, were the makings of the economic component of America's rise to global hegemony.

So far as the morphology of the iron ore market was concerned, the impact of America's rise to number one is therefore best characterised as 'continentalisation'—something of greater scale than localisation, but still less than full globalisation. Being less than fully international, iron ore trade—along with many other mineral trades—therefore failed to attract much attention from American post-war rule-makers in the trade realm. So, in his highly regarded study of world trade patterns, Rowe (1965) classified iron ore as a 'regionally traded' commodity and, on this basis, excluded it from his investigation into 'internationally traded' commodities.

At the same time, precisely because it was made in America, the Carnegie model of vertical market integration, unlike the cheap products it created, did not travel far. American practice had no noticeable impact, for instance, in Australia, where steel production was getting under way more or less as the American model crystallised. The big Australian problem remained, as Hughes (1964) later documented, resolutely traditional: seemingly poor endowments of the necessary natural resources in widely dispersed locations. Over the ensuing decades, a dumbbell pattern of steel production emerged around Port Kembla and Newcastle in New South Wales (NSW) and Whyalla in South Australia (SA), with the iron ore deficiencies in NSW helping counter-balance the shortage of coal in SA. But, all the time, the threat of the imminent exhaustion of iron ore hung over the head of the sector, as exemplified by the well-known embargo on exports on the eve of World War Two (Tsokhas 1995).

In due course, this export restriction survived those hostilities and the immediate post-war decades, only being lifted in 1960 (Lee 2013). By this time, 'localised' European steel producers finally were catching up with the benefits of 'continentalised' American producers through the rollout of the European Coal and Steel Community (Gillingham 1991). However, it was too late: the benefits that once came from continental scale were themselves being overwhelmed by a new and superior market model sporting truly international scale that was driving the reconstituted Japanese steel industry. The implications for Australia, having just 'discovered' the iron mountains of the Pilbara Region, in the Northern part of Western Australia, would prove to be enormous.

Japan: Quantity Plus Quality

When continentalisation was king, Japan was doubly handicapped in that it possessed no significant endowments of either coking coal or iron ore. Not surprisingly, the attempt to relieve these bottlenecks provided an important part of the rationale for its 'imperial experiments' in northern China during the early twentieth century (Rodgers 1948). But even in the narrowest and most instrumental sense, these experiments were not entirely successful. Despite military needs being evermore securely positioned in the driver's seat of the steel industry, production nonetheless peaked well under ten per cent of American levels (Dear 1995), with significant contributions from the melting of pots and pans. Japan's experience was, as Zimmermann (1951) correctly concluded, 'a miracle', albeit one of the less-exalted kind: 'a magician's trick … that shows what a patient long-suffering people can be made to do by unrelenting taskmasters'.

When, following defeat, all these imperial experiments were rolled back, it was hardly surprising that both occupation authorities and the World Bank advised against any Japanese attempt to base post-war reconstruction around steel. However, that advice fell on deaf ears, for 'heavy and chemical industry' already had been preselected by nascent civil governments as the point of focus for a new kind of miracle. In steel, reindustrialisation rested on the newest technology, the Basic Oxygen Furnace, leased from Austria and perfected on an industrial scale with Japan's new steel plants. But, the industry also had a radical solution to the traditional double handicap: to import the necessary ores from the highest quality mines, and to bring the imports home in the new dry bulk carriers of evermore massive scale that were beginning to emerge from Japan's shipyards. As Lundgren (1996) later demonstrated, these ships quickly brought down dry bulk freight rates by nearly two thirds.

Somewhat paradoxically, the first glimpse of the potential advantages in this radical solution was suggested by the facility that grew into the largest American integrated steel works: Bethlehem Steel's Sparrows Point. Its monotonic rise over seven decades was based upon an Atlantic tidewater location that facilitated imports of high-grade ore from Central and later South America—a combination that subverted the continental logic of the US system (Reutter 1988). But, as Bunker and Ciccantell (2005) have documented, Maritime Industrial Development Areas went on to become the Japanese norm rather than the American exception—and global scale therefore was, for the first time, able to trump interior lines of transport in steelmaking. The icing on the cake of Japan's competitive edge came from contrary developments around Lake Superior, where the higher grades of Mesabi ore were by then facing exhaustion and North American steelmakers were being forced to adapt to lower grade taconite (Manuel 2013).

At the point of departure for Japan's industrial rebirth, Australian elites and masses were deeply in the thrall of all things British, and Japan most commonly was remembered as an implacable wartime enemy. National exports were, as usual, concentrated on the UK market half a world away, and were still largely agricultural in composition. Mineral exports represented less than one per cent of total exports, and were clustered around base metals rather than bulk commodities—a sub-sector where national scarcity was thought to prevail (especially in iron ore, as previously noted). So, in 1950, only 3.9% of Australia's exports went to Japan—a very modest one tenth of the UK's share. However, within 20 years, new export-oriented discoveries of Pilbara iron ore, plus the rise of trade in coking coal, not only had made Japan into Australia's premier export market, but had done so to such an extent that it routinely delivered massive 2:1 bilateral trade surpluses in Australia's favour. A decade further on, Australia stood as number one global exporter of iron ore—a status it was about to equal in coal. By that time, the ratio of Japan-bound to UK-bound exports stood nearly 10:1 in the former's favour—a complete inversion of the situation three decades earlier (Lougheed 1987).

Massively impressive though this quantitative shift was, perhaps the main point about Japan's rise as a purchaser of Australian bulk commodities concerned its qualitative impact. This booming bilateral trade induced conservative Australian governments to abandon their strong inclination towards pro-British preferential trade in favour of the principle of non-discrimination. Long memories in Tokyo led their new governments to recall Canberra's infamous trade diversion episode of 1936 (Sissons 1976), as well as the iron ore export embargo—episodes when Japan was on the receiving end of Australian-made commercial discrimination. Consequently, on the cusp of the rise of their resource imports, Japanese authorities initiated commercial talks with Canberra that drew attention to the principle of equality of treatment in trade, the foundation principle of the General Agreement on Tariffs and Trade (GATT). These talks commenced at a time when wheat and agricultural commodities were the backbone of the few existing Australian exports to Japan. So the thin edge of a multilateral normative wedge was delicately inserted into Australian trade policy by the 1957 Treaty on Commerce (Pitty 2001), and later complemented in relation to foreign investment issues by the 1976 NARA treaty. When, therefore, resource trade volumes between Australia and Japan exploded in the 1960s, so, too, did the importance of that wedge of multilateral principles. Hence, in a most unlikely manner, and from a totally unexpected corner began the rise and rise of the Australian government's attachment to commercial multilateralism, an orientation now routinely regarded as Australia's 'first-best' policy in the realm of trade.

But norms alone were insufficient to define exactly how these new markets would be instituted as day-to-day practice. At this lower level of abstraction, two things were clear. First, the GATT offered no risk-minimising protections for iron ore trade, especially after the US washed its hands in 1955 on the liberalisation of international trade in primary products under GATT. Even if GATT had been on the ball, the massive scale of the proposed new Pilbara mines and the lumpiness of bulk commodity inputs into Japan's steel industry suggested that arms-length trade would soon produce unstable market conditions, as Smith (1979) later argued, for resource nationalist gambits from either end of the market. Second, the North American solution of vertical integration was not acceptable in either Japan or Australia, albeit for different reasons. In Japan, post-war governments wanted high-speed growth through maximising the rate of reinvestment in domestic industry—and, to this end, the overseas investment of capital was at this time virtually prohibited. In Australia, prohibition was a conclusion reached by more politically obtuse reasoning, namely, that Australians would not be willing to live with high levels of Japanese foreign investment in their burgeoning resource sector. Memories of Japan's 'southern thrust' in 1942 made bilateral trade contentious enough, but the idea of Japanese ownership of Australian resources was simply beyond the pale.

Given these mutual exclusions, greatness came to be thrust upon long-term contracts as the institutional backbone for the embryonic international trade in bulk commodities. Multi-year commitments to purchase, the first steps of which have been detailed by Boyce (2001), were rolled out for the original Hammersley mine and quickly extended across the other licensed projects in the Pilbara. The long duration of these opening contracts—between 10 and 16 years—allowed mining companies to raise their development costs through debt financing with domestic or international banks. A 'third way' was therefore charted in the Pilbara around the politically dangerous shoals of GATT-style free markets and American-style vertical integration.

As Mohan and Berkowitz (1988) have demonstrated, the high quality of raw material imports was soon an important part of the profitability of Japan's reconstituted steel industry. It therefore was not surprising that the power of Japan's fearsome steel cartel stood resolutely behind the contract pricing system. In the beginning, that cartel had a primary interest in ensuring that high-quality Australian suppliers got up and running; hence the initial contracts fixed both annual tonnages (with a 10% range for annual adjustments) and prices (within a seven and a half percent margin of variation) (Smith 1978). But, from the beginning, it also was clear where the whip hand lay, for the Australian miners negotiating those margins found themselves repeatedly dealing with the very same Japanese official who spoke for the totality of his country's steel industry. By the time the first round of contracts had run their course, Japan's interests in iron ore supplies were shifting to Brazil where, as Byrnes (1994) has argued, possibilities for broadening the market's base were now more seductive to the Japanese than the absolute level of prices. The contract periods available to established Australian mines therefore reduced to one year, creating the benchmark pricing system that has, until recently, been regarded as the norm for the Japanese-origin iron ore export industry.

China: Quantity Minus Quality

By contrast, China's rise within the Japanese-designed international market for iron ore has been, if anything, even more quantitatively spectacular, so much so that analysts regularly have struggled to anticipate its speed. In 1995, to take the first example, the East Asia Analytical Unit (EAAU) in the Australian government's Department of Foreign Affairs and Trade made a comparative study of the Australian and Chinese iron and steel industries, concluding that Australia's iron ore exports to China would grow by 150% over the following decade (EAAU 1995). Being a product of the 1989 Garnaut Report, the EAAU never had been shy about projecting the course of Australian exports to Asia. But in the event, its bullish best nonetheless proved altogether too modest, for by 2005 those exports came in at another 150% greater still (Richmond et al. 2006). By that time, China had executed a 'Japan passing' move to position itself as the world's largest importer of iron ore, not to mention the leading buyer of Australian output.

But unlike Japan, this new number one did not slow down. By 2008, China's iron ore imports from Australia were nearly three times larger than Japan's (Christie et al. 2011). However, arguably, the pinnacle of China's quantitative achievement was reached the following year—the time when red ink from the Global Financial

Crisis (GFC) was depressing the imports of most countries. But not in China, and not in anything related to steel. On the back of Beijing's massive February 2009 stimulus program, recorded iron ore imports surged by more than 40% through the remainder of the year (de Kretser 2010). At this point, China by itself was absorbing some 68% of total global shipments of iron ore, with the lion's share of those marginal spoils accruing to Australian exporters (New 2010). Combined with the high commodity prices, Australia's terms of trade rose to record levels, with iron ore alone accounting for nearly 20% of total exports (Connolly and Orsmond 2011). In a purely quantitative sense, therefore, Canberra's view of the economic rise of China has been even more impressive than its image of the economic rise of Japan four decades earlier.

However, what have not been nearly so impressive are the qualitative changes that China's rise has brought to the iron ore market. On paper, the official Chinese position continued to favour the indefinite continuation of the annual benchmark pricing system, with China simply assuming Japan's mantle of lead negotiator on the purchasing side. In practice, though, China's passing move imprinted certain 'Chinese characteristics', so to speak, on the demand side of that market, which qualified this commitment. The most fundamental of these has been the radical de-concentration associated with China's massively proportioned steel industry.

The root cause of this industry fragmentation is the well-known 'cellular' nature of the Chinese economy, a problem identified by Donnithorne (1972) more than 40 years ago. It remains highly paradoxical that China, the last of the great centrally planned economies, manifests more planning at the provincial level than it does at the central level. So, for instance, the company that has until recently been China's largest steel producer, Baosteel, has a marginally larger absolute output than Nippon Steel, and yet in 2010 it produced only 6% of China's national output as opposed to Nippon Steel's 31% of Japan's (World Steel Association 2010). Beijing's official goal is to have its ten largest producers supplying 60% of the domestic market by 2015 (as opposed to their current share of 44%). But even if this were achieved, China would still feature a low degree of domestic concentration by any reasonable international comparison. Furthermore, Beijing's attempts to create a united front for the industry through the China Iron and Steel Association actively have been resisted by provincial-level planners and entrepreneurs. Compared to Japan's highly cartelised steel industry, it therefore is paradoxical that China has induced a de-concentration of demand-side power as the by-product of its 'peaceful rise' (Pham, 2015).

The combination of rapidly rising Chinese demand with radical demand-side de-concentration has, in turn, enhanced the market power of the major iron ore exporters where, as always, the three largest producing firms routinely control between 60 and 70% of the internationally traded market. This is an impressive figure in a comparative context; in oil, for example, the three largest producers struggle to provide more than 30% of world supply, with the Organization of Petroleum Exporting Countries' (OPEC) share only about 50%. Making matters worse, a string of proposed developments inside Australia—or example, the November 2007 buy-out of Rio by BHP Billiton; and the February 2009 proposal for a joint venture between the same companies in Pilbara-based production—constantly threatened to push effective supply-side concentration higher still (Vivoda and Graetz in this volume). Even though none of these proposals was carried through, the long-lasting residue was a sequence of 'cartel nightmares' for the Chinese steel industry and Beijing authorities alike (Earl 2008). These Beijing nightmares probably assumed surreal qualities at the end of the (GFC), when BHP Billiton, as if confirming its *droit du seigneur* over mining, used a 12 month period of unfulfilled price negotiations to proclaim unilaterally that sales henceforth would be executed either around quarterly indexed prices or daily spot prices.

The cumulative result of this short-term tilt towards 'producer power' in iron ore therefore has been three fold: the destruction of the annual benchmark price system; the provision, under the protective shield of Chinese sovereignty, of an amply-dimensioned speculative market in iron ore; and the quick retreat since 2010 of contract periods, so rapid that the daily spot market price now informs an ever-larger share of transactions.

The Crisis Within the Opportunity: The Re-Emergence of Risk

Over the short-term, Australian governments—perpetually inclined to be great believers in the sanctity of commercial decision-making—quietly were satisfied with the economic opportunities presented by these developments. When the initial boom in iron ore prices was followed by a boom in export volumes, there were dramatic compound effects on the balance of payments and federal budgets from what was eventually judged the most pronounced resource boom in the whole of Australia's economic history. During what Garnaut (2013) later called 'the salad days', the share of iron ore in national exports expanded more than four times in the first decade of the new century (Connolly and Orsmond 2011). Australia's exports to China therefore took off, producing bilateral and, then, multilateral trade surpluses, where there previously had been large deficits (Findlay 2011). Gains from the terms of trade then cut in, delivering sharp rises in national income that outpaced below-par trends in gross domestic product. This income boom, in turn, added windfall gains to the revenue base of the federal and relevant state governments, which allowed them to fund an expansive, but popular, range of so-called 'middle class welfare' programs.

However, the questions that always lurked concerned the longer-term, namely, the sustainability of all these achievements through time. It needs to be remembered that no government from this period ever conceived the possibility of, let alone planned for, a resources boom, let alone a boom of this record magnitude; indeed, quite the contrary, the Howard government had been so driven in its quest for a free trade agreement with the US precisely because it saw in America a 'new economy', the dynamics of which were much less reliant on 'old economy' drivers such as resources, which it regarded as passé. With plans so completely off course during 'the salad days', governments were sleepwalking for more than a decade as crisis brewed inside the opportunity (Cleary, 2011).

The crisis can be thought of as the composite of three types of risks that have been introduced, or reintroduced, into the iron ore market during the boom phase. The first category of risk was created when a large and growing proportion of Australian national exports was allowed to ride the roller coaster of the spot market. The second kind of risk might be regarded as the inversion of sovereign risk, namely, the risk that came out of Canberra's complacent assumption that the corporate interests of the big Anglo-Australian mining houses largely coincided with national interests. The third type of risk centred on the myopia that the Chinese miracle, and Australia's heavy economic reliance upon its continuation, posed for the customary Anglo-centric construct of official strategic calculations in Canberra. A few words on each of these risks are a conclusion to this analysis.

As a country that historically has lived off international sales of primary commodities, Australians understand better than most the cyclical fashion in which these markets move. In the first instance, higher demand provokes a price rise; the price rise then brings forward investment in new capacity; realised new capacity increases the absolute size of physical product entering the market, which in time brings down prices from the dizzy heights recorded at the beginning of the upswing. However, in the case of the China boom in iron ore, the mining companies successfully insinuated a new hypothesis in place of the common-sense understanding, the idea of the 'super-cycle'. According to this hypothesis, Chinese demand for iron ore would continue to expand more or less in line with Chinese city building—at best a half-completed program—and, as a consequence, the end of the cycle would be delayed for decades. If this were true, then self-interest suggested that iron ore companies would benefit through those extra decades by marketing as much of their product as possible, through spot sales rather than lower priced contract sales, which, as previously noted, is precisely where BHP Billiton led the market in 2010. Believing, no doubt, that price declines were a long way off, BHP Billiton in effect suspended the relevance of a second commonplace belief, namely, that spot markets would lead prices up during an upswing, but would later lead prices down during the decline stage.

This downside risk, in fact, began to materialise less than 18 months after BHP Billiton's abandonment of benchmark pricing when iron ore prices showed their first signs of significant retreat, and at the time of writing, prices are now down below the low water mark set during the (GFC). So, although Australian iron ore exports expanded eightfold during the China boom, iron ore nonetheless has proceeded to lose its status as Australia's leading export. Falling terms of trade have, in quick turn, delivered an instant income recession to the Australian economy, which will continue to exact its pound of flesh for a considerable time yet—according to Goldman Sachs, between AUD$500 billion and AUD$1.1 trillion in lost income over the next decade, with the hit to the budget bottom line from the falling iron ore price peaking at AUD$7.5 billion by 2020 (Uren 2015; Greber 2015).

The second category of risk—the belief that what is good for the big iron ore mining companies also will be good for Australia—deeply is embedded in the above analysis. True enough, both BHP Billiton and Rio Tinto have their historic origins in the famous Broken Hill deposit, but both are now listed on the London stock exchange, as well as locally, and neither any longer satisfies the tests laid down by the Foreign Investment Review Board for identification as an Australian company. The operating licence extended by Canberra to these foreign multinational miners is nonetheless massive, with frequent if not quite continuous implications for interstate relations with Beijing. As the world's number one iron ore purchaser, the Chinese state periodically has been vexed by recurrent ideas of formal or informal merger between the world's largest and third largest miners—and perhaps more vexed over Canberra's willingness to regard any such integration as consistent with principles of free competition. Adding insult to injury, China's own attempt to buy into Rio Tinto was crudely rebutted—and this by a company whose own market operations in China were later exposed as corrupt in the Stern Hu affair (Cai 2009).

Two recent developments might now suggest that this extended licence is approaching the end of its tether. After the bursting of the super-cycle bubble, both companies have continued to ramp up their Pilbara production in the face of already falling prices and declining Chinese demand, earning open criticism from the highest levels of government in the pro-mining state of Western Australia (Sprague and Saunders 2014). While the mass dumping of jobs and state royalties have done considerable damage to their standing at the grass-roots level, what has yet to be factored into an already volatile mix is the most obvious implication of dramatic Pilbara expansion, namely, the sharp contraction in the anticipated productive lifespan of the Pilbara province. Before the China boom, this routinely was estimated in the hundreds of years, but by 2010, Canberra already had brought this figure down to 70 years (Commonwealth of Australia 2010). Since aggregate Australian production now is on course to double the 2010 result, the exhaustion of the highest classes of output in the Pilbara now must be capable of expression in terms of generations. Once West Australians began to understand the idea of daily life without Pilbara iron ore, this game of extended licence almost certainly will come to an abrupt halt.

The final class of risk is geo-strategic in the ordinary sense, and arises out of the close economic integration that the China boom has produced between Australia and China. It should be noted that China now dominates both the export and import sides of Australia's trade register in a way that no country has done over the 60 years since Britannia ruled the waves. To many American analysts—most notably, former Secretary of State Hillary Clinton (as reported in McGeough 2014)—this integration suggests Australian governments are 'two-timing' on their alliance commitments to the US, especially when the rhetoric of commitment is not achieving demonstrable expression in increased defence spending.

As has been pointed out (Leaver and Stephens 2009), this kind of dilemma is not without unhappy precedent, albeit on a less portentous scale. During the early decades of the Cold War, conservative Australian governments ignored the desire of their American ally to include food in the network of trade embargoes directed against Mao's China, and wheat sales to China soon were accounting for about five per cent of Australia's total exports. As these sales approached their apogee, those same governments then committed Australian troops to Vietnam, allegedly to contain the regional expansion of Chinese influence. But once the Tet Offensive made it clear to Beijing that this containment was not going to succeed, the Chinese government reduced its Australian wheat purchases in a dramatic fashion, resourcing their import requirements from Canada. One might think that this sorry episode would be closely studied in Australia through the recent China boom, but, in fact, it has been entirely ignored. Perhaps the best that Canberra can hope for is that its own wilful ignorance is being replicated in Beijing.

References

Berglund, A. (1919). The iron-ore problem of Lorraine. *Quarterly Journal of Economics*, 33(3), 531-554.

Boyce, G. (2001). Multilateral contracting in Australian mining: the development of Hamersley Iron, 1961-1966. *Enterprise & Society*, 2(3), 543-575.

Bunker, S., Ciccantell, P. (2005). *Globalization and the race for resources*. Baltimore: Johns Hopkins University Press.

Byrnes, M. (1994). *Australia and the Asia game*. Sydney: Allen & Unwin.

Cai, P. (2009). Secrets, spies and steel: The Rio Tinto case. *East Asia Forum*. http://www.eastasiaforum.org/2010/03/09/secrets-spies-and-steel-the-rio-tinto-case/. Accessed 2 March 2015.

Christie, V., Mitchell, B., Orsmond, D., van Zyl, M. (2011). The iron ore, coal and gas sectors. *Bulletin*, Reserve Bank of Australia, March.

Cleary, P. (2011). *Too much luck: the mining boom and Australia's future.* Melbourne: Penguin Random House Australia.

Commonwealth of Australia. (2010). Budget paper no. 1, statement four, pp. 4–9. http://www.budget.gov.au/2010-11/content/bp1/html/index.htm. Accessed 5 February 2015.

Connolly, E., Orsmond, D. (2011). The mining industry: from bust to boom. In H. Gerard and J. Kearns (Eds.), *Australia in the 2000s*. Sydney: Reserve Bank of Australia.

Dear, I. (1995). *The Oxford companion to World War II*. Oxford: Oxford University Press.

de Kretser, A. (2010, 23 February). Strong ore price tipped for tiger year. *Australian Financial Review*. Accessed 15 March 2014.

Donnithorne, A. (1972). China's cellular economy: some economic trends since the Cultural Revolution. *The China Quarterly*, 52, 605-619.

Earl, G. (2008, October). The new regional reality. *The AFR Magazine*. Accessed 15 March 2014.

East Asia Analytical Unit. (1995). *Iron & steel in China & Australia.* Canberra: Department of Foreign Affairs and Trade.

Findlay, C. (2011). Australia-China economic relations. In J. Golley and L. Song (Eds.), *Rising China: global challenges and opportunities.* Canberra: ANU Epress, June.

Foord, J. (1898). The Great Lakes and our commercial supremacy. *The North American Review*, 167(501), 155-164.

Garnaut, R. (2013). *Dog days: Australia after the boom.* Collingwood, Victoria: Redback.

Gillingham, J. (1991). *Coal, steel and the rebirth of Europe, 1945-1955: The Germans and French from the Ruhr conflict to European Community.* Cambridge: Cambridge University Press.

Greber, J. (2015, 7 February). Politics forces RBA's hand. *Australian Financial Review.* Accessed 7 February 2015.

Hughes, H. (1964). *The Australian iron and steel industry, 1848-1962*. Melbourne: Melbourne University Press.

Irwin, D. (2003). Explaining America's surge in manufactured exports, 1880-1913. *The Review of Economics and Statistics*, 85(2), 364-376.

Leaver, R., Stephens, A. (2009). Somewhere over the rainbow: the Australian Defence White Paper, economic vulnerability, and China. *Austral Policy Forum* 09-15A, Austral Peace and Security Network, Nautilus Institute at RMIT, 4 June http://www.globalcollab.org/Nautilus/australia/apsnet/policy-forum/2009/australian-defence-and-china/. Accessed 12 June 2013.

Lee, D. (2013). Reluctant relaxation: the end of the iron ore embargo and the origins of Australia's mining boom, 1960-1966. *History Australia*, 10(3), 149-170.
Lougheed, A. (1987). International transactions and foreign commerce. In W. Vamplew (Ed.), *Australians: historical statistics*. Broadway, New South Wales.
Lundgren, N-G. (1996). Bulk trade and maritime transport costs: the evolution of global markets. *Resources Policy*, 22(1/2), 5-32.
Mancke, R. (1972). Iron ore and steel: a case study of the economic causes and consequences of vertical integration. *The Journal of Industrial Economics*, 20(3), 220-229.
Manuel, J. (2013). Mr. Taconite: Edward W. Davis and the promotion of low-grade iron ore, 1913–1955. *Technology and Culture*, 54(2), 317-345.
McGeough, P. (2014, 28 June). Time to choose: Clinton warns on ties with China. *Sydney Morning Herald.* Accessed 14 December 2014.
Misa, T. (1995). *A nation of steel: the making of modern America, 1865-1925*. Baltimore: Johns Hopkins University Press.
Mohan, K., Berkowitz, M. (1988). Raw material procurement strategy: the differential advantage in the success of Japanese steel. *Journal of Purchasing and Materials Management*, 24(1), 15.
New, R. (2010). Metals: steel and steel-making raw materials. *Australian Commodities*, 17(1), 168-215.
Pitty, R. (2001). The postwar expansion of trade with East Asia. In D. Goldsworthy (Ed.), *Facing north: a century of Australian engagement with Asia, Volume 1, 1901 to the 1970s*. South Carlton: Melbourne University Press.
Pham, M. (2015). *Energy security and the rise of China: towards a new framework for analysis*. Unpublished Thesis. The University of South Australia.
Reutter, M. (1988). *Sparrows Point: making steel: the rise and ruin of American industrial might*. New York: Summit.
Richmond, S., Millsteed, W., Wilson, R. (2006). Steel and steel making raw materials: prospects for iron ore, steel, metallurgical coal and nickel. *Australian Commodities*, 13(1), 115-135.
Rodgers, A. (1948). The Manchurian iron and steel industry and its resource base. *Geographical Review*, 38(1), 41-54.
Rogers, R. (2009). *An economic history of the American steel industry*. London: Routledge.
Rowe, J. (1965). *Primary commodities in international trade*. London: Cambridge University Press.
Sissons, D. (1976). Manchester v. Japan: the imperial background of the Australian trade diversion dispute with Japan, 1936. *Australian Outlook*, 30(3), 480-502.
Smil, V. (2005). *Creating the twentieth century: technical innovations of 1867-1914 and their lasting impact*. Oxford: Oxford University Press.
Smith, B. (1979). Security and stability in minerals markets—the role of long-term contracts. *The World Economy*, 2(1), 65-78.
Smith, B. (1978). The Japanese connection: negotiating a two-way street. In P. Hastings and A. Farran (Eds.), *Australia's resources future: threats, myths and realities in the 1980s*. West Melbourne: Nelson in Association with the Australian Institute of International Affairs.
Smith, C., Forbes, R. (1957). Metallurgy and assaying. In C. Singer, E Holmyard, A. Hall, & T. Williams. (Eds.), *A history of technology: volume III, from the renaissance to the industrial revolution 1500-1750*. Oxford: The Clarendon Press.
Sprague, J-A., Saunders, A. (2014, 10 October). WA turns blowtorch on Rio, BHP. *Australian Financial Review*. Accessed 10 October 2014.
Tsokhas, K. (1995). Trouble must follow: Australia's ban on iron ore exports to Japan in 1938. *Modern Asian Studies*, 29(4), 871-892.
Uren, D. (2015, 9 February). No political will to cut spending as falling commodity prices starve Canberra of revenues. *The Australian*. Accessed 9 February 2015.
World Steel Association. (2010) World steel in figures 2010. http://www.worldsteel.org/?action=newsdetail&id=302. Accessed 13 July 2015.
Zimmermann, E. (1951). *World resources and industries*. New York: Harper & Row.

Chapter 10
Coal

Philip Crowson

Abstract This chapter looks at coal's changing contribution to primary energy supplies in the Asia-Pacific Region, most notably in China. It then examines how coal's uses in the region have evolved in recent decades and how they might change in the future, paying particular attention to steam and coking coals. Attention then moves to a discussion of the region's domestic coal production, with a brief examination of the region's coal reserves. Although the region, as a whole, is largely self-sufficient, imports and exports in the coal trade are important. The main geographical origins and destinations of trade are described, before a discussion of the constraints, risks, and challenges faced both by domestic and foreign producers, as well as end-users of the product.

Abbreviations

BTU	British Thermal Unit
EIA	Energy Information Administration
FOB	Free on Board
GDP	Gross Domestic Product
GW	Gigawatt
IEA	International Energy Agency
MT	Million Tonnes
MTCE	Million Tonnes of Coal Equivalent
MTOE	Million Tonnes of Oil Equivalent
PPP	Purchasing Power Parity
TWH	Terrawatt Hour
UCG	Underground Coal Gasification
US	United States

P. Crowson (✉)
Centre for Energy, Petroleum and Mineral Law and Policy, University of Dundee, Nethergate, Dundee DD1 4HN, UK
e-mail: p.c.f.crowson@dundee.ac.uk

T. O'Callaghan, G. Graetz (eds.), *Mining in the Asia-Pacific*, The Political Economy of the Asia Pacific, DOI 10.1007/978-3-319-61395-6_10

Introduction

Coal is the collective description for a wide variety of organic fuels contained in sedimentary rocks. Sometimes, appropriately referred to as Old King Coal, as it is one of the first minerals to power the modern international economy. These combustible materials cover a spectrum of different qualities and characteristics. They are categorised in different ways by national governments and various international organisations. The International Coal Classification of the Economic Commission for Europe has developed a standard based on the work done by the EIA (2013). This classification is now widely used for international comparisons of the intensity of coal, and serves as the guiding definition for this chapter.

In general, the classification of coal is based on the inherent quality of each material, rather than its final use value. The quality of coal, based on its calorific value, content of volatile matter and fixed carbon, and its caking and coking properties, can vary widely, even in the same deposit. The relative value of each coal product within a particular category depends on its ash and moisture content. It also depends on the contamination by other undesirable elements, such as sulphur, chlorine, and phosphorus. These elements do not determine where coal fits within the various categories. However, it does impact on the calorific value of coal, and within each category.

Given these caveats, coal is divided into two main categories: divided according to the calorific value of the mineral. The first is hard coal, which has a gross calorific value of 5700 kcal/kg.[1] Hard coal has less ash and is more moist than other forms of coal. It has a mean random reflectance of vitrinite of at least 0.6.[2] This is subdivided into anthracite (generally less than 10% volatile matter and a high carbon content) and bituminous coal, which is further sub-divided into coking coal and other bituminous coal. The second category is brown coal, which is non-agglomerating coal with a gross calorific value of less than 5700 kcal/kg and containing more than 31% volatile matter on an ash free, but moist basis. This is then sub-divided into sub-bituminous coal (gross calorific value between 4165 and 5700 kcal/kg) and lignite (less than 4165 kcal/kg). In some countries, including Australia, Japan, South Korea, and the United States (US) sub-bituminous coal is included with hard coal. Coking coal, which is sometimes described as metallurgical coal, is bituminous coal of a quality that allows the production of a coke suitable to support a blast furnace charge. Anthracite, as well as other bituminous and sub-bituminous coals, variously are known as steam or thermal coal. Bituminous coal, used in pulverised (or granular) fuel injection blast furnaces, is a substitute for coke and is included with coking coal in Japan and South Korea.

This chapter first looks at coal's changing contribution to primary energy supplies in the Asia-Pacific Region, most notably in China. Then it examines how coal's

---

[1] This also is referred to as black coal, stone coal, coffee coal, and dark coal.

[2] Vitrinite is a shiny glass-like material composed of the woody tissue from which coal was formed. Its reflectance is a measure of the percentage of incident light reflected from a polished surface. The higher the reflectance, the higher the thermal quality of coal.

uses in the region have evolved in recent decades and how this might change in the future, paying particular attention to steam and coking coals. The paper then moves to a discussion of the region's domestic coal production, with a brief examination of its coal reserves. Although the region, as a whole, is largely self-sufficient, imports and exports in the coal trade are important. The main geographical origins and destinations of the trade are then described, before a discussion of the constraints, risks, and challenges faced both by domestic and foreign producers, and by the region's leading coal users.

Primary Energy Supplies

Four the world's top ten coal producers are located in the Asia-Pacific Region. With the exception of the United States, China India, Australia and Indonesia are the largest producers in the world. The only producer in Western Europe is Germany. However, this is dwarfed by China's output. Indeed, China produces more coal, than the other nine producers combined. Table 10.1 demonstrates this imbalance in production between the leading producers.

The Asia-Pacific Region's consumption of primary energy of all forms has risen much faster in recent decades than usage in the rest of the world. This mainly is due to rapid expansion in China, particularly during the past decade. China's share of the global total increased from 6.1% in 1980 to 8.3% in 1990 and 10.7% in 2000, before shooting up to 19.8% in 2010 and 22.1% in 2012. The shares of other Asia-Pacific states, excluding Japan, increased more rapidly in the 1980s and 1990s, from 2.2% in 1980 to 6.1% in 2000, but more slowly in the subsequent decade, reaching 7.1% in 2012. In the meantime, Japan's share rose slightly from 5.4% in 1980 to a peak of 5.8% in 1995, before dropping to 3.8% in 2012 (BP 2013).[3] Table 10.2 highlights the almost doubling of coal production since 1990. Table 10.3 shows the changes in consumption patterns from 2010 to 2015.

Chinese consumption grew substantially faster than elsewhere in the 2000s. The faster rates of growth of energy consumption in the Asia-Pacific Region (other than

Table 10.1 Top ten coal producers, 2014 (million tonnes)

China	3747.5 mt	Russian Federation	334.1 mt
United States	916.2 mt	South Africa	253.2 mt
India	668.4 mt	Germany	186.5 mt
Australia	491.2 mt	Poland	137.1 mt
Indonesia	470.8 mt	Kazakhstan	115.5 mt

Source: World Coal Association (2015)

[3] The BP Statistical Review expresses primary energy consumption in million tonnes of oil equivalent (mtoe). In this chapter, all comparisons are given in million tonnes of coal equivalent (mtce). The conversion used is based on the relative calorific values of oil and coal given in IEA (2013), namely, oil at 41.868 gigajoules per tonne and coal at 29.3076 gigajoules per tonne.

Table 10.2 Total global coal production in 2013, 2014 and 2015, compared to 1990

2015	3830.1 mt
2014	8022.5 mt
2013	8075.5 mt
1990	4677.0 mt

Source: World Coal Association (2015)

Table 10.3 Coal consumption in the Asia-Pacific Region, 2010–2015

	2010	2011	2012	2013	2014	2015
Australia	50.6	50.2	47.3	45.0	44.7	46.6
China	1743.4	1899.0	1923.0	1964.4	1943.3	1920.4
China SAR	6.2	7.4	7.3	7.8	8.1	6.7
India	292.9	300.4	330.0	355.6	388.7	407.2
Indonesia	39.5	46.9	53.0	57.6	69.8	80.3
Japan	115.7	109.6	115.8	120.7	118.7	119.4
Malaysia	14.8	14.8	15.9	15.1	15.4	17.6
New Zealand	1.4	1.4	1.7	1.5	1.5	1.4
Philippines	7.0	7.7	8.1	10.0	10.6	11.4
Singapore				0.3	0.4	0.4
South Korea	75.9	83.6	81.0	81.9	84.6	84.5
Taiwan	37.6	38.9	38.0	38.6	39.0	37.8
Thailand	15.5	15.8	16.4	15.8	17.9	17.6
Vietnam	14.0	16.5	15.0	15.8	19.3	22.2

Japan) largely reflect the region's much faster rate of economic growth, as measured by its gross domestic product (GDP). Japan's energy usage rose less rapidly than its GDP for most of the period since 1980, although the disparity was not as large as in the rest of the world. By contrast, energy usage in other Asia-Pacific countries rose more rapidly than their real GDP in the 1980s and 1990s, before following the general trend. China's performance was very different. Its usage of primary energy grew much more slowly than real GDP between 1980 and 2002, so that it became a more efficient user of energy. The declining trend in its energy intensity was only temporarily reversed in 2003 and 2004. China uses substantially more energy per dollar of GDP than other countries in the region.

Coal's Contribution to Primary Energy Usage

Until well into the twentieth century, coal was the world's most important primary source of energy. Its leading role has been superseded by petroleum and natural gas in many countries, but not in China, where it accounts for over two thirds of primary energy supply. Figure 10.1 brings out coal's dominance in China's energy balance in comparison with other Asia-Pacific countries and with the rest of the world.

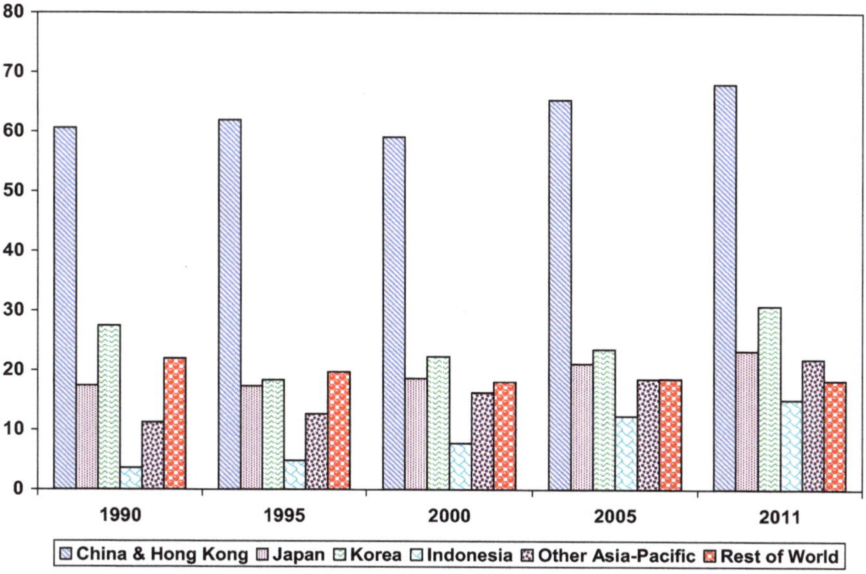

Fig. 10.1 Coal's percentage share of primary energy supply in selected countries, 1990–2011. Source: IEA (2013a)

Coal's share of China's energy supply dipped in the late 1990s, but rose in the subsequent decade to reach 68% in 2011. In the latter regard, China's trend has been typical of the region generally, where coal's share has been rising since the mid-1990s. The increases in share have been most marked in Indonesia and in the region's smaller countries. By contrast, coal's share gradually has declined in the rest of the world over the past two decades, continuing a longer-term decline. Compared with its 18.3% share of primary energy supply outside the Asia-Pacific Region, coal accounts for nearly 31% of energy supply in South Korea, 23% in Japan, 15% in Indonesia, and 22% in the other Asia-Pacific countries.

China's dependence on coal to meet its energy needs means that its usage has come to dwarf that of all other countries. This dominance is shown in Fig. 10.2 which allocates global usage of coal and coal products between selected countries of the region and the rest of the world.

In 2011, China consumed 2.7 billion tonnes of the global usage of 5.5 billion tonnes (coal equivalent) of coal and coal products, its share having more than doubled since 1990 when it used almost 0.8 billion tonnes. The other Asia-Pacific countries together used 0.45 billion tonnes, with Japan accounting for one third of that total. Table 10.4 shows how usage moved within the Asia-Pacific Region between 1980 and 2011. This shows actual volumes, regardless of calorific values, rather than coal equivalent tonnages.

Domestic coal supplies embrace local production, net trade, and changes in stocks. As I discuss later, most countries in the region, other than China, Indonesia, and, to a lesser extent, Thailand, depend on imports to meet their rising demand. Indonesia

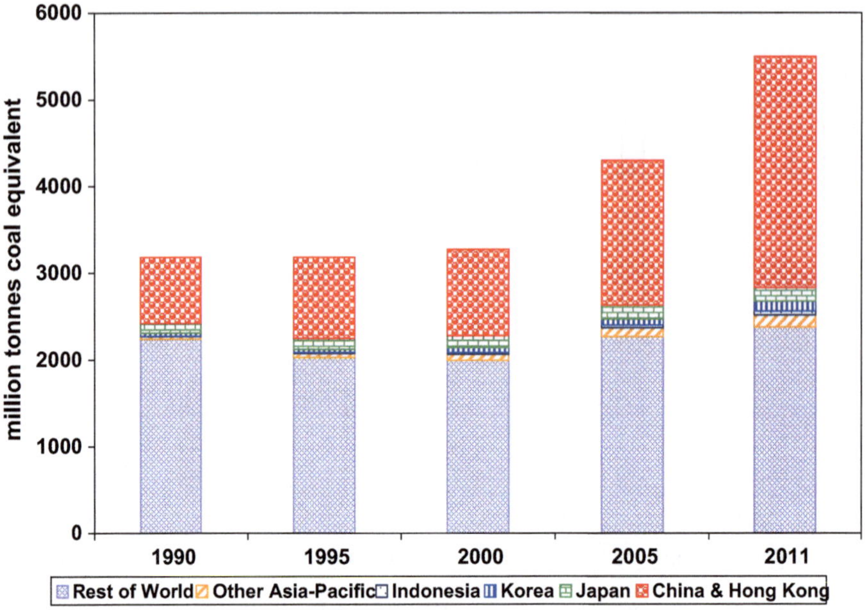

Fig. 10.2 Supplies of coal and coal products, 1990–2011 (mtce). Source: IEA (2013a)

Table 10.4 Domestic coal supplies in the Asia-Pacific Region, 1980–2011 (million tonnes)

	1980	1990	2000	2011
China	626.0	1049.6	1337.5	3501.3
Hong Kong	0.0	8.9	6.1	12.5
Indonesia	0.1	4.2	22.0	59.7
Japan	87.7	115.0	153.0	174.1
Malaysia	0.1	2.2	3.7	24.7
Myanmar	0.0	0.0	0.0	0.7
The Philippines	0.6	2.6	8.6	14.6
South Korea	27.8	44.8	71.8	130.9
Taiwan	6.0	17.2	46.8	66.5
Thailand	1.6	12.7	21.3	39.7
Vietnam	4.1	4.0	7.8	27.7
Total	753.9	1261.1	1678.7	4052.6

Source: IEA (2013b)

relies entirely on domestic production, much of which is classified by the IEA as sub-bituminous coal, and China largely has been self-sufficient until recent years. Demand has risen strongly throughout the region. Indeed, the region's rate of growth of demand slowed in the 1990s, but accelerated in the 2000s, largely because of developments in China. The rate of growth of Japanese and Taiwanese usage slackened considerably in that decade. China's usage has continued to grow since 2011. The mainsprings of the region's growing demand are examined in the next section.

Types of Coal used in the Asia-Pacific Region

The intrinsic properties of the different types of coal determine their suitability for different uses, the main distinction being between coking (metallurgical) coal and other types. Table 10.5 illustrates how supplies of the main types of coal moved in China, Japan, and South Korea between 1980 and 2011.

Usage of coking coal shrank in Japan from 49% of the region's total in 1980 to some 8% in 2011, whilst China's share rose strongly from 47% to 85%. Its demand grew at an annual average rate of 4% in the 1990s and 14.8% in the 2000s. South Korea and Taiwan are the region's only other users. As its name implies, coking coal mainly is used in coke ovens for the manufacture of coke, which is, in turn, predominantly used in the iron and steel industry. However, coking coal is not the only type of coal used in making coke, nor is coke the only coal-based product used in the iron and steel industry.

Consumption of anthracite, the highest quality grade of coal, is constrained by availability and accordingly accounts for less than 2% of the region's total domestic supplies. The region's demand for all other types of bituminous coal, including small quantities of lignite in Thailand, grew at an annual average of 8% in the 2000s, a significant acceleration on the 2.8% per annum of the 1990s, and above the 6% per annum of the 1980s.

Table 10.5 Domestic supplies of coal by type, 1980–2011 (million tonnes)

Raw material	1980	1990	2000	2011
Coking coal	142.5	161.0	200.8	633.5
Anthracite	4.1	6.2	17.3	75.5
Other bituminous coal	605.1	1073.8	1402.1	3240.5
Sub-bituminous coal	0	0.9	10.3	82.8
Lignite	1.6	12.5	17.6	20.3
China				
Coking coal	66.8	80.1	119.0	541.1
Anthracite	0	0	0.2	36.1
Other bituminous coal, including sub-bituminous and lignite	559.0	978.4	1224.3	2936.5
Japan				
Coking coal	70.2	64.9	57.1	53.8
Anthracite	0	2.2	3.8	5.6
Other bituminous coal	17.5	47.8	92.3	114.7
South Korea				
Coking coal	4.0	11.7	19.4	32.6
Anthracite	0	0	5.3	6.3
Other bituminous coal	23.8	32.2	41.5	87.8
Sub-bituminous coal	0	0.9	5.6	4.2

Source: IEA (2013b)

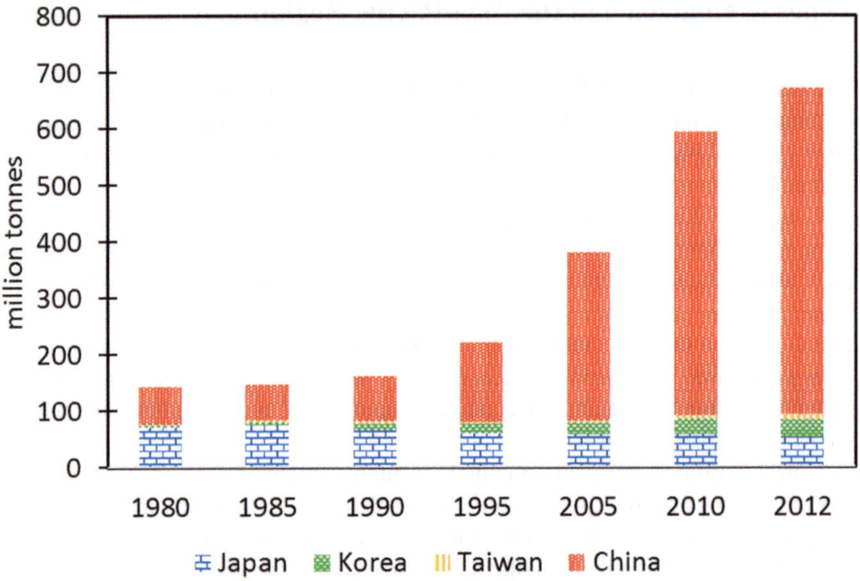

Fig. 10.3 Domestic supplies of coking coal, 1980–2012 (million tonnes). Source: IEA (2013)

Demand for non-coking or steam coal grew more strongly during the past decade than in the 1990s in most countries of the region, except Japan. Once again, China accounted for the lion's share of total usage (88%) and of the decade's increase (over 89%). Its demand has continued to grow rapidly since 2011.

Figure 10.3 shows how consumption of coking coal moved between 1980 and 2012 in the different countries of the region, emphasising the importance of Chinese demand, while Fig. 10.4 contains similar information for steam coal, excluding anthracite, but including Indonesia's sub-bituminous supplies.

The next sections examine how the uses of coal have developed in the region and discuss their likely future trends.

Coke Ovens and the Iron and Steel Industry

A coke oven produces coke, a bi-product of coal. By mixing bitumous coal at very high temperatures (between 1000 and 2000 °C), coke is the solid remainder of what is left of the burnt carbonaceous material (coal). Coke is a solid product, low in moisture content and volatile matter, which is obtained from the carbonisation of coal at high temperature. Its value lies in its use as a fuel in iron ore furnaces. The end product of this process is the manufacture of steel.

As already noted, most coking coal used in the region is for the manufacture of coke in coke ovens. The coke ovens consumed 93% of China's total domestic

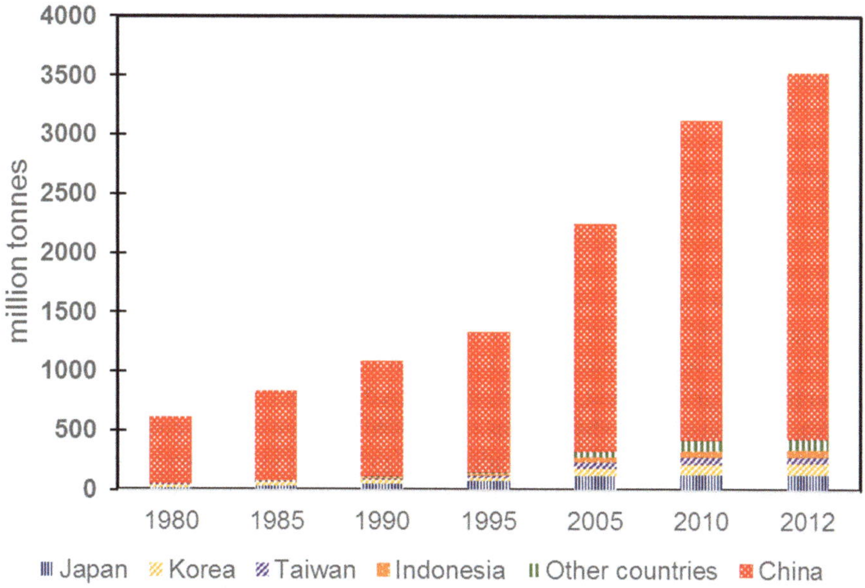

Fig. 10.4 Domestic supplies of steam coal, 1980–2012 (million tonnes). Source: IEA (2013)

supplies of coking coal in 2011. Supplies of coking coal increasingly have been inadequate, or too costly, to fulfil the needs of the coke ovens that supplement supplies of coking coal with other grades of coal. Usage in Japan's coke ovens has contracted from its level in the 1980s, whereas China's demand has grown strongly, largely to meet the needs of its iron and steel industry. Japan, South Korea, and Taiwan produce all the coke they use, whereas China has traded small quantities. It has moved from being a net exporter in the early 1990s to become a small net importer.

Burning off moisture and volatile constituents means that the tonnage of the coke ovens' output substantially is less than the tonnage of raw materials they consume. In China, the yield of coke from the coal consumed has risen from 65% in 1990 to about 75% in 2013. This is comparable to the yields in the other countries of the region, although yields fluctuate from year to year in all countries.

There is a wide range of mainly industrial uses of coke in China outside of the steel industry. Their share of consumption rose from 30% in 1980 to 43% in 1990, but since have contracted to 27% in 2000 and 17% in 2011. The growth rate of these other uses was 8% per annum in the 2000s compared with 14% per annum for consumption in the iron and steel industry. Demand from the iron and steel industry accounts for the balance of consumption. Industrial uses make up 19% of Japan's coke consumption, but in South Korea and Taiwan the iron and steel industry is essentially the coke ovens' only consumer.

Modest quantities of coking coal are burnt in a range of Chinese industries, as discussed in the next section. These quantities rose from 7 million tonnes (8.4% of total Chinese usage of coking coal) in 1990 to 17 million tonnes (3.1 per cent) in 2011. The remainder of the region's coking coal consumption is in the iron and steel industry, either as a replacement for coke, or as a direct charge. Increasingly, it is supplemented by lower grade bituminous coals, which are injected in pulverised form. Coal and coke are used to support the blast furnace charge, to provide heat and carbon for the reduction of iron ore, and for the production of oxygen steel furnace gas.

In 2011, the iron and steel industry accounted for some 15% of China's total consumption of coal, down on the 17% of 1980, but an increase on the 9% of 1990 and 12% of 2000. In Japan, the industry's share has fluctuated around a declining trend to 47% in 2011, and the share has halved to about 12% in Taiwan. In South Korea, iron and steel production has maintained its share of total coal consumption at about 26% over the past two decades. Steel producers in the other countries of the region depend exclusively on electric furnaces, using scrap and imported raw materials.

Although the steel output of electric furnaces has increased, their share has fallen, except in Taiwan. Nearly 90% of China's production comes from oxygen blown converters; with the figures being 77% in Japan and 61% in South Korea. These two processes now are the only two routes to steel production. Over the past 30 years, less efficient, and more coal intensive, processes have been phased out, and the average size of converters has risen, most notably in China. This has meant that the amount of steel produced per tonne of coal consumed in China has risen closer to the Japanese and South Korean levels.

The US Energy Information Administration (EIA 2013) forecasts that the gross output of China's iron and steel industry will more than double between 2010 and 2030, before beginning to decline. There still is scope for further improvements in the efficiency of China's industry so that its coal consumption will continue to grow less rapidly than steel production. The coal-based steel production of the other countries will increase far more slowly, if at all. The EIA (2013) forecasts that Japan's steel output will decline after 2020, due to weakening domestic demand.

Other Non-Power Generating Uses

Whilst the bulk of coal used in the region goes either to the production of iron and steel or is used for electrical power generation, there is a range of other uses. Outside China, the production of cement accounts for more than two fifths of the total, spread across several countries. The remaining uses are diverse; for example, providing heat and power for industrial processes and raw materials for chemical products. Anthracite accounts for less than one-fifth of the total, mainly in Vietnam, and the balance is of steam coal of varying grades. Other non-power uses of coal include rail transport, residential uses, the production of non-metallic minerals, in heat plants, and a range of other nonspecific industrial uses.

Industrial usage contracted sharply in South Korea, and stagnated in Japan, but rose strongly in Indonesia, Thailand, and Vietnam during the 2000s. That was on the back of increased domestic coal production, and, in some cases, was encouraged by government policy. Indonesia's actions in that regard are discussed in a later section.

The cement and chemical industries are China's two largest industrial users after iron and steel. China's cement output increased from 80 Mt in 1980 to 210 Mt in 1990, 597 Mt in 2000 and 2099 Mt in 2011 (USGS 2013). Comparison with the industry's coal usage implies that it has become a much more efficient user of coal over the years. Further increases in coal usage are likely with rising production of cement. In the chemical industry, coal is used for the production of ammonia and methanol, which is encouraged by government policy. The other industrial uses are mainly for the provision of heat for industrial processes. The EIA (2013) expects industrial gross output to quadruple between 2010 and 2040, but industrial coal consumption will expand less rapidly. The Chinese government is encouraging increased efficiency in coal consumption, a shift towards less energy-intensive industries, and reductions in emissions, all of which is reducing coal consumption per unit of output. Furthermore, coal consumption is moving from industry to power generation as a result of electrification. Such a trend is witnessed by the declining use of coal in rail transport over the past twenty years. Residential use of coal also is falling as coal is replaced by more convenient and less polluting fuels for cooking and space heating. Countering that trend has been a strong rise in coal burning in district heating plants.

The EIA (2013) forecasts that total consumption of coal in China's industrial sector, including coke ovens and the iron and steel industry, will increase from 1114 mtce in 2010 to 1658 mtce in 2020 and 1908 mtce in 2030, before falling back to 1778 mtce in 2040.[4] This is based on an assumed 5.7% annual average growth rate of GDP from 2010 through 2040. The implied annual average rates of change for industrial coal consumption are increases of 4.1% per annum between 2010 and 2020 and 1.4% per annum in the 2020s, followed by a decline of 0.7% per annum in the 2030s. Industrial uses of all types, including space heating, are expected to take a slightly rising share of China's total coal consumption in the present decade, but to account for a smaller share in 2040 (40.7% compared with 44.6% in 2010).

Electrical Power Generation

Electrical power generating stations are the largest users of coal in the Asia-Pacific Region and, in some countries, are the only users. Including combined heat and power plants, their share of the region's total coal consumption has risen from 18% in 1980 to 28% in 1990, 44% in 2000 and 52% in 2011. In China, the share was just

[4]These forecasts are expressed in quadrillion British Thermal Units (BTUs) and here, have been converted into mtce for convenience.

Table 10.6 Usage of coal in electricity generation in Asia-Pacific countries, 1980–2011 (million tonnes)

	1980	1990	2000	2011	1980–1990	1990–2000	2000–2011
					Percent per annum increases		
China	120.0	281.8	563.9	1766.2	8.9	7.2	12.1
Indonesia	0.0	4.4	16.0	41.6	–	13.7	10.0
Japan	9.6	31.0	64.9	86.2	12.4	7.7	2.9
Malaysia	0.0	1.3	2.4	20.6	–	6.3	24.2
Myanmar	0.0	0.0	0.0	0.3	–	–	–
Philippines	0.1	1.1	7.2	11.0	26.4	21.1	4.3
South Korea	1.9	7.7	39.0	81.6	15.2	17.6	7.7
Taiwan	2.4	8.0	25.9	36.1	12.6	12.5	3.4
Thailand	1.3	9.9	14.6	24.0	22.3	4.0	5.1
Vietnam	1.4	1.6	2.1	9.0	1.6	2.6	15.9

Source: EIA (2013)

over 50% in 2011 compared with 55% in the rest of the region. Table 10.6 shows how the electrical generation industry's coal consumption has developed since 1980.

Coal's share of China's installed electricity generating capacity has been declining from over 75% of the total. At the end of 2012, coal fired power stations still accounted for 66% of China's 1145 gigawatts (GW) of electrical capacity. The country's net power generation grew by some 11% per annum between 2005 and 2011 to 4476 terawatt-hours (TWh), with the industrial sector consuming three-quarters of produced electricity. The EIA (2013) projects a 3.7% per annum growth in electricity demand up to 2040 on the basis of an assumed 5.7% per annum rate of growth of GDP. That will take net generation to 7295 TWh by 2020 and 11,595 TWh by 2040.

Coal will account for a declining share of generating capacity as the Chinese government expands nuclear power capacity, the use of renewables (mainly wind power), and natural gas (based on imports and the development of domestic shale gas resources), with the objective of reducing carbon emissions and unacceptable levels of air pollution in urban areas. The EIA predicts that the share of nuclear generating capacity will rise from one per cent in 2010 to seven per cent by 2040, whilst the share of gas will grow from 3% to 5%. The fastest growth in generating capacity will be from renewable sources whose share is forecast to increase from 6% in 2010 to 21% in 2040. Wind turbines will be the major source of renewable energy. Conversely, hydropower capacity will lag, and its share will drop from 22% to 18%.

Coal accounted for 79% of the energy used in 2010 and its share will fall to 62% in 2040. Even so, coal-fired generating capacity is expected to increase by almost 530 GW between 2010 and 2040. By then, total generating capacity will grow to 2265 GW, with coal-fired stations accounting for 52% of the total. Not all installed capacity can operate continuously. Coal supplies a higher proportion of the energy used to generate electricity than its share of installed capacity. Coal consumption for electricity generation will rise from 1250 mtce in 2010 to 1860 mtce in 2020 and 2470 mtce in 2040 (EIA 2013).

The Chinese government strongly is encouraging improved efficiency in coal usage, both through directives and economic incentives. That partly is to reduce costs, but also to combat climate change by reducing CO^2 emissions. Less efficient and small coal-fired plants are being scrapped, and advanced coal-burning systems, such as supercritical and ultra-supercritical pulverised coal-fired generation, are being adopted in new power stations. These advanced systems emit almost 40% less CO^2 than conventional plants. One percentage point improvement in the efficiency of a conventional pulverised coal combustion plant results in a 2–3% reduction in CO^2 emissions. Some 80 GW of small and inefficient coal-fired plants were closed between 2005 and 2010, and further replacements of small units by much larger modern plants are progressing.

The main growth in China's electricity generating capacity will be to supply the large urban areas in the south and east of the country, where demand for power is increasing most rapidly. Whereas coal remains the cheapest fuel for power generation in most of the country, transport costs from the mines reduce coal's competitiveness in coastal cities. Here, nuclear power and natural gas-based capacity will replace some coal-fired capacity. The constraints on coal supplies are discussed in a later section. In the longer-term, the Chinese government plans to invest heavily in more gas-fired power plants. The country can obtain gas from growing domestic sources, as well as from imports. Some new efficient gas-fired units are being developed in conjunction with new LNG terminals in Guangdong and Shanghai. Shifting away from coal to gas-fired generation depends on the country's ability to increase domestic production through shale gas and offshore reserves, as well as imported sources.

The pressure to switch away from coal is given added impetus by the need to improve air quality in the major cities. The government's Airborne Pollution Prevention and Control Action Plan 2013–17 aims to reduce coal's share of total energy consumption to 65% by 2017. Companies are prohibited from building new coal-fired power plants around three major cities, Beijing, Shanghai, and Guangzhou, in order to counter pollution. Beijing has gone further, replacing all its coal-fired facilities (about 2.4 GW of capacity) with gas-fired plants by the end of 2014. There also are proposals to restrict imports of lower grade coals with high ash and sulphur contents. Figure 10.5 summarises the US Energy Administration's projections of China's total demand for all types of coal in all uses. It shows index numbers of total demand based on 2010 as 100.

The rate of growth is projected to slacken considerably over coming decades, with demand peaking around 2035. Whilst other commentators may be more sanguine about growth in the coming decade, the overall picture of demand eventually levelling out is realistic.

The rate of growth of Japanese consumption of coal for power stations has slackened over the past three decades, but the electrical sector still took a rising share of total usage, up to almost 50% in 2011. Coal was then given a temporary fillip by the nuclear power plant shutdowns that followed the Fukushima disaster. However,

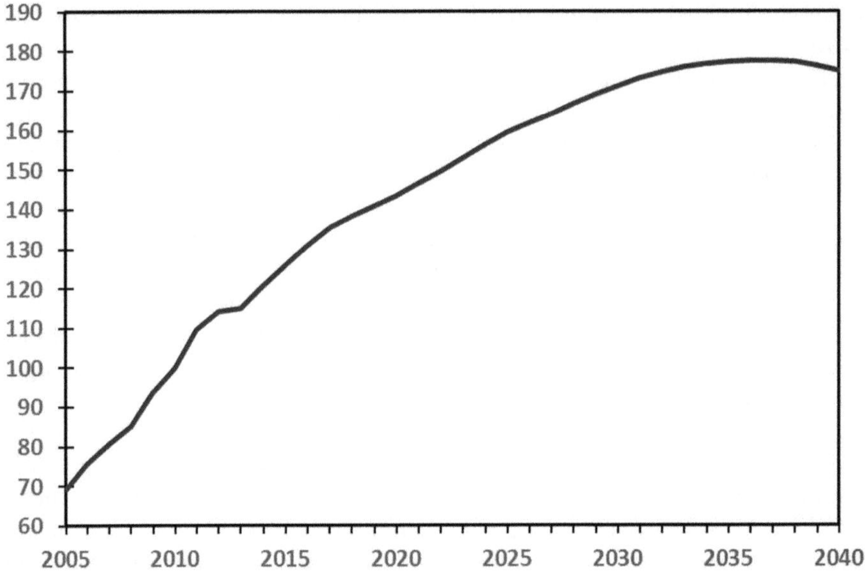

Fig. 10.5 EIA projections of Chinese coal demand to 2040. Index numbers based on consumption expressed in BTUs (2010 = 100). Source: EIA (2013)

over the longer-term, a shift toward renewable energy and natural gas for electricity generation will weaken the electric power sector's demand for coal.

South Korea's electricity industry took 62% of coal usage in 2011 compared with 7% in 1980, but the period of rapid growth has ended. Although the South Korean government has greatly downgraded its plans for nuclear power, it will remain important. Furthermore, changes in relative tax burdens are expected to favour natural gas at the expense of coal in order to reduce coal consumption. Coal's share of generating capacity is expected to fall from 44% in 2010 to 27% in 2040. This implies that the sector's consumption will flatten off in coming decades.

Demand for coal for power generation has developed strongly in most other Asia-Pacific countries. Malaysia's demand has risen the most rapidly in the past decade, with the country relying on imports. Multiple coal-fired power plants of 1 GW capacity or greater either are entering commercial service, are under construction, or are in an advanced development stage. Indonesia, Thailand, and Vietnam depend on domestic production. The continuation of recent government policies will ensure further growth of their coal-fired power capacity. Vietnam plans to build substantial new coal-fired electric generating capacity over the next 20 years, mainly in the south, whilst the country's major coal deposits are located in the north.

In all countries, and in all uses besides power generation, demand for coal is not totally independent of supply. Countries with domestic mines have tended to burn coal in preference to other fuels whose supply may be costlier or less assured. Trends in domestic mine production are discussed in the next sections.

Sources of Coal

Japan, South Korea, and Taiwan depend entirely on imports of coal, and the smaller consumers of the region, such as Malaysia and the Philippines, rely mainly on imports. Vietnam is a modest net exporter. Figure 10.6 shows how the production and consumption of all types of coal, in terms of coal equivalents rather than actual tonnages, have moved in the region since 1981, excluding China and Indonesia.

The main producers, who largely have been responsible for the increased output from 2000 onwards, are Mongolia, Thailand, and Vietnam. Much smaller quantities come from the Philippines and Malaysia. The fall in demand in 2009 was concentrated in Japan, with only slight reductions elsewhere. In 2012, Japan accounted for 38% of the consumption, while South Korea and Taiwan accounted for 25% and 13% respectively.

Until the 2000s, China was self-sufficient, with a modest two-way trade. However, its production has been unable to keep pace with demand; accordingly, imports have risen strongly since 2008. Fig. 10.7 shows how China's production and consumption of all types of coal, again in terms of coal equivalents rather than actual tonnages, have moved since 1981. There was a brief period in the late 1980s when consumption outstripped production, but for most of the period until 2009 there were annual export surpluses. The acceleration in the rates of increase of both consumption and output from 2000, after stagnation in the late 1990s, is readily apparent.

Figure 10.7 shows how Chinese production, trade, and consumption have moved in selected years since 1990 for different types of coal. The table also contains data on Indonesia, which has become an important coal producer and exporter since the early 2000s. It now is the world's largest exporter of steam coal, and also produces small quantities of coking coal for export (Table 10.7).

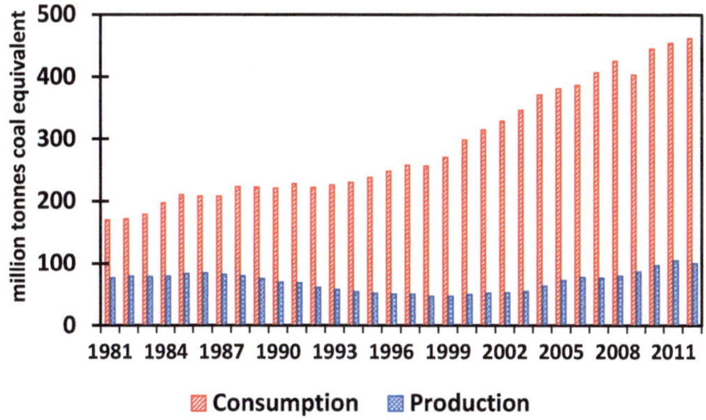

Fig. 10.6 Asia-Pacific coal production and consumption, 1981–2012, excluding China and Indonesia. Source: BP (2013)

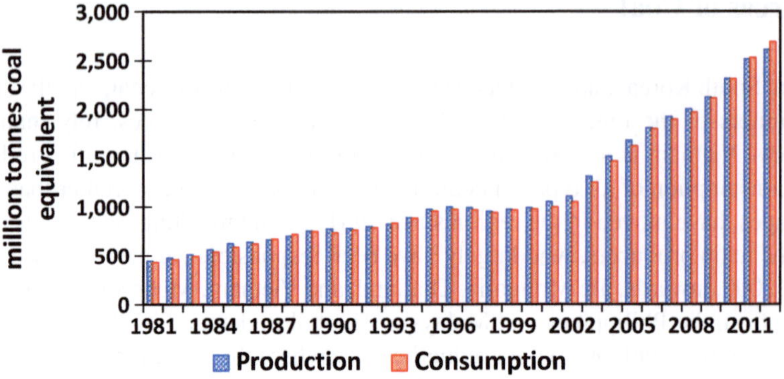

Fig. 10.7 China's coal production and consumption, 1981–2012. Source: BP (2013)

Table 10.7 Production and consumption of various coal types in selected countries and years, 1990–2012

	1990	1995	2005	2006	2007	2008	2009	2010	2011	2012
China: coking coal										
Production	86	147	297	339	360	385	416	459	509	510
Imports	0	0	7	5	6	7	34	47	45	71
Exports	−3	−7	−5	−4	−3	−3	−1	−1	−4	−2
Domestic supply	80	140	298	340	362	388	439	502	541	579
China: steam coal										
Production	909	1142	1910	2036	2161	2230	2348	2538	2771	2895
Imports	11	11	30	45	57	45	104	126	150	230
Exports	−14	−22	−67	−59	−50	−42	−21	−20	−18	−8
Domestic supply	979	1182	1932	2132	2248	2307	2509	2709	2973	3099
Indonesia: hard coal										
Production	5	21	94	136	140	129	150	170	191	182
Exports	−3	−20	−94	−136	−140	−128	−149	−169	−190	−181
Domestic supply	2	1	0	1	0	1	1	1	1	1
Indonesia: sub-bituminous coal										
Production	5	21	76	97	108	120	141	155	169	260
Exports	−2	−11	−35	−48	−55	−72	−84	−98	−110	−201
Domestic supply	4	11	42	49	54	48	57	57	59	59

Source: IEA (2013)
Note: There may be double counting in the IEA's Indonesian figures in the earlier years, as sub-bituminous coal may then have been included with hard coal

Table 10.8 Estimated Asia-Pacific coal reserves, end 2011 (million tonnes)

	Bituminous, including anthracite	Sub-bituminous	Lignite	Total
Australia	37,100	2100	37,200	76,400
China	62,200	33,700	18,600	114,500
Indonesia	–	28,017	–	28,017
Japan	337		10	347
Malaysia	4	–	–	4
Philippines	41	170	105	316
South Korea	–	126	–	126
Taiwan	1	–	–	1
Thailand	–	–	1239	
Vietnam	150	–	–	150

Source: World Energy Council (2013).

The rise in China's imports largely reflects bottlenecks in its production and transport systems, rather than any overall shortage of reserves. There is limited reliable information about China's coal resources and reserves, partly because of problems of terminology and definition. Published estimates often confuse in-situ deposits with recoverable amounts, which can be considerably less. Conversely, there often is little economic incentive to prove reserves well in advance of mining. China's coal resources are estimated at 988 billion tonnes and its proved recoverable reserves at 114.5 billion tonnes (WEC 2013). Table 10.8 compares China's estimated coal reserves with those of other Asia-Pacific countries. Australia's reserves are included in the table for comparative reference.

China's proved reserves dwarf those of other countries, apart from Australia. Coal deposits exist in most of China's regions. Twenty-eight provinces produce coal, but three-quarters of proved recoverable reserves are in the north and northwest, particularly, in the provinces of Shanxi, Shaanxi, Xinjiang, and Inner Mongolia. The geological conditions in those provinces allow larger, lower cost mining, and more potential for future expansion compared with deposits elsewhere. Coking coal reserves are found mostly in central and coastal districts, where the mining conditions are more difficult, and the remaining coal is less accessible. The Chinese Government is encouraging imports of coking coal with the aim of extending the life of its domestic reserves.

Proved reserves of all grades of coal will last 34 years at 2011 production rates. This does not account for the probable movement of resources into proved reserves. Nor does it allow for the extraction of coal by new technologies, such as underground coal gasification (UCG). This is a method of converting coal in the ground into a gas for use in industrial heating, power generation, or the manufacture of hydrogen, synthetic natural gas, or diesel fuel. China has about 30 projects using UCG in different phases of preparation.

Indonesia's total coal resource base is estimated at nearly 120 billion tonnes, with measured resources of 24.1 billion tonnes. Whilst most of its coal has medium to low calorific value and relatively high percentages of volatile matter, its low ash and sulphur content provide important competitive offsets.

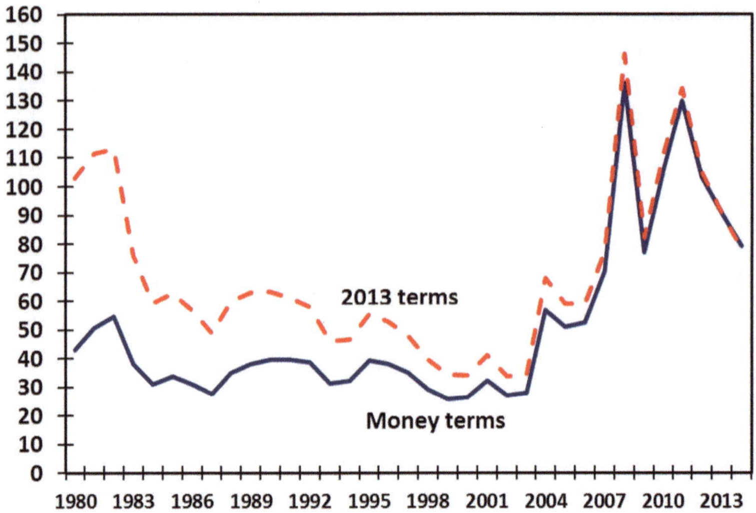

Fig. 10.8 Annual prices of Australian steam coal in money and in 2013 terms since 1980. Sources: IMF (2014); BEA (2014) (US$ per metric tonne FOB Newcastle/Port Kembla, 1200 btu/pound, less than 1% sulfur, 14% ash.)

Thailand has large reserves of lignite, which are mined for domestic power generation, notwithstanding their adverse environmental impacts. Mongolia is the only other country in the region with significant coal reserves being developed for export. The published estimates probably understate Vietnam's reserves, although Vietnam never is likely to be more than a modest exporter. Indeed, its exports will decline with the expansion of its coal-fired generating capacity. Whilst Table 10.8 reports reserve estimates for Japan and South Korea, production has ceased in both countries, so that the figures refer more to resources than to economically viable reserves.

Australia is endowed with substantial coal resources, mainly in Queensland and New South Wales. About 36% of the reserves of bituminous coal are of coking quality. All the lignite, and more than half the bituminous coal is extractable by surface mining. There are vast resources of both bituminous coal and lignite in addition to the proved in-situ reserves listed in Table 10.8, many of which potentially are recoverable (WEC 2013).

Coal Prices

The wide variety of different grades produced in a great number of mines for many end-use markets means that there is no standard world price for coal. Nor is there a common method of price determination. Mines supplying nearby power stations may have long-term cost-related contracts, whilst internationally traded coal increasingly is priced on a market basis. Annual price negotiations have

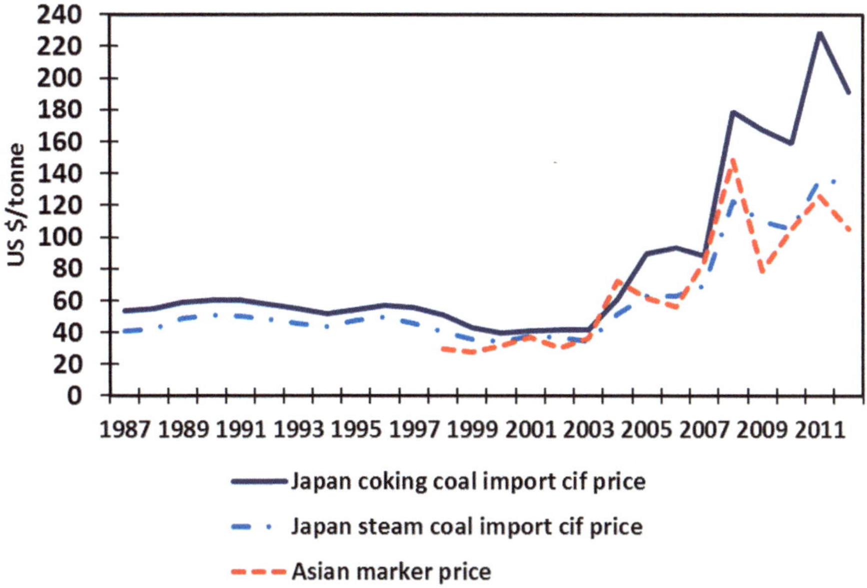

Fig. 10.9 Japanese coal prices and an Asian marker price, 1987–2012. Source: BP (2013)

tended to give way to shorter-term index-based pricing, a trend that has gone much further in the pricing of coking than of steam coal. The result is that price volatility has increased for internationally traded coal in recent years. Fig. 10.8 demonstrates this by reference to the development of the free on board (FOB) price of Australian steam coal since 1980. Prices are shown in US dollars in both money and 2013 terms.

Prices drifted downwards in the two decades to 2002, but then rose sharply, with burgeoning demand running up against capacity constraints. The peak came in 2008, and prices then plummeted in response to the global financial crisis (GFC), before rising again in 2010–2011, when China became a net importer. Prices of steam coal since have fallen back to 2009s levels, but those of coking coal have remained buoyant. Figure 10.9 compares the landed US dollar prices of coking and steam coal in Japan between 1987 and 2012, and also shows the trend in an Asian marker price. The latter reflects movements in spot prices of steam coal, and is heavily influenced by Chinese sales, whereas the Japanese imports partly are based on contractually fixed prices. With more suppliers and a wider range of qualities, there is more competition in the markets for steam than for coking coal. Its price always has exceeded that of steam coal, but the margin was relatively small until 2007, apart from a blip in 2005–2006. China's switch from market balance to net imports of coking coal after 2008 created shortages that forced up, and then sustained, its price, so that it did not fall back as much as the price of steam coal in 2009. The Queensland floods of 2011 exacerbated the tightness of the market and pushed prices higher. Subsequently, prices of coking coal eased. Those of a premium grade more than halved in China between early 2011 and late 2013.

Finally, it is worth noting the US EIA's comments about the uncertainties involved in predicting trends in international trade in coal apply with even more force to prices. Ultimately, prices are driven by the balance between supply and demand, but this is subject to a wide range of often unpredictable forces. Perhaps the one certainty is that the relative stability of prices of internationally traded coal that characterised the 1980s and 1990s is unlikely to return.

Conclusion

In the 2000s, regional demand for both coking coal and steam coal rose strongly, in part due to Chinese economic growth. The outlook is for much slower growth in the coming decade, followed by stagnation and/or decline. Concern about climate change, carbon emissions, and atmospheric pollution is leading to a reappraisal of earlier plans for dependence on coal-fired electricity generating capacity in a number of countries. Whilst the prospects for a massive expansion of nuclear power have been weakened by Japan's Fukushima disaster, the emphasis is shifting towards liquified natural gas and renewables, most notably wind power (Vivoda 2014). Efficient existing coal-fired stations will continue to operate for many years and new stations will be constructed. The less efficient stations will be phased out over time. The emphasis will be on technological innovation to reduce, or even capture, carbon emissions. Where possible, power utilities will minimise their consumption of lower-quality coal with lower calorific values and high ash and sulphur contents.

On the supply side, China is rationalising production into larger and more efficient operations, and Indonesia's exports will be constrained by central government policies. Both developments offer potential scope for alternative suppliers of steam coal to Asian markets, although China will favour domestic supplies over imports in order to ensure security of supply. The main impediments to increasing supply capacity, apart from the uncertainties created by measures to combat carbon emissions, are the capital costs of new mines and associated transport infrastructure, and the lengthy lead times involved. Those lead times are being extended in some coal producing areas by planning delays and public objections to new developments. All producers face rising operating costs and, in exporting countries, the ever-present threat of higher taxes and royalties.

In conclusion, the coal industry faces major challenges both to demand and supply. Lower cost suppliers of higher grades of coal should continue to thrive, but higher cost and smaller operations could be threatened. The future is as uncertain as it always has been.

References

Bureau of Economic Analysis. (2014). *National income and product account tables*. http://www.bea.gov. Accessed 8 May 2014.
BP. (2013). *BP statistical review of energy 2013*. BP plc, London. http://www.imf.org/external/np/res/commod/pdf/ppt/BP0613.pdf. Accessed17 September 2013.
Energy Information Agency. (2013). *International energy outlook 2013*. Report Number: DOE/EIA-0484. Washington: Energy Information Administration. Accessed 25 July 2013.
International Energy Agency. (2013a). *Coal information 2013*. Paris.
International Energy Agency. (2013b). *World energy balances 2013*. Paris.
International Energy Agency. (2013c). *World energy statistics 2013*. Paris.
International Monetary Fund. (2014). *World economic outlook database*. Washington. http://wwwimf.org. Accessed 8 May 2014.
United States Geological Survey. (2013). *2011 minerals yearbook*: *Cement*. .Washington. http://minerals.er.usgs.gov. Accessed 14 June 2014.
Vivoda, V. (2014). *Energy security in Japan*. London: Ashgate.
World Coal Association. (2015). Coal Facts 2015. http://www.worldcoal.org/sites/default/files/Coalpercent20Factspercent202015.pdf. Accessed 15 April 2015.
World Energy Council. (2013). Coal. *World energy resources: 2013 survey*. .London. http://www.worldenergy.org/publications/2013/world-energy-resources-2013-survey/. Accessed 14 November 2014.

References

Chapter 11
Rare Earth Elements

Ming Hwa Ting

Abstract This chapter investigates the current rare earth elements trade in the Asia-Pacific Region. It begins by noting the elements that fall within this category of minerals, noting their rarity. It then looks at the trade in the commodity and some of the issues associated with its supply chain. In the last part of the chapter I examine the tension between Japan and the United States over China's restriction of exports to these countries. The chapter concludes with some speculation about the future of the trade.

Abbreviations

CCP	Chinese Communist Party
DSB	Dispute Settlement Body
GATT	General Agreement on Tariffs and Trade
GFC	Global Financial Crisis
HHI	Herfindahl-Hirschman Index
OPEC	Organization of Petroleum Exporting Countries
US	United States
USGS	United States Geological Survey
USD	US Dollar
WTO	World Trade Organization

Introduction

There are 15 rare earth elements in the periodic table. Scandium (21 Se) and Yttrium (39 Y) are the lowest two elements. The other 13 occupy a higher position on the periodic table. In order of their position on the table, they include: Lanthanum (57 La); Cerium (58 Ce); Praseodymium (59 Pr); Neodymium (60 Nd); Promethium

M.H. Ting (✉)
University of Adelaide, Adelaide, SA 5005, Australia
e-mail: ming.ting@adelaide.edu.au

T. O'Callaghan, G. Graetz (eds.), *Mining in the Asia-Pacific*, The Political Economy of the Asia Pacific, DOI 10.1007/978-3-319-61395-6_11

(61 Pm); Samarium (62 Sm); Europium (63 Eu); Gadolinium (64 Gd); Terbium (65 Tb); Dysprosium (66 Dy); Holmium (67 Ho); Erbium (68 Er); Thulium (69 Tm); Ytterbium (70 Yb); and, Lutetium (Lu).

These rather obscure elements only have come into prominence in the past few years, mainly as a consequence of China's suspension of rare earth exports to Japan in September 2010.

The commercial exploitation of rare earths began in Treibach, Austria in 1903, even though the elements were discovered by Johan Gaddin in Ytterby, Sweden, in 1794. The gap of over a 100 years between initial discovery and commercial exploitation provides a clear indication of the historical difficulties of working with such elements.

Arguably, the sudden prominence of rare earth elements in recent years is attributable to the simmering geopolitical tensions between China and Japan over contested maritime claims in the East China Sea, not because of their elemental qualities *per se*. Even though their elemental properties make them increasingly important in the green energy market, rare earth elements still do not, as yet, constitute a major share of the global commodities market when compared with mainstream resources, such as oil and coal (Ting 2010).

Why Focus on Rare Earth Elements Now?

Arguably, state and non-state actors value rare earth elements for their potential to disrupt activity in major industries, such as the automotive and green power generation sectors; not because of their current economic value. Hence, these actors are concerned with any developments or policies introduced by major producers, China being the principle example, that artificially may affect the supply and price of such resources. However, since reaching a peak in 2011, the demand for rare earths has fallen. This, in turn, has put downward pressure on the high prices that rare earths previously attracted. This declining demand has resulted in less international attention on the trade and supply of rare earths.

China currently is the largest producer of rare earths in the world, and its operations in Baotou, Inner Mongolia, provide the vast majority of rare earths available in the global market. In recent years, China has calibrated global rare earths supply through the implementation of supply chain innovations, such as the introduction of variable quota limits and an export licencing system (Ting and Seamen 2013). With the introduction of these measures and others, the Chinese government has been able to influence the production and supply chain of rare earths to the global market. For example, between 2008 and 2009, China's production of rare earths accounted for 240,000 tonnes out of the global total of 248,000 tonnes. The Herfindahl-Hirschman Index (HHI), a measure used by the United States (US) Justice Department to quantify monopolistic behaviour, provides insight into the extent of China's influence over the sector. According to the Justice Department, when the HHI is between 1000 and 1800, the particular industry is regarded as 'concentrated'.

Table 11.1 OPEC basket price, 2005–2014

Year	Yearly basket price (USD)
2005	50.64
2006	61.08
2007	69.08
2008	94.45
2009	61.06
2010	77.45
2011	107.46
2012	109.45
2013	105.87
2014	96.29

Source: OPEC (2015)

Based on figures from the United States Geological Survey (USGS), the HHI for the rare earths sector is above 9000 (Ting 2011).

The high level of Chinese government control of the industry may serve as a deterrent to private commercial investment in rare earths mining in other countries. Rare earths mining, as with other extractive sectors, is a costly business, and requires a substantial investment of time and financial resources across the mine life-cycle. Moreover, given the time lag between prospecting and supplying the market, it is probable that market conditions will fluctuate, potentially affecting investment, mining, and trade activities. For instance, at the height of the rare earths dispute between China and Japan, commodity prices spiked, and commercial interest in rare earths mining increased.

However, in recent months, the prices of rare earth elements have been falling. In a way, the general decline in rare earths prices mirrors that of the oil market. Indeed, between 2003 and 2008, oil prices were on an upward trajectory due to the Organization of Petroleum Exporting Countries' (OPEC) limited spare capacity, which restricted the organisation's ability to respond to spikes in global oil demand. However, in the past few years, demand has eased, and prices have decreased (Bousso and Schneyer 2014). These annual price fluctuations can be seen below in Table 11.1.

In Table 11.1, oil prices have been on an upward trajectory, reaching a peak of US$109.45 in 2012. However, the average price of oil in February 2016 dropped to US$29.00 a barrel. Accordingly, motorists now are paying less when filling up at service stations, and air travellers are benefiting from the drop in oil prices. Businesses also have made changes to maximise the benefits of lower oil prices. Such changes especially are evident in the commercial passenger aircraft industry, in which fuel prices constitute a significant proportion of operating costs. With the recent decrease in fuel prices, airlines have begun deferring purchases of more fuel-efficient aircraft (Clark 2015). For example, Thai Airways recently announced the company's intention to hedge its fuel prices to take advantage of the prevailing low prices (Reuters 2015a). Australian airline, Qantas, also has indicated that it will remove fuel surcharges, as fuel prices come down (Reuters 2015b). These examples demonstrate that industries seek to maximise profits, and that commercial decisions are driven principally by the 'bottom line'. More importantly, the example of oil demonstrates the vagaries of international trade in oil, gas, and mining; no commodity, even the rare earth elements that are fuelling growth in green technology industries, is immune to fluctuating price cycles.

Rare Earth Elements a Driver of New Technologies

Due to growing concerns over the impacts of climate change, much attention has been given to exploring the use of green, renewable energies (Ting 2013). Rare earth elements are playing a crucial role in the green energy sector. For example, neodymium and dysprosium are used in the manufacture of magnets used in wind turbines and hybrid drive electric motors. Depending on the generative capacity of wind turbines, between 300 and 500 kg of rare earth elements are used in the manufacturing process. Similar efforts have been made in the automotive sector, as manufacturers look for ways to reduce dependence on fossil fuels. Hence, hybrid and electric vehicles increasingly are becoming common. However, as rare earth elements are used in the production of engines and batteries for these vehicles, any disruption in the supply chain, in terms of access or price, is likely to become a roadblock both to production and adoption by the public (Bourzac 2011). After all, if the cost of production increases, the prices of rare earths will increase accordingly. Naturally, this would put upward pressure on the ticket prices of vehicles, thereby reducing the extent of adoption. As an indication of the significance of rare earth elements to the automotive industry, major companies, including Ford, have begun to focus on the place of rare earths in their supply chains, paying specific attention to sustainable development implications (Ford 2012). The inclusion of a section on rare earths in Ford's 2012/2013 sustainability report provides *prima facie* evidence that these elements increasingly are assuming a significant role in the manufacturing process of one of the major industries in the world.

The Rare Earths Supply Chain

The thrusting of rare earths into the international spotlight in the wake of China's suspension of supply to Japan highlighted the main vulnerability within the global rare earths supply chain. Indeed, only then did rare earths appear as a major blip on the radar of the major consumers, Japan and the US. The US formerly was the largest rare earths producer. However, the country relinquished its position when it became cheaper to import materials from China, rather than mining and processing them locally. With growing international recognition of the critical role that rare earths play in growth industries discussed above, countries increasingly are paying attention to how they might ensure that the global rare earths supply chain remains stable and equitable. To reduce reliance on foreign-sourced rare earths, the US has been researching and developing new technologies that make less use of these resources. For example, the country's Department of Energy has established the 'Rare Earths Alternative in Critical Technologies' project to drive innovations in the research and development of the next generation of permanent magnets and batteries.

As with any research and development process, results are not guaranteed, and a long gestation period sometimes is necessary before evaluation can occur. Hence, there also are increasing efforts aimed at recycling electronic devices, such as smartphones and tablets. This practice, often referred to as 'urban mining', is the extraction of economically valuable materials, such as gold, silver, and rare earth elements, from discarded electronic waste (e-waste). Urban mining is attractive, because it involves a much lower level of financial risk. This is because the commodities sought are present in e-waste in a known quantity and, more importantly, of a certain purity level. This is to be expected, as e-waste comprises formally fully functional high-tech products (Clancy 2014). Large scale e-waste recycling therefore can help to mitigate the need to undertake traditional forms of mining that can have significant environmental externalities. There are two caveats, however. First, the weight of commodities contained within typical e-waste is quite low. Thus, for e-waste recycling to be economically viable, significant economies of scale need to be realised. This would require substantial financial investment. Second, if not done in an environmentally sensitive manner, the recycling of e-waste can result in negative environmental externalities. For example, the burning of rubber and plastic wire casing in order to extract the copper contained within emits hazardous fumes that can harm both the individual performing the operation, as well as the environment. However, at present, e-waste recycling principally occurs in developing nations, where environmental standards and regulations tend to be lax (Acaroglu 2013). Consequently, even though such activities potentially can reduce the need for rare earth mining, there are potentially significant impediments that must be considered.

International Disputes Over Rare Earths and Rising Nationalist Sentiment in China

Given the aforementioned limitations associated with securing access to rare earths supplies from alternative means, on 13 March 2012, the US brought China's restrictions on rare earths exports to the attention of the World Trade Organisation (WTO). In its complaint, the US argued that China's control of the exportation of rare earths contravened its obligations upon ascension to the WTO. Summarised briefly, the US argued that China's regulation of the rare earths sector through export restrictions, export quotas, and licencing served as a *de facto* trade barrier that distorted the global trade of such resources. On the 22 of March of that year, the European Union and Japan, both being major consumers of rare earths, requested to be added to the complaint. On the 26 March, Canada joined as well. The crux of the dispute was that upon the country's entry into the WTO, China was expected to abolish all tariffs on goods and services, except those listed in Annex 6. Eighty-four items are listed in the Annex; rare earths do not appear on the list of exclusions. However, China argued that even though rare earths were not included in Annex 6, it still had the right to impose duties through the 'General Exceptions' clause in Article XX of the General Agreement on Tariffs and Trade (GATT). Specifically, Article XX states that member states are able to impose restrictions on trade in certain commodities on the basis that they are 'necessary to protect human, animal or plant or health', especially since such restrictions are based on the conservation of exhaustible minerals. In its response to the claim, China argued that rare earths mining imposed significant environment impacts (Kilby 2014).

The mining and refining processes associated with rare earth elements, like other commodities, inherently present significant negative environmental costs. However, in the case of rare earths, the processes arguably entail more environmental externalities. For rare earths to be used for industrial manufacturing processes, the elements need to be of a high purity level. However, the elements generally are not found in nature it their pure state; they nearly always are found with other elements. Consequently, they have to undergo a process of 'liberation'. This is a complex undertaking.

Given the significant environmental externalities associated with the processing of rare earth elements, it is unsurprising that there is much opposition by local communities when rare earth operations are proposed. For example, when an Australian rare earths mining company Lynas, established its processing plant in Kuantan, Malaysia, there was a great deal of community disquiet. Environmental activists brought legal action and sought an injunction to stop the company's operations. Nevertheless, their actions were unsuccessful, and Lynas was able to commence operations in 2012.

Rare earths processing in Malaysia is not new. In 1992, Mitsubishi closed its rare earths refinery in Bukit Merah, after its operations were alleged to have resulted in a higher-than-usual prevalence of birth defects and leukemia in the small community of 11,000 people (Bradsher 2011). Even though there was no clinical evidence indicating that the rare earths processing activities caused the health problems, the experience heightened the sense of unease among the Malaysian public, which was ignited again when Lynas commenced operations in Kuantan.

Rare earths mining and refining activities impose significant environmental costs. Hence, it is understandable that China is concerned about protecting its own environment. However, this line of defence is not convincing given China's resistance to the capping of its greenhouse emissions, or its continued reliance on generating energy from coal-fired power plants. The country's resistance largely is due to its concerns that any significant lowering of its emission levels would have to come from a corresponding reduction in its industrial activity. Accordingly, despite China's best efforts at defending itself against the complaint, the WTO ruled against the country in March 2014. China appealed against the decision, but the original verdict was upheld by the Dispute Settlement Body (DSB) on 29 August 2014. China accepted the decision on 26 September 2014, but indicated that it needed time to implement the recommendations of the DSB. In January 2015, the Chinese government announced that its first step in implementing the DSB's decision would be its removal of export quotas for rare earths from May 2015 (Ministry of Commerce, People's Republic of China 2015). According to Zhanheng, deputy secretary of the China Rare Earths Industry Association, the removal of government restrictions on rare earths exports likely will make it easier for new mining companies to enter the industry (Wang 2015).

At present, the absolute economic value of the global rare earths trade is low compared with more common commodities, such as iron ore and crude oil. However, it must be acknowledged that even though the present economic value of rare earths may not be that high, it is impossible to place a figure on the *future* economic and strategic values of rare earth elements. Just as the humble silicon chip has revolutionised the computing industry and produced exponential growth in downstream and previously non-existent industries, rare earths hold the potential to change the way our vehicles are powered and electricity is generated. Similarly, their elemental properties lend themselves well to the defence industries, which means that these elements likely will become increasingly important over time.

As early as the 1980s, the Chinese government recognised the strategic importance of rare earth elements. Lobbying by Deheng, Ganchang, Jiachi, and Fangyun convinced former Chinese leader Deng Xiaoping to hasten Chinese efforts with regard to the scientific development of rare earths. This is acknowledged in the establishment of the '863 Program'. One of program's primary aims, according to the Chinese Ministry of Science and Technology, was 'to achieve breakthroughs in key technologies for environmental protection, resources and energy development to serve the sustainable development of [Chinese] society' (Ministry of Science and Technology, the People's Republic of China n.d.). With strong, ongoing support from the Chinese government, Chinese expertise in the rare earths sector increased rapidly. The implementation of 'Program 973' in March 1997 provided stronger evidence of the Chinese government's commitment to strengthening its position in the sector. Like his predecessor, Deng Xiaoping, former President Jiang Zemin recognised the economic and strategic value of rare earths. Visiting Baotou in 1999, Jiang stated that it was important for China to '[improve] the development and applications of rare earth[s], and change the resource advantage into economic superiority' (Baotou National Rare-Earth Hi-Tech Industrial Development Zone 2015).

In China, the political stability and legitimacy of the ruling Chinese Communist Party (CCP) largely is dependent on the Party's ability to manage citizens' rising expectations with regard to socio-economic development, which were precipitated by Deng Xiaoping's introduction of economic reforms in 1978. In a way, the Chinese citizenry can be conceived as having entered into a pseudo-Faustian pact with the CCP: indeed, so long as there is economic growth and the perceived 'trickling down' of benefits, there is minimal incentive for citizens to seek a change to the political status quo. However, in order to maintain the economy's growth trajectory, it is necessary for the CCP to ensure that the country can leverage its natural resource endowment, including rare earths, which arguably will be pivotal to the development of new industries.

Rare earth elements are regarded as crucial to driving the next wave of industrialisation, especially in China (Baldi et al. 2014). Accordingly, since September 2012, plans have been underway to establish an 18,000 tonne strategic rare earths stockpile. There also have been growing calls from within China to increase rare earths imports, especially of the medium and heavy varieties, so as to reduce depletion of domestic supplies. In contrast, some stakeholders in China have raised concerns that rare earth mining has gone on for too long with little, if any, regulation. According to the Chinese Ministry of Land Resources, the over-exploitation of rare earths between 1996 and 2009 resulted in Chinese reserves declining by 37 per cent. As a consequence, Chinese state media outlets, such as *Xinhua* and *People's Daily*, have been vocal in advocating the nationalist stance that China, as an aspiring global superpower, should not accede to external pressure from other states to remove the country's restrictions on the rare earths supply chain. Given the tight control the Chinese government has over the media, it can be assumed that such strident posturing has the tacit or explicit approval of the Chinese government.

Commercial Considerations

What is clear from the above discussion is that the international competition for rare earth elements, and the ensuing price fluctuations are not new. What has changed, though, is the type of commodities driving commercial and geo-political contests. Being finite by definition, the concern that dwindling oil, gas, and mineral resources eventually will run out has resulted in the establishment of regulations to control the extraction, use, and sale of such commodities. Governments also have begun to think more strategically about how rare earth elements may be used to serve wider geo-political purposes. As an example, when China suspended rare earths supplies to Japan, former Australian Prime Minister Kevin Rudd said, 'The Australian government understands the significance of rare earths globally. Australia stands ready to be a long term, secure, reliable supplier of rare earths to the Japanese economy' (Taylor 2010). However, since former Prime Minister Rudd's comments, the prices and geo-political importance of rare earth elements have fallen, as demand has waned. Accordingly, despite strong political support, the commercial fortunes of rare earths mining companies also declined. For example, Lynas had to defer some debt repayments and sell AUD$40 million in shares after rare earth prices fell sharply from their 2011 peak (Stringer 2014).

The lifting of export quotas (discussed above) ironically also may put unintended downward pressure on rare earths prices in the long-term. Given that prices have been on a downward trajectory, the removal of the export quotas and a probable commensurate increase in supply, likely will result in the further diminution of prices. Such a development would place further commercial pressure on rare earth mining companies. It is worth recalling that the low prices of the 1980s and 1990s constituted the primary reason that American rare earths companies ceased their operations. This paved the way for Chinese companies, with lower operating costs, to dominate the rare earths market. Misplaced confidence in free market principles to regulate the demand and supply of rare earths led to China's dominance of the industry, and, subsequently consolidating its position in the global rare earths supply chain. Similarly, the restarting of rare earths mining by Molymer Corporation at its Mountain Pass operation in California in the early 2000s coincided with renewed understanding of the importance of reducing American dependency on Chinese-sourced rare earths; this also coincided with rising market prices for rare earth elements. With rare earths prices having fallen significantly from their 2011 peak, the extent to which non-Chinese commercial entities will persist in undertaking mining activities in light of the lack of attractive commercial returns remains to be seen.

Moreover, even though the Chinese government has undertaken to abolish its export quotas for rare earths, any potential increases in supply, for which Japan and the US have been lobbying, may well have the unintended impact of driving non-Chinese miners out of business. This is because the operating costs for Chinese mines remain much lower than those for non-Chinese mines, which means that they potentially can weather the negative impacts of lower prices for a longer period. Hence, the lifting of quotas, rather than leading to a diminution of Chinese influence on the global rare earths supply chain, may have the unintended consequence of enabling Chinese mines to supply the global market at a time non-Chinese mines are reducing their mining and refining capacity. Should such a development occur, other producing states may have little recourse available to them. After all, China has acceded to international pressure by implementing policy changes that directly address the judgement of the DSB.

At this juncture, it ought to be noted that the removal of export quotas does not mean that China will have relinquished all control over the trade in rare earths, nor that domestic mining companies will have unfettered access to the global market. This is because export licences still are required, and the export of oil, gas, and mineral resources only can take place through nine designated custom ports.[1] Any disruptions in the operations of these ports likely will affect global rare earths supplies in the short-term. Export licences henceforth will be handled by the Special Commissioner's Office within the Chinese Ministry of Commerce. Given that these regulations only were announced in January 2014, it remains to be seen if the application and approval processes for export licences will serve as pseudo-export quotas, albeit under a different guise, in order to restrict and/or exert China's influence over the global rare earths supply chain. Likewise, given that global demand for rare earths has decreased in the last two years, the impact of the existing Chinese quotas arguably have not been felt by the major consumers of rare earths, as Chinese mining companies have exported less than is permitted under the quota system. Thus, the jury still is out as to what, if any, long-term effects this new Chinese policy will have on the supply and prices of rare earth elements.

[1] The nine designated custom ports are located in Tianjin, Shanghai, Qingdao, Huangpu, Hohhot, Nanchang, Ningbo, Nanjing, and Xiamen Customs.

Conclusion

It is clear that the global rare earths supply chain is dominated by China. Even though China does not have a geological monopoly on rare earth elements, the costs of mining and refining, both financially and environmentally, deterred many states from undertaking mining activities in the 1980s and 1990s. Consequently, China displaced the United States, with the former assuming the mantle of being the world's largest rare earths producer by default. Despite the myriad of uses in global growth industries, rare earths did not enter international society's collective consciousness until China suspended exports to Japan over maritime disputes in the East China Sea in September 2010. That episode demonstrated China's ability unilaterally to disrupt the global supply chain, and highlighted the imbalance that had formed in the industry over the preceding 20–30 years. At the time of writing, the global demand for rare earths had fallen in tandem with the falling oil prices and slower global economic activity, which is yet to fully recover from the worst effects of the recent Global Financial Crisis (GFC). Consequently, international attention on rare earth elements has receded. However, given the long period between prospecting and refining these commodities, it therefore is necessary for major consumers to adopt a long-term perspective on this issue. Research and development into rare earths extraction and recycling need to continue apace, and alternative suppliers need to be identified. Arguably, prices and demand for rare earths will at some point increase. Should the political and regulatory status quo remain, the same vulnerabilities no doubt will occur again.

References

Acaroglu, L. (2013, May 4). Where do old cellphones go to die? *New York Times*. http://www.nytimes.com/2013/05/05/opinion/sunday/where-do-old-cellphones-go-to-die.html. Accessed 30 December 2013.

Baldi, L., Peri, M., Vandone, D. (2014). Clean energy industries and rare earth materials: economic and financial issues. *Energy Policy*, 66, 53-61.

Baotou National Rare-Earth Hi-Tech Industrial Development Zone. Rare earth: an introduction. http://www.rev.cn/en/int.htm. Accessed 6 January 2015.

Bourzac, K. (2013). The rare-earth crisis. *MIT Review*. http://www.technologyreview.com/featuredstory/423730/the-rare-earth-crisis/. Accessed 3 December 2013.

Bousso, R., Schneyer, J. (2014, October 13). Exclusive: privately, Saudis tell oil market - get used to lower prices. *Reuters*. http://www.reuters.com/article/2014/10/13/us-oil-saudi-policy-idUSKCN0I201Y20141013. Accessed 15 October 2014.

Bradsher, K. (2011, March 8). Mitsubishi quietly cleans up its former refinery. *New York Times*. http://www.nytimes.com/2011/03/09/business/energy-environment/09rareside.html. Accessed 3 December 2013.

Clancy, H. (2014, February 22). Rare earths recycling takes on new Luster. *Forbes*. http://www.forbes.com/sites/heatherclancy/2014/02/25/rare-earth-recycling-takes-on-new-luster/. Accessed 3 March 2014.

Clark, N. (2015, January 13). Falling oil prices, a boon to airlines, pose a challenge for Airbus and Boeing. *New York Times.* http://www.nytimes.com/2015/01/14/business/international/falling-oil-prices-a-boon-to-airlines-pose-a-challenge-for-airbus-and-boeing.html?_r=0. Accessed 19 September 2015.

Ford. (2012). Sustainability 2012/2013. http://corporate.ford.com/microsites/sustainability-report-2012-13/supply-materials-elements. Accessed 3 December 2013.

Freese, B. (2003). *Coal: a human history.* London: Arrow Books.

Kilby, C. (2014). China's rare earth trade: health and the environment. *The China Quarterly*, 218, 540-550.

Ministry of Commerce, People's Republic of China. (2015). Official of Chinese Ministry of Commerce Comments on cancelling rare earth quotas. http://english.mofcom.gov.cn/article/newsrelease/significantnews/201501/20150100874624.shtml. Accessed 15 January 2015.

Ministry of Science and Technology, the People's Republic of China (n.d.). National high-tech R&D program (863 Program). http://www.most.gov.cn/eng/programmes1/200610/t20061009_36225.htm. Accessed 15 January 2015.

Organization of Petroleum Exporting Countries. (n.d.). OPEC basket price. http://www.opec.org/opec_web/en/data_graphs/40.htm?selectedTab=annually. Accessed 1 February 2015.

Reuters. (2015a, January 27). Update 2 - Asian Airlines slash fuel surcharges but fares likely to rise. http://www.reuters.com/article/2015/01/27/australia-qantas-idUSL4N0V600F20150127. Accessed 27 January 2015.

Reuters. (2015b, January 8). Thai Airways plans to fully hedge fuel costs as oil prices skid. http://www.reuters.com/article/2015/01/09/thai-airways-oil-hedging-idUSL3N0UO1IG20150109. Accessed 23 January 2015.

Stringer, D. (2014, May 6). Lynas to sell shares, defer debt after rare earths slump. *Bloomberg*. http://www.bloomberg.com/news/articles/2014-05-05/lynas-to-sell-shares-defer-debt-payments-after-rare-earth-slump. Accessed 6 May 2015.

Taylor, R. (2010, November 23). Australia commits to rare earths supply to Japan. *Reuters.* http://ca.reuters.com/article/topNews/idCATRE6AM1BF20101123. Accessed 5 November 2015.

Ting, Ming Hwa. (2013). Rare metals and green technologies. In White, K., Durham, L. (Eds), *America Goes Green: An Encyclopaedia of Eco-Friendly Culture in the United States.* Santa Barbara: ABC-Clio, pp. 552-553.

Ting, Ming Hwa. (2011). China's rare earth motives. *The Diplomat.* http://thediplomat.com/2011/06/chinas-rare-earth-motives/ Accessed 5 May 2013.

Ting, Ming Hwa. Seaman, J. (2013). Rare earths: future elements of conflict in Asia. *Asian Studies Review*, 37(2), 234-252.

Wang, Z. (2015). Export quota system for earths abolished. *China Daily*, 6 January. http://usa.chinadaily.com.cn/business/2015-01/06/content_19245940.htm. Accessed 15 January 2015.

World Trade Organization. Article XX: general exceptions. http://www.wto.org/english/res_e/booksp_e/gatt_ai_e/art20_e.pdf, p. 565. Accessed 5 December 2015.

Chapter 12
Natural Gas

Vlado Vivoda

Abstract This chapter analyses change and continuity in the regional approach to LNG markets since 2010, focusing, in particular, on Japan, the world's largest LNG importer. The primary focus is on natural gas markets in the region as opposed to extraction, production and/or domestic regulation in producing states in the Asia-Pacific Region. This approach is driven by the fact that the regional gas market is in flux and the evolution of regional gas pricing will have a significant effect on the economics of natural gas production worldwide. Japan's approach to LNG markets is evaluated in the context of its interaction with other formal institutional actors and is couched within the broader regional institutional setting. Consequently, the chapter documents recent developments in Japan's LNG policy and summarises its activities in the LNG market since 2010.

Abbreviations

APEC	Asia-Pacific Economic Cooperation
ARF	Asian Regional Forum
ASEAN	Association of South East Asian Nations
CBM	Coal Bed Methane
CEO	Chief Executive Officer
CNOOC	Chinese National Offshore Oil Corporation
CNPC	Chinese National Petroleum Corporation
CSG	Coal Seam Gas
EAS	East Asian Summit
EIA	Energy Information Administration
ENI	Ente Nazionale Idrocarburi
G7	Group of 7
GAIL	Gas Authority of India Limited
GAZPROM	Gazovaya Promyshlennost

V. Vivoda (✉)
Centre for Social Responsibility in Mining, University of Queensland,
St Lucia, QLD 4072, Australia
e-mail: v.vivoda@uq.edu.au

T. O'Callaghan, G. Graetz (eds.), *Mining in the Asia-Pacific*, The Political Economy of the Asia Pacific, DOI 10.1007/978-3-319-61395-6_12

GIIGNL	International Group of Liquefied Natural Gas Exporters
HH	Henry Hub
IEA	International Energy Agency
IEEJ	Institute for Energy Economics, Japan
JBIC	Japan Bank for International Cooperation
JCC	Japan Crude Cocktail
JKM	Japan South Korea marker
JOGMEC	Japan Oil, Gas and Metals National Corporation
KOGAS	Korea Gas Corporation
LNG	Liquefied Natural Gas
METI	Ministry of Economics, Trade and Industry
MOF	Ministry of Finance
MT	Million Tonnes
MTPA	Million Tonnes per Annum
MMBTU	Million British Thermal Units
NSW	New South Wales
OPEC	Organization of Petroleum Exporting Countries
POSCO	Pohang Iron and Steel Company
SCO	Shanghai Cooperation Organization
TEPCO	Tokyo Electric Power Company

Introduction

Global trade in liquefied natural gas (LNG) has undergone significant change over the past decade with a proliferation of new market entrants on both the producer and consumer sides. However, despite significant change in the global marketplace, and the growing importance of natural gas as a relatively clean energy source, natural gas prices in Asia have remained linked to crude oil prices and thus do not reflect regional supply/demand fundamentals. Historically, satisfied with secure supplies, Asian importers have not pursued a policy to abandon oil-indexation. For as long as they were willing and able to afford LNG under long-term contracts and pass the costs onto customers, oil-linked pricing remained unchallenged. With energy security prerogatives high on the agenda, long-term contracts also guaranteed secure supply of LNG for 15–25 years, without the unforeseen price and supply volatility that has characterised the oil market. However, LNG market developments in recent years have changed this perspective. Given that the international LNG market is regionally fragmented, there is a significant price differential among the three major basins (Fig. 12.1). Most importantly for Asian importers, since 2010, the price in Asia has been considerably higher than in North America or Europe.

Globally, the price of natural gas is benchmarked against competing fuels, mainly, coal, fuel oil, and uranium; increasingly, it also is being benchmarked against renewable energies. In addition, the price of LNG is benchmarked against

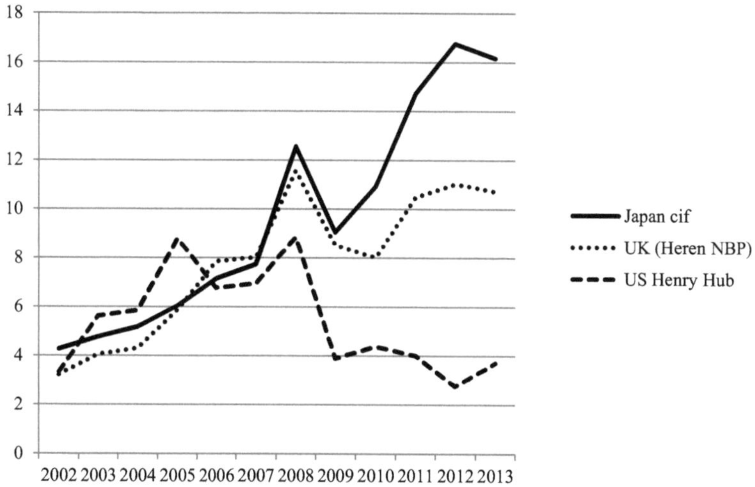

Fig. 12.1 LNG price in the three basins, 2002–2013 (US$ per million Btu). Source: BP (2014)

pipeline natural gas.[1] There are three separate and relatively independent regional natural gas markets. In the United States (US), LNG competes with pipeline natural gas and is benchmarked against the Henry Hub (HH) price for domestic spot and short-term transactions. In Europe, the LNG price is benchmarked against fuel oil and natural gas spot prices. The LNG price in Asia is benchmarked against the average monthly price of crude oil imported into Japan. This benchmark is known as the Japan Crude Cocktail (JCC). As is evident in Fig. 13.1, since 2010, Asian importers have been paying a large premium on LNG prices in other regional markets.

Crude oil parity provides for LNG prices that are linked to cost-insurance-freight crude oil prices on the calorific value equivalent basis (Langton 1994). In other words, the JCC price is an indicator of when the price of LNG is in parity with the price of crude oil on an energy equivalent basis (Miyamoto et al. 2009). The pricing of long-term LNG contracts in Asia generally reflects the fact that one million British thermal units (MMBtu) of natural gas contains one-sixth (16.67 per cent) of the energy content of a barrel of oil, a relationship which is referred to as oil parity. This linkage of LNG to oil prices also is referred to as oil-indexation. In simplified terms, the pricing formula for the cost of Asian LNG imports is (JCC x price slope) + a negotiated premium. The slope defines the relationship between oil and LNG prices and, when the energy-equivalent parity is used, the slope is 16.67 per cent. However, slopes can be lower or higher than 16.67 per cent, depending on

[1] Natural gas transportation costs between producer and consumer countries are an important aspect of competitiveness of gas supplies. In general, liquefying natural gas and shipping it via ocean transport becomes cheaper than transporting natural gas in offshore pipelines for distances of more than 1000 km or in onshore pipelines for distances greater than 3500 km (Mokhatab and Purewal 2006). However, the actual break-even point depends on numerous parameters, which, besides distance, include volumes, project financing, market size, and geo-politics.

whether the buyer agrees to pay a premium over the energy-equivalent oil price. A negotiating premium is the constant, or an element of the price that is independent of movement in oil prices. Most LNG contracts include a premium that corresponds to average costs of shipping EY (2013). For example, a crude basket price of $100/barrel with a 15 per cent slope and $2.00/MMBtu premium would yield an LNG import price of $17.00/MMBtu.[2]

Oil-indexation for long-term LNG contracts has remained the industry standard in the region because initial contractual arrangements, signed between Japan and various exporters during the 1970s and 1980s, have used the price of crude oil (the most widely traded global energy commodity) as a benchmark (Standard & Poor's 2012). In 1977, Japan commenced LNG imports from Abu Dhabi and Indonesia. Initially, while fixed price was used for Abu Dhabi LNG, a formula that linked LNG to crude oil prices was used for imports from Indonesia. In 1979, the price of Abu Dhabi LNG was linked to Japan's crude oil import prices, which was also reflected in the Organization of Petroleum Exporting Countries (OPEC) policy after the Second Oil Crisis. In 1983, Japan commenced LNG imports from Malaysia and a new Indonesian project based on an oil price-linked formula (Flower 2008; Miyamoto et al. 2009). By the time that South Korea (1986) and Taiwan (1990) joined Japan as LNG importers, this pricing principle was well established and suppliers were reluctant to accept other mechanisms.

Since 2010, the continued prevalence of this pricing structure has had severe consequences for Asian importers. The adverse effect has been most noticeable in Japan, the world's largest LNG importer, particularly following the nuclear shutdown after the March 2011 Fukushima earthquake and tsunami. According to data from Japan's Ministry of Finance (MoF), the average cost per unit of imported LNG increased by 84 per cent between 2009 and 2013, with increases recorded in every year (MoF 2014). As a consequence of the Fukushima disaster and the necessity to replace 25 to 30 per cent of electricity supply provided by nuclear power, according to the International Group of Liquefied Natural Gas Exporters (GIIGNL 2011, 2014), Japan's LNG imports increased from 70.87 million tons (mt) in 2010 to 87.98 mt in 2013. At the same time, Japan's LNG import bill doubled from ¥3.5 trillion ($39.9 billion) in 2010 to ¥7.0 trillion ($71.8 billion) in 2013 (MoF 2014).[3] Due to the unresolved status regarding the future of nuclear power, the rapid increase in Japan's LNG demand has led to higher prices under newly entered long-term contracts (Miyamoto et al. 2012). Given that oil-indexation has imposed a signifi-

[2] To further add complexity, some contracts will have varying slope percentages used at different oil price levels. Broadly speaking, there can be four basic forms: the simplest is a straight-line constant slope that exposes both the buyer and seller to adverse price movements. A second type is the so-called 'S-curve', which will have a flatter slope at low oil prices to protect sellers and a flatter slope at high oil prices to protect buyers. The other two types are variations on the S-curve, where either only sellers have some protection (an oil-linked contract with a floor) or only buyers have protection (an oil-linked contract with a ceiling) (EY 2013).

[3] The effects have not been isolated to Japan. South Korea is the world's second largest LNG importer. Although not affected by a nuclear disaster, the cost of LNG procurement has increased from US$13.9 billion in 2009 to US$30.6 billion in 2013, with the average cost per unit increasing by 43 per cent during the same period.

cant additional cost for Japan and other regional LNG importers since 2010, establishing a more flexible pricing system that at least partially reflects the regional supply/demand balance has become an energy policy priority.

High transaction costs in the region in recent years suggest that the Asian LNG market is sub-optimally organised. The challenge is that major regional importers historically have been unable to move away from competitive institutional structures, which favour the unilateral pursuit of supply security. Historically, regional energy trade has been dominated by statist approaches and energy has been 'securitised' (Phillips 2013; Hancock and Vivoda 2014). The main assumption is that, if transaction costs for regional LNG trade are to be reduced and oil-indexation challenged, it is necessary for key regional buyers to transform their energy policy approaches away from these historically dominant state-centred structures toward a market-based approach.

Against this backdrop, this chapter analyses change and continuity in the regional approach to LNG markets since 2010, focusing, in particular, on Japan, the world's largest LNG importer. The primary focus is on natural gas markets in the region as opposed to extraction, production and/or domestic regulation in producing states in the Asia-Pacific region. This approach is driven by the fact that the regional gas market is in flux and the evolution of regional gas pricing will have a significant effect on the economics of natural gas production worldwide. Japan's approach to LNG markets is evaluated in the context of its interaction with other formal institutional actors and is couched within the broader regional institutional setting. Consequently, the chapter documents recent developments in Japan's LNG policy and summarises its activities in the LNG market since 2010. Japan's main objective has been to challenge oil-indexation in the region. Its recent efforts have included a greater emphasis on LNG procurement under spot and short-term contracts, sourcing new supplies from the US under alternative pricing arrangements and driving regional buyer cooperation. The chapter evaluates the potential effects of Japan's LNG strategy on regional pricing in the broader political-economic context. Due to continued dependence on long-term oil-indexed contracts that allow limited scope for renegotiation, LNG pricing in the region only partially will shift away from oil-indexation by the end of the decade. The extent of the shift will be affected by market conditions, such as HH prices relative to JCC-indexed prices, gas-to-gas competition in China, as well as political developments, which include regional buyer cooperation, domestic gas market reforms, Japan's nuclear politics, and US domestic politics. The chapter concludes by arguing that entrenched traditional informal institutional structures will continue to hinder market integration, transaction cost reduction, and the potential for greater regional cooperation.

Natural Gas Trade in Asia: Institutional Setting

Over a decade ago, natural gas was described as Asia's 'fuel of the future' (Fesharak et al. 2000). A particular appeal of natural gas is that it is 30 per cent less carbon-intensive than oil and 50 per cent less intensive than coal (Stevens 2010). As such, natural gas is regarded as the energy source that will be Asia's bridging fuel to a sustainable energy future. While Asia's gas consumption has doubled over the past decade, the relative share of natural gas in regional energy consumption has remained at only 11 per cent (BP 2014). The high price of natural gas has been the key hindrance to its growth relative to other energy sources, particularly since 2010.

The Asian natural gas market is complex and fragmented. Unlike its European and North American counterparts, it is not well connected by pipelines, and is dominated by LNG trade. In 2013, Asian states imported 75 per cent of internationally traded LNG, but only eight per cent of piped gas. Approximately 81 per cent of natural gas was imported as LNG and the remainder via pipelines (BP 2014). Politics notwithstanding, structural issues, such as regional geography and long distances between major producers and consumers, often make it prohibitively expensive to build gas pipelines across Asia and the Pacific. Regional pipeline infrastructure includes China's import pipelines from Myanmar and Central Asia and several international pipelines in South East Asia. Pipeline trade is forecast to increase as China starts importing natural gas from Russia when the Power of Siberia pipeline links Eastern Siberia with North Eastern China is constructed in 2020.

As noted above, Japan is Asia's largest LNG importer, accounting for 49.4 per cent of regional LNG imports. Japan is followed by South Korea 22.7 per cent), China (10.4 per cent), India (7.3 per cent), and Taiwan (7.1 per cent). Several other countries recently have become LNG importers, but their shares are below one per cent (GIIGNL 2014).[4] In terms of suppliers, regional natural gas producers provide for 46 per cent of LNG and 51 per cent of piped gas consumed in Asia (BP 2014). Regional producers exported natural gas exclusively to regional buyers. However, since regional producers have been unable to keep pace with growing demand, LNG increasingly has been sourced from non-regional exporters. As a consequence, with a 31.3 per cent share, Qatar is the largest source of Asia's LNG imports, and is followed by Malaysia (14.1 per cent), Australia (12.6 per cent), Indonesia (10.2 per cent), and Russia (six per cent) (GIIGNL 2014).

Australia is the fastest growing regional LNG supplier. With numerous new conventional and unconventional gas projects currently under development, by 2020, the Energy Information Administration (EIA 2013a) forecasts that Australia will replace Qatar as the world's largest LNG exporter. Most of Australia's traditional gas resources are located in the North West Shelf offshore in the Carnarvon, Browse, and Bonaparte basins. Coal bed methane (CBM) economic resources, located in the Bowen and Surat Basins in Queensland and New South Wales (NSW) have doubled since 2010, implying that the resource distribution of gas likely will shift away from offshore traditional gas production to CBM or other sources over the next two decades.[5] Australia also has significant technically recoverable shale gas reserves dispersed throughout the country (EIA 2013a). Australia's new gas field developments are tied to liquefaction projects that will facilitate a greater export potential.

---

[4] In the context of this chapter, a reference to Asia includes East Asia, South Asia, Southeast Asia, Oceania, and Russia's far eastern regions.

[5] In Australia, CBM is referred to as CSG.

Specifically, several major new LNG projects, including Gladstone, Gorgon, Ichthys, Queensland Curtis, Pluto, and Wheatstone, are under construction or in advanced planning stages to support Asia's increasing appetite for natural gas.

However, for Australia's export potential to materialise, the industry and the government will need to overcome community opposition to unconventional gas projects. Investors face challenges with project delays based on greater public resistance to potential environmental impacts (Mercer et al. 2014). In response, the New South Wales and Queensland state governments, and the Federal government, have adopted more stringent environmental regulations, most notably those related to water use and disposal and land rights in CBM and shale gas projects (Lacey and Lamont 2014). In addition, Australia's growing LNG industry faces acute capital cost escalation requiring greater investment for new greenfield projects, which places some of the proposed projects at risk of delay or cancellation. Recently publicised cost overruns are attributed to labour shortages and resultant high wages, appreciation of the Australian dollar in relation to the US dollar since 2009, greater environmental hurdles due to more stringent regulations, and the remote locations of some projects (Balfe 2013; EIA 2013a).[6]

After accounting for more than a third of regional LNG supplies in 2002, Indonesia's share of the Asian market declined to 10 per cent in 2013. Natural gas shortages caused by production problems and rising demand (23 per cent since 2007) have, in recent years, forced Indonesia to procure LNG spot cargoes in order to meet contractual export obligations. ExxonMobil and Statoil relinquished deepwater blocks in 2013 after failing to discover economically viable reserves (EIA 2014). The government began constructing new LNG receiving terminals and gas transmission pipelines to address demand issues, and this likely will reduce the volume of natural gas available for export. While the government is committed to exporting part of its national gas production in the future, the rapidly growing domestic market likely will become the main priority (Mujiyanto and Tiess 2013; Pramudatama 2012).

In 2013, Malaysia was the second largest global LNG exporter after Qatar (GIIGNL 2014). Despite its status, Malaysia experiences a geographic disparity of natural gas supply and demand among its regions. Natural gas demand primarily is from the power and industrial sectors in Peninsular Malaysia, while gas supply is in the eastern states of Sarawak and Sabah on Borneo. In order to meet the gas needs in Peninsular Malaysia, the government-owned Petronas is developing various regasification terminals to secure supply from the global gas market (EIA 2013b). Malaysia's natural gas production has increased over the past two decades to serve growing domestic demand and export contracts. Most of Malaysia's natural gas production is offshore of Sarawak and supports LNG exports from Bintulu. When it

[6] The following LNG projects have experienced cost overruns of between 12 and 32 per cent: Gladstone, Gorgon, Ichthys, Queensland Curtis, Pluto, and Wheatstone. Ichthys LNG, sanctioned in 2012, currently is the world's most expensive liquefaction project on a per unit basis, and Chevron's Gorgon LNG project cited cost increases of over 40 per cent in US dollar terms from US$37 billion to US$52 billion.

started in 1983, Bintulu LNG was the world's first large scale LNG project (Memon et al. 2014). Recent foreign investment in deep-water and technically challenging fields primarily in Sarawak and Sabah provides impetus to maintain production levels for the remainder of the decade (EIA 2013b).

Governments and energy companies (state-owned or private) of major LNG-importing and exporting countries are the most important participants in regional natural gas trade. The inflexibility of the LNG supply chain historically has provided significant impetus to Asian governments to procure LNG under long-term contracts through their state-owned or state-controlled companies (International Energy Agency (IEA) 2013). High capital intensity and risks associated with development times that can stretch for up to a decade historically have deterred private oil and gas companies from substantial project participation (Langton 1994). Instead, there has been a high component of government participation, particularly from importing Asian states. The large capital costs and the inherent inflexibility in the value chain have required contractual arrangements to protect both the supplier and buyer. As a consequence, the international trade in LNG has been dominated by long-term contracts with strict destination clauses. These arrangements have offered little opportunity for flexibility in delivery without financial penalty.

The structure of domestic natural gas markets across Asia limits competition due to the preference for regulated monopolies that deliver LNG under long-term oil-indexed contracts (Rogers 2012). State involvement in domestic natural gas markets also is evident in the ownership structure of regasification terminals, which are predominantly owned and operated by vertically integrated state-owned companies (IEA 2013). Security of supply policy is the primary objective for most regional LNG importers, and state-owned companies play a key role in this policy. As a consequence, historically, regional natural gas importers have engaged in a competitive pursuit of bilateral deals with suppliers. A lack of buyer cooperation in LNG procurement negotiations with suppliers has endowed the latter with excessive bargaining power to influence negotiation outcomes in their favour, thus perpetuating the existing contractual and pricing arrangements.

The interaction between governments and oil and gas companies in the Asian natural gas trade is embedded within the broader regional institutional setting. While regional institutions coordinate some aspects in the energy relations between states and companies, these institutions often have less resources and a weaker actor status and, at best, enjoy delegated powers from states. Over the past decade, Asian states have created a complex web of regional, sub-regional, and extra-regional fora.[7] This complex web of overlapping and multifaceted policy-making institutions is characterised by flexibility, informality, non-binding and voluntary commitments,

[7] The most important regional policy-making institutions include the East-Asia Summit (EAS), the Association of South-East Asian Nations (ASEAN), ASEAN+3 (China, Japan, and South Korea), ASEAN's Regional Forum (ARF), the Asia-Pacific Economic Cooperation (APEC), and the Shanghai Cooperation Organization (SCO).

and consensual decision-making.[8] Thus far, regional institutions have failed to establish the necessary degrees of trust and solidarity to credibly support energy market integration. While major Asian economies are dependent on uninterrupted energy supplies, and share common interests as major energy consumers, the rules and mechanisms for governing energy have been sub-optimal, and regional cooperative mechanisms remain underdeveloped. There is no evidence of any substantive cooperative outcomes beyond information sharing, confidence building measures, or the setting of aspirational targets (Ravenhill 2013).

The interaction between countries, companies, and regional institutions is embedded in a deeper structure of informal institutions, which represent well-established rules, norms, and practices. As such, they can set formidable constraints upon the behaviour of market participants and regional institutions. They create path-dependencies where numerous choices made create an institutional matrix resisting radical change, instead favouring incremental development (North 1990). Since the early 1990s, the informal institutional setting in Asia has been dominated by state sovereignty, a strong preference for policy autonomy, the persistence of bilateral patterns of energy diplomacy, and a lack of management among the region's great powers (Andrews-Speed 2014; Wilson 2014). Given that energy has been 'securitised' across the region (Phillips 2013), Asian LNG importers have practised neo-mercantilist protectionism of domestic markets, which have been dominated by state-owned or state-controlled companies (IEA 2013).[9] Moreover, major regional LNG importers have had no concrete form of policy adjustment toward energy suppliers at the bilateral level and trade has been dominated by bilateral relationships with suppliers. For example, since the mid-2000s, Japan, South Korea, and China have used free trade agreements with LNG exporters in order to improve their energy security at the expense of other regional states. This tendency has been reflective of their desires to secure access to scarce resources in a competitive environment (Wilson 2012).

The competitive pursuit of security of long-term supplies through bilateral energy deals and energy diplomacy has been the *modus operandi* of Asia's LNG markets. State-market interaction and regional institutional frameworks have been dominated by traditional structures that emphasise sovereignty and great power management. The unilateral pursuit of secure supplies under long-term contracts has implied that major market participants have been unwilling or unable to chal-

[8] In January 2007, representatives of the EAS member states met in the Philippines and adopted the Cebu Declaration on East Asian Energy Security. However, the declaration offered nothing more than a non-binding agreement based on a series of non-binding principles. The APEC Energy Security Initiative is another example of a non-binding, regional initiative.

[9] A neo-mercantilist approach to energy implies that states should not rely on energy markets to provide optimal energy outcomes, but rather should steer markets to benefit the state. In this context, national energy policy is considered a security challenge and, as such, is 'securitised', as opposed to being conceived as largely an economic issue. State intervention via subsidies to specific energy sources, diplomatic activity designed to enhance access to energy resources and energy security (energy diplomacy), and support for national energy champions exemplify this approach (Phillips 2013; Stoddard 2013; Hancock and Vivoda 2014).

lenge the prevailing LNG market structure. While the statist pursuit of energy security has produced a degree of balance between security of demand and supply availability, since 2010, this market structure has resulted in sub-optimal market outcomes and significantly has increased transaction costs for regional LNG importers. If transaction costs for regional LNG trade are to be reduced and oil-indexation challenged, it is necessary that key regional buyers transform their energy policy from state-centred to market-based approaches.

By evaluating the prospects for such a transformation to materialise, this chapter contributes to the ongoing political economy debate about the desired extent of state intervention in energy markets (Phillips 2013; Hughes and Lipscy 2013; Hancock and Vivoda 2014). Moreover, recent studies stress the critical importance of evaluating change and continuity in LNG market structure and pricing mechanisms in the region (Aguilera et al. 2014; Inchauspe 2014). Given that the global LNG market remains regionally segmented, analysing recent developments in the evolution of the LNG market in Asia contributes to the literature on LNG pricing and the prospects for the Asia-Pacific Region, and potential global market integration (Moryadee et al. 2014; Wu 2014).

Japan as a Case Study

The chapter evaluates the change in the regional approach to LNG markets by focusing on Japan as a case study. A pioneer in regional LNG trade, Japan is the world's largest LNG importer, accounting for 37.1 per cent of global LNG imports (GIIGNL 2015). Given its market share, Japan is a key player and its actions are crucial in influencing Regional LNG pricing. If the region is to challenge oil indexation, Japan's efforts need to be supported by South Korea, due to its relative market share, and China, due to the significance of its pipeline imports and projected increase in LNG imports (Table 12.1).

Japan's LNG imports are dominated by long-term contracts, with Japanese buyers obliged to import a fixed quantity of LNG over a 15–25 year period regardless of supply/demand fluctuations. Such inflexible arrangements have been necessary in order to amortise the initial project investment and secure project finance. Most long-term LNG contracts include a destination clause, which restricts buyers from reselling purchased cargo. Long-term contracts historically have been favoured by Japanese utilities and gas companies due to their end-user delivery obligations and also have been in alignment with Japan's energy security policy (Vivoda 2014a) (Table 12.2).

Although most of Japan's entities are not state-owned (Table 12.2), they operate internationally under governmental policy guidelines, and the government is involved in the supply of natural gas to Japan across the entire value chain. The vast majority of LNG supply projects in which Japanese companies have participated principally have been supported by the Japan Bank for International Cooperation (JBIC). Cofinancing from JBIC has played an important role in securing the participation of

Table 12.1 Basic natural gas data

	Japan		South Korea		China	
	2009	2013	2009	2013	2009	2013
Share of natural gas in energy demand	17.0 per cent	22.2 per cent	12.8 per cent	17.4 per cent	3.7 per cent	5.1 per cent
Natural gas net import dependence	100 per cent	100 per cent	100 per cent	100 per cent	4.7 per cent	27.5 per cent
Pipeline imports (billion m^3)	–	–	–	–	–	24.5
LNG imports (billion m^3)	85.9	119.0	34.3	54.2	7.6	24.5
Average LNG import prices ($/mmBtu)	$7.40	$16.00	7.20 per cent	$14.90	$3.60	$10.90
Share of spot and short-term LNG imports	9.0 per cent	24.7 per cent	10.0 per cent	27.1 per cent	17.0 per cent	21.1 per cent
Number of natural gas suppliers	13	18	13	17	11	15
Largest LNG supplier (share)	Indonesia (20.0 per cent)	Australia (20.5 per cent)	Qatar (27.0 per cent)	Qatar (33.8 per cent)	Australia (62.3 per cent)	Qatar (37.6 per cent)

Table 12.2 Key market participants in the LNG trade

	Japan	South Korea	China
Major market participants (excluding distribution and retail ownership)	Chubu, Chugoku, Kansai, Kyushu, Shikoku, Tohoku (utilities; private) Tokyo Electric Power Company (TEPCO) (utility; state-owned) Hiroshima Gas, Osaka Gas, Saibu Gas, Shizuoka Gas, Toho Gas, Tokyo Gas (city gas; private) INPEX, JX Nippon, JAPEX (oil and gas; private) Itochu, Marubeni, Mitsubishi, Mitsui, Sojitz, Sumitomo, Toshiba (trading houses; private)	South Korea Gas Corporation (KOGAS) (state-owned) POSCO (private)	China National Offshore Oil Corporation (CNOOC) China National Petroleum Company (CNPC) Sinopec (all state-owned)

Japanese commercial banks in project financing. Additional government support also is evident in the form of insurance provided by the Ministry of Economy, Trade and Industry (METI), and equity financing, guarantees, and other support provided by the state-owned Japan Oil, Gas and Metals National Corporation (JOGMEC). Reflecting the government's supply security policy, JOGMEC's core function is to secure a stable supply of oil and natural gas for Japan's domestic consumption. As of April 2014, JOGMEC has provided equity capital in 44 projects and liability guarantees for project development in 13 oil and gas projects worldwide (Ichikawa 2014). Electric utilities (~60 per cent) and city gas companies (~35 per cent) dominate LNG imports. Domestically, these entities have enjoyed monopoly over their respective regional franchise service areas and also own and operate gas transportation and distribution infrastructure, with limited third party access.. Internationally, the government has encouraged utilities and gas companies to leverage their purchasing power through joint LNG procurement.

Japan's 2010 *Strategic Energy Plan* aimed to reduce natural gas demand by 25 per cent of 2007 levels by 2030, based on an assumption that the share of nuclear power would increase. However, after Fukushima, this policy became obsolete as LNG came to be seen as a vital fuel in replacing power produced by idled nuclear reactors. Increasing the share of LNG was a quick and politically easy solution. LNG became the principal fuel for thermal power generation, because burning coal and crude oil was not perceived to be a long-term solution given environmental regulations and supply limitations. In 2012, the government announced that it will build ten LNG import terminals within a decade.

Satisfied with secure supplies, historically, the Japanese government has not pursued an effective policy to scrap oil-indexation. However, with growing demand for increasingly expensive LNG under oil-indexed long-term contracts after Fukushima, establishing a more flexible pricing system that, at least partially, reflects a regional supply/demand balance has become an energy policy priority. According to analysis from the Institute for Energy Economics, Japan (IEEJ), a METI-linked think-tank, increasingly it has become important for Japan to revise the oil-indexed pricing system for LNG procurement (Yanagisawa 2013). Consequently, Japanese policy has focused on the pricing of internationally procured LNG and aimed at lowering prices toward European or HH levels. Japanese politicians, increasingly supported by electric utilities, publicly have challenged the continuation of oil-indexation in long-term LNG contracts. In September 2012, METI Minister Yukio Edano stated that Japan was experiencing an 'outflow of national wealth' as a result of high priced, oil-indexed LNG import contracts (Rogers and Stern 2014). Japan's most recent *Strategic Energy Plan* (April 2014) promotes the establishment of a stable and flexible LNG supply-demand structure in Asia and a relaxation of destination clauses in long-term contracts. The government supports and encourages movements by Japanese companies to relax the destination clause and eventually abolish them from LNG contracts (METI 2014). In mid-2014, Japanese officials made reference to the Rome Group of 7 (G7) Energy Initiative (May 2014) and the Brussels G7 Summit Declaration (June 2014), where the G7 countries agreed on the 'promotion of flexible gas markets, including relaxation of destination clauses and producer-consumer dialogue' (European Commission 2014).

Japan's Challenge to Oil-Indexation

Japan has challenged oil-indexation in two ways. First through diversification. Second, by pursuing buyer cooperation.

(a) Diversification

When energy supply sources are diversified, the power controlled by a single supplier is diminished (Vivoda and Manicom 2011; Vivoda 2009, 2014b). Diversification of supply sources has been a key strategy employed by the Japanese government to improve energy security, enhance bargaining power in relation to individual producers, and secure more competitively priced gas. Between 2009 and 2013, Japan increased the number of LNG suppliers from 13 to 18. At the same time, the country's reliance on the largest supplier has remained constant at 20 per cent. Since 2009, Japan has begun to implement its diversification strategy by increasing its participation in upstream development and LNG projects across the Asia-Pacific and elsewhere, including Australia, Canada, Russia, Papua New Guinea, Mozambique, and, most importantly, the US. While LNG from these countries largely will be sourced through long-term contracts based on oil-indexation, the US is an important exception.

Since 2012, Japanese companies have committed to US upstream and liquefaction projects that would provide approximately 17 million tonnes per annum (mtpa), or 20 per cent of Japanese total gas demand, with first gas deliveries to Japan expected in 2017 (Fensom 2014). KOGAS and the Gas Authority of India Limited (GAIL) also are participating in US projects. Long-term contracts have been signed with complete destination flexibility under HH pricing and, as such, have the potential to challenge the traditional contractual arrangements for LNG procurement.[10] Following the substantial increase in shale gas production, the US has been undergoing a transformation from an importer to a future exporter of natural gas, with first LNG exports from the Sabine Pass LNG project in Louisiana commencing in 2015. With the opening of the widened Panama Canal project in June 2016, Gulf Coast LNG terminals will see their distance disadvantage to growing Asian LNG markets reduced by ten days (IGU 2014). The widened Panama Canal locks will be able to accommodate at least 90 per cent of the LNG fleet, compared to seven per cent before the project had begun (GIIGNL 2014). The HH price is determined by supply and demand for natural gas in the US. Given the large discrepancy between HH gas prices and Asian LNG prices during 2013 (see Fig. 13.1), LNG from the US would cost 20 to 30 per cent less than LNG imported under existing long-term contracts (Koyama 2013). HH prices of $3–5/MMBtu would produce a cost range of $10–12/MMBtu delivered to Japan (Argus Media 2014). Consequently, the strategy to secure upstream and liquefaction agreements in the US has been driven by the belief that importing high volumes of LNG from the US may put pressure on oil-indexation in Asia.

[10] Most liquefaction projects in the US have been signed on a free on board basis, where Asian companies agree to pay the project developer a $2.50–$3.00/MMBtu liquefaction fee to lift the volumes from the Gulf of Mexico facilities, along with 115 per cent of the HH price to source the volumes from the US natural gas pipeline system.

Since 2010, Japan and other Asian buyers increasingly have relied on spot and short-term LNG procurement.[11] With growing international trade in terms of volume, accompanied by an increase in the number of exporters, Japan's spot and short-term LNG imports (defined as LNG traded under contracts with duration of four years or less) increased from nine per cent in 2009 to 24.7 per cent of the overall LNG imports in 2013 (GIIGNL 2010, 2014). Under tight market conditions, between 2007 and 2012, 77 per cent of spot and short-term cargoes predominantly purchased by Japan were more expensive than oil-indexed cargoes procured under long-term contracts (IEA 2013). Driven by increased demand and limited availability of long-term cargoes, particularly after Fukushima, Japanese buyers accepted the higher prices. While Japanese spot and short-term LNG purchases have increased significantly over the past several years, the price differential between these and long-term contract LNG deliveries has, until recently, remained minimal. In fact, although the price for spot and short-term LNG cargoes is an outcome of negotiation, it has remained linked to the JCC. However, in the first half of 2014, Japan's push for increased spot and short-term purchases finally started paying off in terms of de-linking them from the JCC-indexed price. The Platts' Japan South Korea Marker (JKM) price of spot and short-term LNG has dropped by 42.7 per cent, from their record high monthly average of $19.419/MMBtu in March 2014 to $11.125/MMBtu in early July 2014. This is the lowest level since the March 2011 Fukushima disaster.[12] The significant drop in spot prices is in stark contrast to the relatively unchanged prices of LNG delivered to Japan and South Korea under oil-indexed long-term contracts during this period ($15–17/MMBtu).

(b) Buyer Cooperation

Over the past several years, it has become apparent to Asian LNG importers that increased spot and short-term LNG purchases, while improving supply diversification, have not had an immediate effect on regional pricing. Similarly, the flurry of contracts that promise the flow of significant volumes of HH-priced North American LNG to Asia from 2015 has not led to long-term contract renegotiations with regional suppliers, nor has it challenged oil-indexation. While this may materialise in due course, from 2013, Japan has employed another measure that may assist in challenging oil-indexation in the region, namely buyer cooperation.

---

[11] LNG for spot and short-term trading is available due to excess production (above the fulfilment of the producers' supply obligations), or is available from plants with marginal capacity obtained by de-bottlenecking. Such LNG often is an outcome of the time lag between the production and shipping starts of LNG projects. Spot and short-term LNG also becomes available as a result of conflict between parties to contracts, amortised plants, or expired contracts. Factors that free-up spot and short-term LNG are technical or contractual matters, which are part of individual projects (Namikawa 2003).

[12] Platts' JKM is a daily assessment of LNG prices for spot cargoes delivered to Japan and South South Korea based on the most recent trades and/or bids and offers from buyers and sellers in the spot market. Although not a physical spot market, JKM serves as a proxy for all Asian prices. It is the first index that reflects regional LNG supply and demand fundamentals. JKM reached an historic high of $20.20/MMBtu on 14 February 2014 (Rogers et al. 2014).

As the regional LNG market expands, there is much scope for buyer cooperation and, more specifically, for joint participation in overseas projects and cooperation during contractual negotiations. Collaboration between Japan and other regional LNG importers is essential for de-linking LNG from oil prices in the region (Koyama 2013; Vivoda 2014b). It also may provide for a small, but important, step toward enhancing energy security cooperation in a region where such cooperation has been missing and, consequently, a signal of a possible shift away from traditional structures centred on the competitive pursuit of supply security.

The first round of LNG purchase talks between Japan and South Korea in November 2011 was the first indication of an intention to enhance buyer cooperation. The second meeting between the two countries that jointly account for half of global LNG imports was held in April 2012 (Sato and Okada 2013). In October 2013, KOGAS President and Chief Executive Officer (CEO) Jang Seok-hyo announced that South Korea was involved in the third round of talks with several Japanese companies jointly to purchase gas at less expensive prices (Choi 2013). Demonstrating evidence of a concerted buyer effort, which materialised between the second and third round of the Japan-South Korea talks in January 2013, Japan's Chubu Electric Power Company concluded a memorandum of understanding with South Korea's KOGAS for the joint procurement of LNG from ENI.[13] This was the first case of joint LNG procurement by companies from different Asian states. From May 2013 to December 2017, Chubu and KOGAS will purchase 1.7 mt of LNG (Natural Gas Asia 2013). In March 2014, Hirobumi Kawano, President of JOGMEC, stated that KOGAS and an unnamed Japanese company had purchased LNG jointly for the first time and that such cooperative purchases will become more common (Chung 2014).

In a further demonstration of regional buyer cooperation, in September 2013, the energy ministers of the EAS member-states launched a joint inquiry into LNG markets with the aim of securing affordable regional LNG supply. The study was initiated by Japan's METI Minister Kazuyoshi Akaba and was adopted by all 18 member states (Kumagai 2013). In the same month, a joint statement was released by Japan and India to address higher prices of LNG in Asia compared with Europe and North America. The statement stressed that Japan and India would work together to develop a market environment that would enable effective, stable, and globally competitive LNG procurement (Amaha and Kumagai 2014). As a first step, in January 2014, Indian state-owned energy firm Oil and Natural Gas Corporation signed a memorandum of understanding with Mitsui for cooperation in pursuing petroleum and natural gas opportunities in India and third countries (Natural Gas Asia 2014). Moreover, in February 2014, Japan's Chubu and GAIL signed a memorandum of understanding to jointly procure LNG, with a view to leveraging their combined purchasing capacity to reduce prices (Amaha and Kumagai 2014).

In December 2013, Japan's largest oil and gas explorer, INPEX, established an office in Singapore to trade LNG.[14] With a trading post in Singapore, it will be easier

[13] ENI originally was called Ente Nazionale Idrocarburi.

[14] Originally referred to as the 1966 as North Sumatra Offshore Petroleum Exploration Co., Ltd.

for INPEX to purchase LNG to feed its new Naoetsu receiving terminal (Tan and Tsukimori 2013). The announcement by INPEX followed news regarding Singapore's intention to become Asia's LNG trading hub, as it is expanding its LNG importing capacity to 9 mt by 2017 (Platts 2013). Regional LNG importers expressed support for the development of a natural gas trading hub in Singapore (Corbeau 2014). Finally, in January 2014, TEPCO proposed to invite other Japanese and foreign firms to jointly procure up to 40 mtpa of LNG to reduce costs. In what would be the first concrete move toward forming a buyers group for LNG, TEPCO started talks in 2014 with other LNG buyers, in order to establish a joint purchasing company to negotiate with suppliers and to invest in gas projects. The purchases covers both spot and contractual volumes and are aimed at negotiating better contract terms and pricing (Sheldrick and Topham 2014). While regional LNG buyers were scheduled to meet in February 2014 to discuss the club and other ways of reducing prices, the meeting was delayed due to Japan's political tensions with China and South Korea (Chung 2014).

Oil-Indexation in the Asia-Pacific Region?

Investments in new capacity in the US increases the share of spot and short-term purchases, and recent evidence of Japan-led buyer cooperation in Asia likely will enhance flexibility in the supply chain, and is indicative of regional LNG importers' desire to move away from oil-indexation. Since March 2014, a significant price gap has emerged between the JKM price of spot purchases by Japan and South Korea and the JCC-indexed price of LNG delivered to Japan under long-term contracts. A rapid drop in JKM prices (over 40 per cent in four months) provides an early indication that an oversupply may be emerging in the market and that the price is partly influenced by supply/demand in the region and is no longer exclusively referenced to the oil-indexed gas price. The potential significance of such a rapid drop in spot prices is highlighted by the fact that it coincided with Japanese utilities using the highest volume of LNG on record in June 2014 (Amaha 2014). The seemingly ample supply has offered Japanese utilities an opportunity to balance demand and supply through cheaper spot purchases. Spot purchases increase importers' choices, add liquidity to markets, and allow importers to hedge financially and physically (Ernst and Young 2013). An oversupplied market limits the bargaining power of exporters, and thus their ability to raise prices as demand gains, particularly in the context of coordinated purchases by Asian buyers (Russell 2014). Moreover, the shale gas revolution in the US may provide Asian LNG importers with negotiating leverage and a new degree of pricing flexibility. Policy-makers in Japan and other Asian countries believe that the liquefaction tolling arrangements that their companies have negotiated at several LNG export projects in the US may provide a new model of LNG procurement, based on the HH-linked pricing system rather than traditional oil-linked LNG pricing. Two additional factors also may work in their favour: the growth in China's coal-indexed pipeline imports, and the planned restart of 31 nuclear reactors in Japan.

China is the only major regional LNG importer that has significant domestic gas production and pipeline connections to major gas exporters. In China, LNG competes with piped natural gas and coal, the predominant power source. As a consequence, LNG demand in China is more price sensitive than in Japan and South Korea. In fact, China's LNG price levels have been lower than those of Japan and South Korea due to favourable long-term price formulas for its Australian and Indonesian supply contracts. In order to enhance supplier diversification, China has pursued long-term pipeline supplies from Turkmenistan (since 2010), Myanmar (since 2014) and Russia (from 2020) at prices lower than for LNG imports (Russell 2014). Moreover, since 2011, imported coal has been considerably cheaper than LNG sourced from Qatar, its major supplier. As a consequence, China has been able to leverage its price sensitivity in negotiations with potential natural gas suppliers. For example, most recently, coal-indexation has been used in Sino-Russian negotiations over long-term pipeline imports and China also has renegotiated a favourable price for LNG imports from Indonesia, well below the regional average.[15] While China's coal-indexed pipeline imports from Russia will contribute to regional diversification away from oil-indexed contracts, they also are likely to curb the growth of LNG demand in China and subsequently ease the balance between regional LNG supply and demand. The Sino-Russian agreement is likely to precipitate the development of natural gas fields in eastern Siberia, and open the way for LNG projects, such as Vladivostok LNG, which could make additional supplies available to the region. The IEEJ (2014) views the Sino-Russian gas deal as a positive development in regional contract diversification.

Challenges

Although future US LNG exports appear promising for Asian LNG importers in relation to both supply and price diversification, they raise risks with regard to HH pricing. While HH prices have been in the range of $2–5/MMBtu since 2010, there is a considerable risk that if, over the duration of new contracts, HH prices rise to (or above) $7/MMBtu, importers could find themselves paying prices in excess of JCC levels, particularly if oil prices drop below $100/barrel. A potential increase in HH prices that coincides with first US LNG exports also may lead to domestic political opposition in the US against exports in favour of keeping low energy costs for US

[15] In May 2014, China and Russia (through China National Petroleum Company (CNPC) and Gazovaya Promyshlennost (Gazprom) signed a $400 billion natural gas deal. Russia will supply 38 billion cubic meters (28 mtpa LNG-equivalent) of natural gas per annum via pipeline for 30 years. The cost is estimated at $10/MMBtu (Russell 2014). In June 2014, China National Offshore Oil Corporation (CNOOC) agreed to pay a ceiling price of US$8/MMBtu for LNG imported from the Tangguh project in Indonesia (Cahyafitri 2014).

manufacturers. Some petrochemical producers, manufacturers, and industrial gas users maintain that future LNG exports likely will remove the competitive advantage provided by cheap domestic gas (Ebinger and Avasarala 2013). Consequently, the extent to which US projects are able to compete for markets in Asia is yet to be demonstrated and will be affected by HH prices relative to JCC-indexed prices and domestic political machinations in the US.

Moreover, Japan's LNG contracts with major suppliers are dominated by long-term contracts. In 2013, Australia, sold only 2.9 per cent of its LNG under spot and short-term contracts (GIIGNL 2014). Since 2010, suppliers have benefited from the prevailing pricing mechanism. For example, the value of Australian LNG exports has increased by 76 per cent (in Australian dollar terms) between financial year 2009–2010 and 2012–2013 (Bureau of Resources and Energy Economics 2014). Consequently, traditional LNG suppliers, such as Australia, prefer the status quo in which LNG prices remain indexed to crude oil. Australia accounts for over 60 per cent of the global planned LNG liquefaction projects. These projects are backed by \$200 billion of largely Asian-financed upstream projects that either are planned, awaiting regulatory approval and licencing, or are under construction. As much as 62 mtpa of new supply capacity is due to come online in Australia between 2014 and 2017, all of which is destined for Asian markets. Australia is on track to overtake Qatar as the largest global LNG exporter by the end of the decade (Bloxham and Hartigan 2012).

In 2013, Australia was Japan's largest LNG supplier with 18.16 mt, or 20.6 per cent of Japan's imports (GIIGNL 2014). However, between 2009 and 2013, Japanese companies committed to an additional 15.51 mtpa of LNG from future Australian projects, including Queensland's Curtis, Gorgon, Wheatstone, and Ichthys projects, all under long-term oil-indexed contracts. With only 4.96 mtpa of the existing contracts expiring before the end of the decade (Fig. 12.2), there is limited scope for negotiating more flexible ricing and supply arrangements (GIIGNL 2010, 2011, 2012, 2013, 2014). Australia's increasing market share and preference for long-term contracts due to high project costs limits the availability of spot and short-term LNG in the region. There also is limited scope for contract renegotiations. According to the IEA (2013), while long-term contracts comprise price re-negotiation clauses that allow for minor price adaptation to changes in the market, volume or destination clauses are not subject to re-negotiation. A re-negotiation of price terms will be very difficult because Asian LNG contracts contain little in the way of enforceable price re-opener provisions, and sellers that have made enormous investments (particularly in new projects that have yet to start production) will be resistant to any change in pricing.

Regional LNG producers, including Australia, thus far have resisted the idea of joint LNG purchases and demands for price renegotiation. They have pointed to the illegality of joint purchasing due to restricted destination clauses in the existing supply contracts. They also have insisted that long-term supply contracts in their present form are necessary to ensure they can take on the risk of developing capital-intensive projects (Sheldrick and Topham 2014). LNG projects require

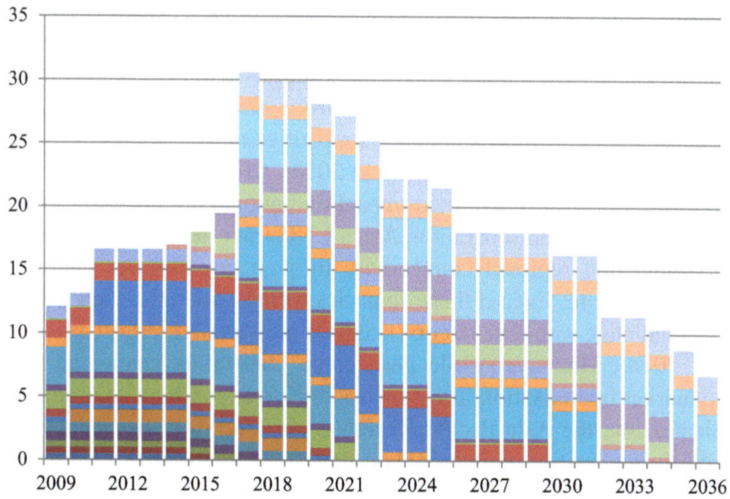

Fig. 12.2 Japan's long-term LNG contracts in Australia, 2009–2036 (mtpa). Source: GIIGNL (2009, 2010, 2011, 2012, 2013, 2014).

security of contracted volumes in order to be assured of commercial viability. Without this security, projects may not be developed in line with market demand. As a result, regional LNG suppliers warn that the wave of supply expected to come online before the end of the decade would be delayed and LNG procurement costs would rise (WoodMackenzie 2013). Consequently, while a shift from oil-indexation in the region would represent a significant risk for regional suppliers, particularly in terms of commercial viability of future projects, it also would undermine the security of supplies for regional importers.

A move toward LNG contracts that allow for greater volume and destination flexibility will be predicated on a change in the global markets that will increase these features in newly-signed contracts, such as the expected destination-free volumes from the US. However, LNG exports from the US to Asia unlikely will be sufficient fundamentally to affect oil-indexation in the region. Most of the new supply contribution will come from Australia, where the HH-linked formula does not support project economics. Even if all US LNG were to be HH-linked, this likely would represent no more than 10 per cent of the total Asian demand in 2020. Given that imports from the US under HH pricing will not rival oil-indexed import volumes from Australia by the end of the decade, it is unlikely that regional LNG prices will be de-linked from oil.

An additional constraint in challenging oil-indexation in the region is that none of the Asian LNG importers have liberalised gas markets. Instead, domestic markets are dominated by state-owned or state-controlled companies, so there is limited third party access to pipelines and other infrastructure, and governments are able regulate domestic gas prices (IEA 2013). With the exception of Japan, Asian importers have not demonstrated any determination at the policy level to achieve market liberalisation before the end of the decade.[16] Some have suggested that for de-linking LNG prices from crude oil to materialise, it is essential that regional governments adopt a hands-off approach to their domestic gas markets. Market liberalisation would set the foundation for a move away from currently monopolised domestic markets in the region, toward a mature, competitive market across the entire value chain (Gao 2010; IEA 2013; Chang and Li 2014). Critics argue that domestic market liberalisation may have a limited impact on regional pricing (Rogers and Stern 2014). Due to geographic separation, Asian LNG importers do not have any prospect of establishing regional connectivity through cross-border interconnecting pipeline infrastructure. Even if these countries were to liberalise their domestic gas markets, a challenge would be to correlate their individual reference prices (Rogers and Stern 2014). However, while geographic constraints cannot be overcome, domestic market liberalisation in Japan, South Korea, or China would be a significant indicator that the Asia-Pacific Region is shifting away from the historically dominant informal institutions centred on state sovereignty and supply security.

Given that geography prevents one or more of the large Asian markets from acting as the 'host' liberalised gas market to provide the regional price reference for LNG spot cargoes, Rogers and Stern (2014) argue that regional importers should establish a regional liquid trading hub. In time, an Asian trading hub could function as a centre for price discovery of physically traded spot cargoes (based on JKM) and pave the way for the adoption of an exchange-based price for LNG futures (Wan 2011). According to the IEA (2013) and Rogers and Stern (2014), Singapore is best suited to develop a trading hub as the government has a hands-off approach to the market. Such a trading hub would facilitate the exchange of natural gas that reflects prices that correspond to regional supply and demand, and is priced against coal, nuclear power, and other competitors (Stern 2014).[17] The opening of the Singapore LNG terminal in 2013, and its future expansion, may provide the foundation for such a regional market hub. If it were to materialise, the dynamics of the demand side would be driven by the quantities of LNG (in excess of other long-term LNG contract supplies and, in the case of China, domestic and pipeline supplies) required

[16] In November 2012, the Japanese government proposed to establish an LNG futures market in Japan that sets a price based on domestic supply/demand factors. The aim was to launch the market in 2015. Critics have suggested that, given that the electricity sector represents approximately two-thirds of total natural gas demand, deregulation or liberalisation in the gas market would be hindered if competition is not introduced in the power-generation sector, which has been dominated by ten regional monopolies (IEA 2013). In response, in November 2013, the Japanese Diet passed the first of three electricity market reform packages. One of the aims of these reforms is to encourage efforts for the reduction of LNG procurement costs through competition between electric utility monopolies. In 2014, the 'top-runner approach' has been under METI's consideration. If adopted, for long-term LNG contracts reviewed after 2015, METI may use LNG price-estimates, which partly reflect linkage with gas prices (mainly HH) for the rate assessment (Kihara 2014).

[17] The key requirements for a hub are: a defined physical location (from which shipping costs to importing centres could be assessed by buyers considering trades); and development (and impartial enforcement) of a trading code, which would apply to all participants, whether conducting bilateral, over-the-counter, or (later) exchange-based trades. On both these points, Singapore currently appears to be a viable option (Rogers and Stern 2014).

to meet demand at the Asian hub price. Supply side dynamics would be driven by LNG availability (at the Asian hub price) in competition with other LNG consuming regions (Rogers and Stern 2014).

However, the likelihood of a regional hub developing the required level of liquidity to serve as a reference price for long-term contracts seems overly optimistic as well, unless volumes can expand sufficiently to become a credible indicator of regional LNG supply and demand fundamentals. While potential market oversupply in tandem with the arrival of US supplies under HH pricing, and the implicit challenge (in volumetric terms) to other suppliers, may help in creating change in this direction toward the end of the present decade, this probably is the earliest date a liquid traded market for Asian LNG is likely to materialise. This prognosis is based on an unlikely scenario in which regional LNG buyers move from a state-centric approach to energy policy toward a shared commitment for a market-based solution to regional pricing. However, even a move from the securitisation of energy toward greater cooperation would not guarantee that there will be sufficient liquidity to allow for a reliable reference price. As argued above, until the end of this decade, Japan and other LNG importers will remain reliant on inflexible long-term contracts with little provision for price review/renegotiation.

Conclusion

Given that the competitive relationship between oil and natural gas has weakened over the past four decades, Japan and other Asian LNG importers believe that LNG prices no longer should be linked to crude oil import prices. As this chapter demonstrates, such a sub-optimal market outcome in Asia has led Japan to implement several measures in order to challenge oil-indexation and ultimately reduce transaction costs. First, in order to enhance market flexibility, since 2010, Japan significantly has increased spot and short-term LNG purchases. Initially, and under tight market conditions, this proved to be an expensive measure that nevertheless helped Japan to replace idled nuclear power post-Fukushima. However, with market oversupply emerging in 2014, the liquidity that was made available by hitherto expensive spot purchases enabled Japan to negotiate lower prices for spot cargoes and to de-link them from JCC-indexed long-term prices. Second, since 2012, Japanese companies have concluded several long-term contracts with US suppliers that promise the flow of significant volumes of HH-priced LNG with free destination clauses in the second half of this decade. Finally, in late 2011, Japan initiated a cooperative effort with other regional LNG importers. Initially, the discussions were limited to annual meetings between Japanese and South Korean energy ministers and gas companies. In 2013, other regional importers became involved and discussions regarding concrete proposals, such as joint bidding and the establishment of a trading hub, intensified.

Toward the end of this decade, oil-indexation in Asia will be challenged by greater competition between sellers, more price-sensitive buyers, increasing gas-on-gas competition from new pipeline infrastructure in China, increasing spot market liquidity, and the supply of HH-priced exports from the US (EY 2013). Arguably, US LNG exports will be a major catalyst for a shift from exclusive reliance on oil-indexation. By 2020, we are likely to witness a gradual and partial migration away from oil-indexation toward a mixture of contract pricing approaches, hybrid oil/coal/HH indexation, and potential regional gas hub indexation (Choi and Robertson 2013).

These developments notwithstanding, oil-indexation is expected to remain the predominant pricing approach given that Japan unlikely will be able to renegotiate the price formation terms in the existing long-term contracts with Australia and other traditional sellers. The contractual conditions in existing JCC-linked contracts for all Asian buyers (with the notable exception of China) suggest limited scope for radically changing national average LNG import prices. However, the introduction of the additional source of LNG under a new pricing system, a high and growing share of spot and short-term contracts, and enhanced buyer cooperation likely will improve importers' bargaining power in contract negotiations for expired legacy projects and future projects. Their bargaining power likely will be enhanced by the looming market oversupply. By 2018, the global supply of LNG is expected to increase by 25 to 30 per cent, while China's increased pipeline imports and Japan's gradual nuclear restarts may limit regional LNG demand growth (IGU 2014). Market oversupply may enable Asian buyers to negotiate more flexible pricing arrangements in expired legacy projects, as well as upgrades to existing projects and potential new projects. Asian buyers, in their negotiations with upstream sellers, will need to insist on a move away from exclusive oil-indexation to a price mechanism that reflects anticipated market fundamentals. In fact, they should refuse any suggestions that they should pay the JCC-related price for new supplies even if this were to result in the deferral of projects.

Security of supply has been the historic justification for oil-indexation in Japan. However, with increasing market liquidity, Asian security premiums have become more difficult to justify, and Japan has promoted buyer cooperation in order to reduce transaction costs in the regional LNG market. If other importers follow Japan's lead, and an informal LNG buyers' club is formalised, we may witness fierce competition with suppliers over pricing in expired legacy contracts and future projects. An unprecedented cooperative effort in the region indicates that there is willingness in Tokyo to move away from competitive institutional structures based on energy policy approaches that emphasise supply security. Development of an LNG trading hub as a reference price for Asian LNG would be a logical next step for Japan and other regional buyers, particularly during this period of high market liquidity. However, this would require a deeper level of regional cooperation and, significantly, a different approach to energy markets across the region. Historically, security concerns have created a culture of early contracting of long-term LNG supplies in order to cover possible future demand requirements. Notwithstanding the increased importance of price considerations in recent years, across Asia, with high (and growing) energy import dependencies, security of supply considerations continue to exert significant influence over energy policy-making processes. The role of the government as a market participant - as opposed to a regulator - through vertically integrated gas companies limits competition and market efficiency. Continued domination of domestic gas markets by vertically integrated state-owned or quasi-government companies, with their own supply infrastructure that suits specific long-term, oil-indexed contracts, unlikely will stimulate further market integration.

While Japan may have taken a small, but important, step away from a securitised approach to energy markets, it would be naïve to suggest that security of supply priorities no longer are significant and that Japan is moving toward a 'hands-off' approach at a fast pace. After all, the core function of JOGMEC is to secure a stable supply of oil and natural gas for Japan's use. Similarly, security of supply remains a top priority for the Chinese and South Korean governments and their state-owned companies (IEA 2013). Distrust and unresolved historical issues continue to plague relationships between Asia's major powers, which often escalate into diplomatic rows over, for example, maritime territorial disputes in the East China Sea (Manicom 2014). Entrenched structures hinder deeper cooperation, and path dependencies imply that change only can be incremental. A specific challenge for governments will be to overcome the security of supply rationale and to relinquish control of what frequently is regarded as a strategic sector of the economy.

References

Aguilera, R., Inchauspe J., Ripple, R. (2014). The Asia-Pacific natural gas market: large enough for all? *Energy Policy,* 65, 1-6.

Amaha, E. (2014, July 11). Major Japan utilities' Jun LNG use hits record high for the month. *Platts.* http://www.platts.com/latest-news/natural-gas/tokyo/update-major-japan-utilities-jun-lng-use-hits-26832220. Accessed 27 February 2015.

Amaha E., Kumagai, T. (2014, January 27). Japan's Chubu Electric, India's GAIL to ink MOU on LNG procurement this week. *Platts.* http://www.platts.com/latest-news/natural-gas/tokyo/japans-chubu-electric-indias-gail-to-ink-mou-27867393. Accessed 27 February 2015.

Andrews-Speed, P. (2014). A regional public goods approach towards energy market integration in East Asia. In Y. Wu, F. Kimura, X. Shi (Eds.), *Energy market integration in East Asia: deepening understanding and moving forward* (pp. 25-55). New York: Routledge.

Argus Media. (2014, June). Global LNG. LNG markets, projects and infrastructure, volume X, issue 6.

Balfe, P. (2013). Australia's LNG boom: opportunities and challenges. ACIL Allen Consulting, Insights. http://www.acilallen.com.au/insight/4/australias-lng-boom-opportunities-and-challenges. Accessed 25 September 2015.

Bloxham P., Hartigan L. (2012). Downunder digest: Australia's gas-fired mining boom. HSBC Global Research, 3 May. https://www.hsbc.com.au/1/PA_ES_Content_Mgmt/content/australia/about/research/archive/2012/OZ120504.pdf. Accessed 21 April 2015.

BP. (2014). Statistical review of world energy. http://www.bp.com/en/global/corporate/energy-economics/statistical-review-of-world-energy.html. Accessed 4 October 2014.

BREE. (2014). Bureau of Resources and Energy Economics, Australian Government. 2014 Australian energy statistics. Canberra: BREE.

Cahyafitri R. (2014, July 1). Tangguh LNG price raised after deal. *The Jakarta Post.* http://www.thejakartapost.com/news/2014/07/01/tangguh-lng-price-raised-after-deal.html. Accessed 1 December 2015.

Chang Y., Li, Y. (2014). Towards an integrated Asia-Pacific natural gas market. In Wu, Y., Kimura F., Shi, X. (Eds.), *Energy market integration in East Asia: deepening understanding and moving forward* (pp.178-202). New York: Routledge.
Choi, K. (2013, October 15). Korea, Japan to cooperate on gas purchase. *Korea Times*.
Choi T., Robertson, P. (2013). Exporting the American renaissance: global impacts of LNG exports from the United States. a report by the Deloitte Center for Energy Solutions and Deloitte MarketPoint LLC. https://www2.deloitte.com/content/dam/Deloitte/global/Documents/dttl-er-exportingamericanrenaissance-08072013.pdf. Accessed 10 October 2014.
Chung, J. (2014, March 25). KOGAS, Japan firm buy gas together; more joint deals to come – JOGMEC. *Kuwait Times*. http://news.kuwaittimes.net/pdf/2014/mar/25/p25.pdf. Accessed 14 April 2015.
Corbeau, A-S. (2014, July 15). A way to reprice Asia's costly gas. IEA Journal, issue 6. http://www.iea.org/ieaenergy/issue6/a-way-to-reprice-asias-costly-gas.html. Accessed 14 April 2015.
Ebinger C., Avasarala, G. (2013). The case for U.S. liquefied natural gas exports. *Oxford Energy Forum,* February.
Energy Information Administration. (2014). Indonesia: analysis. 5 March. U.S. Department of Energy. http://www.eia.gov/countries/cab.cfm?fips=id. Accessed 27 August 2014.
Energy Information Administration. (2013a). Australia: analysis. U.S. Department of Energy, 21 June. http://www.eia.gov/countries/cab.cfm?fips=as. Accessed 27 August 2014.
Energy Information Administration. (2013b). Malaysia: analysis. U.S. Department of Energy, 3 September. http://www.eia.gov/countries/cab.cfm?fips=my. Accessed 27 August 2014.
European Commission. (2014). The Brussels G7 Summit Declaration. MEMO/14/402, 5 June, http://europa.eu/rapid/press-release_MEMO-14-402_en.htm?locale=FR. Accessed 14 July 2014.
Ernst & Young. (2013). Global LNG: will new demand and new supply mean new pricing? *Ernst & Young Global Limited*. http://lngbunkering.org/lng/sites/default/files/2013per cent20EYper cent20impactper cent20ofper cent20globalper cent20supplyper cent20patterensper cent20on-per cent20LNGper cent20price.pdf. Accessed 27 August 2014.
Fensom, A. (2014, July 2). Asia's gas blowout? The Diplomat. http://thediplomat.com/2014/07/asias-gas-blowout/. Accessed 21 September 2014.
Fesharak,i F., Wu, K., Banaszak, S. (2000). Natural gas: the fuel of the future in Asia. East-West Center, Asia-Pacific Issues, 44.
Flower, A. (2008). Appendix 2, LNG pricing in Asia- Japan Crude Cocktail (JCC) and S-curves. In J. Stern (Ed.), *Natural gas in Asia: the challenges of growth in China, India, Japan, and Korea* (pp. 405-409). Oxford: Oxford University Press.
Gao, AM-Z. (2010). Regulating gas liberalisation: a comparative study on unbundling and open access regimes in the US, Europe, Japan, South Korea and Taiwan. Netherlands: *Kluwer Law International.*
International Group of Liquefied Natural Gas Exporters. (2015). The LNG industry in 2014. Paris: GIIGNL.
International Group of Liquefied Natural Gas Importers. (2014). The LNG industry in 2013. Paris: GIIGNL.
International Group of Liquefied Natural Gas Importers. (2013). The LNG industry in 2012. Paris: GIIGNL.
International Group of Liquefied Natural Gas Importers. (2012). The LNG industry in 2011. Paris: GIIGNL.
International Group of Liquefied Natural Gas Importers. (2011). The LNG industry in 2010. Paris: GIIGNL.
International Group of Liquefied Natural Gas Importers. (2010). The LNG industry in 2009. Paris: GIIGNL.
Hafner, M,, Tagliapietra, S. (2013). *The globalization of natural gas markets: new challenges and opportunities for Europe*. Deventer: Netherlands: Claeys & Casteels.
Hancock, K., Vivoda, V. (2014). International political economy: a field born of the OPEC crisis returns to its energy roots. *Energy Research & Social Science*, 1, 206-216.

Henderson, J. (2012). The potential impact of North American LNG exports. Oxford: Oxford Institute for Energy Studies, NG68, October.
Hughes, L., Lipscy, P. (2013). The politics of energy. *Annual Review of Political Science*, 16, 449-469.
Ichikawa, M. (2014). Offshore oil, gas and metals development. Paper presented at the SEA Japan 2014 conference. Tokyo, Japan, 9 April.
International Energy Agency. (2013). Developing a natural gas trading hub in Asia. Paris: OECD/IEA.
Institute of Energy Economics, Japan. (2014). IEEJ e–newsletter, No. 40, 27 June.
International Gas Union. (2014.) IGU World LNG report - 2014 edition.
Inchauspe, J. (2014). Natural gas markets in Asia: pricing, integration, regionalism, and energy security issues. In Z. Zhu (Ed.) *Globalization, development and security in Asia* (pp. 9-39). Singapore: World Scientific.
Kihara, S. (2014). Japan's LNG policy. Paper presented at the 35th Australia-Japan High Level Group on Energy and Minerals Consultations. Brisbane, Australia, 23 June.
Koyama, K. (2013). Japan's post-Fukushima energy policy challenges. *Asian Economic Policy Review*, 8(2), 274–293.
Kumagai, T. (2013). Asia-Pacific producers, consumers to study 'affordable' LNG supply: EAS. Platts, 27 September. http://www.platts.com/latest-news/natural-gas/tokyo/asia-pacific-producers-consumers-to-study-affordable-27463148. Accessed 21 April 2014.
Lacey, J., Lamont, J. (2014). Using social contract to inform social licence to operate: an application in the Australian coal seam gas industry. *Journal of Cleaner Production*, 84, 831-839.
Langton, T. (1994). LNG prospects in the Asia-Pacific Region. *Resources Policy*, 20(4). 257-264.
Manicom, J. (2014). *Bridging troubled waters: China, Japan, and maritime order in the East China Sea*. Washington, DC: Georgetown University Press.
Memon, M., Tunio, S., Memon, K., Lashari, A. (2014). A comparative study of liquefied natural gas: an overview. *Research Journal of Applied Sciences, Engineering and Technology*, 7(17), 3522-3528.
Mercer, A., de Riyke, K., Dressler, W. (2014). Silences in the midst of the boom: coal seam gas, neoliberalizing discourse, and the future of regional Australia. *Journal of Political Ecology*, 21, 279-302.
Ministry of Economy, Trade and Industry, Government of Japan. (2014). Strategic energy plan. Tokyo, Japan.
Miyamoto, A., Ishiguro, C., Nakamura, M. (2012). A realistic perspective on Japan's LNG demand after Fukushima. Oxford: The Oxford Institute for Energy Studies, NG 62, June.
Miyamoto, A., Ishiguro, C., Yamada, T. (2009). Irrational LNG pricing impedes development of Asian natural gas markets: a perspective on market value. Paper presented at the 24th World Gas Conference, Buenos Aires, Argentina, 5-9 October.
Ministry of Finance, Government of Japan. (2014). Trade statistics of Japan. http://www.customs.go.jp/toukei/info/index_e.htm. Accessed 14 February 2015.
Mokhatab, S., Purewal, S. (2006). Letter to the editor: is LNG a competitive source of natural gas? *Petroleum Science and Technology*, 24(2), 243-245.
Moryadee, S., Gabriel, S., Avetisyan, H. (2014). Investigating the potential effects of U.S. LNG exports on global natural gas markets. *Energy Strategy Reviews*, 2(3/4), 273-288.
Mujiyanto, S., Tiess, G. (2013). Secure energy supply in 2025: Indonesia's need for an energy policy strategy. *Energy Policy*, 61, 31-41.
Namikawa, R. (2003). Take-or-pay under Japanese energy policy. *Energy Policy*, 31(13), 1327-1237.
Natural Gas Asia (2013) Eni inks LNG supply deal with South Korea, Japan. 11 January. http://www.naturalgasasia.com/eni-inks-lng-supply-deal-with-south-korea-japan. Accessed June 30 2015.
Natural Gas Asia. (2014). ONGC, Mitsui to cooperate in exploring unconventional, conventional oil, gas. 26 January. http://www.naturalgasasia.com/ongc-mitsui-to-cooperate-in-exploring-unconventional-conventional-oil-gas-11621. Accessed 26 January 2014.
Neumann, A. (2009). Linking natural gas markets: is LNG doing its job? *The Energy Journal, special issue*, 187-199.

North, D. (1990). *Institutions, institutional change and economic performance*. Cambridge: Cambridge University Press.
Phillips, A. (2013). A dangerous synergy: energy securitization, great power rivalry and strategic stability in the Asian century. *The Pacific Review*, 26(1), 17-38.
Platts, (2013). Singapore set to take a role as Asia's LNG trading hub: executive on Platts Energy Week. 2 December. http://www.platts.com/pressreleases/2013/120213. Accessed 5 December 2015.
Pramudatama, R. (2012, July 13) RI to keep exporting LNG, despite local need. *The Jakarta Post*. http://www.thejakartapost.com/news/2012/07/13/ri-keep-exporting-lng-despite-local-need.html. Accessed 5 September 2014.
Ravenhill, J. (2013). Resource insecurity and international institutions in the Asia-Pacific Region. *The Pacific Review*, 26(1), 39-64.
Rogers, D., White, M., Phua, D. (2014). Prospects for development of an Asian LNG trading hub. King & Spalding, Energy Newsletter, February. http://www.kslaw.com/library/newsletters/EnergyNewsletter/2014/February/article1.html. Accessed 1 February 2015.
Rogers, H., Stern, J. (2014). Challenges to JCC pricing in Asian LNG markets. The Oxford Institute for Energy Studies, NG 82, February.
Rogers, H. (2012). The impact of a globalising market on future European gas supply and pricing: the importance of Asian demand and North American supply. The Oxford Institute for Energy Studies, NG 59, January.
Russell, C. (2014, March 25). Cheaper Asian LNG depends on coal, Japan nuclear. *Reuters*.
Sato, S., Okada, Y. (2013, March 6). Japan said to plan talks with South Korea on joint LNG purchases. *Bloomberg*.
Sheldrick A, Topham J (2014, January 17). Japan's TEPCO plans for LNG buyers' group, with 40 mln t/yr target. *Reuters*. http://www.reuters.com/article/japan-lng-procurement-idUSL3N0KR21M20140117. Accessed 3 June 2014.
Siliverstovs, B., L'Hégaret, G., Neumann, A., von Hirschhausen, C. (2005). International market integration for natural gas? A cointegration analysis of prices in Europe, North America and Japan. *Energy Economics*, 27(4), 603-615.
Standard & Poor's. (2012). What's behind the boom in global liquefied natural gas development? Global Credit Portal, 20 April.
Stern, J. (2014). International gas pricing in Europe and Asia: a crisis of fundamentals. *Energy Policy*, 64, 44-48.
Stern, J. (Ed.). (2012). The pricing of internationally traded gas. Oxford: The Oxford Institute for Energy Studies.
Stevens, P. (2010). The 'shale gas revolution': hype and reality. A Chatham House Report, September.
Stoddard, E. (2013). Reconsidering the ontological foundations of international energy affairs: realist geopolitics, market liberalism and a politico-economic alternative. *European Security*, 22(4), 437-463.
Tan, F., Tsukimori, O. (2013, December 11). Japan's Inpex to make Singapore a hub for trading oil, LNG-sources. *Reuters*.
Vivoda, V. (2014a). *Energy security in Japan: challenges after Fukushima*. Aldershot: Ashgate.
Vivoda, V. (2014b). LNG import diversification in Asia. *Energy Strategy Reviews*, 2(3/4), 289-297.
Vivoda, V. (2009). Diversification of oil import sources and energy security: a key strategy or an elusive objective? *Energy Policy*, 37(11), 4615-4623.
Vivoda, V., Manicom, J. (2011). Oil import diversification in Northeast Asia: a comparison between China and Japan. *Journal of East Asian Studies*, 11(2), 223-254.
Wan, K. (2011). Asian LNG price-index gains traction. Petroleum Economist, 29 September.
Wilson, J. (2014). Northeast Asian resource security strategies and international resource politics in Asia. *Asian Studies Review*, 38(1), 15-32.
Wilson, J. (2012). Resource security: a new motivation for free trade agreements in the Asia-Pacific Region. *The Pacific Review*, 25(4), 429-453.

Wood Mackenzie. (2013, September 13). Uncertainty in Japan and Korea's LNG markets will prolong tight Asian market beyond 2020. http://www.energy-pedia.com/news/general/new-155924. Accessed 1 March 2014.

Wu, Y. (2014). Gas market integration: global trends and implications for the EAS region. In Wu, Y., Kimura, F., Shi, X. (Eds.) *Energy market integration in East Asia: deepening understanding and moving forward* (pp.159-177). New York: Routledge.

Yanagisawa, A. (2013). The burden reduction effects of importing U.S. LNG for Japan. Institute for Energy Economics, January.

[illegible]

Chapter 13
Copper

Philip Crowson

Abstract This chapter examines the demand for copper in the Asia-Pacific Region. Like steel, most refined copper is processed into a range of semi-fabricated products before it is embodied in finished goods. Often it is alloyed with other metals, such as zinc, tin, and silver. Data on how copper is used only is available for semi-manufactured products, whether or not these have been made solely from refined copper or from copper alloys. Accordingly, the uses and production of, and trade in, copper are analysed here, before a discussion of trends in the region's demand for copper. In the final section, I examine the future of the copper trade and the political risks that may weaken its commercial value.

Abbreviation

AFC	Asian Financial Crisis
APEC	Asia-Pacific Economic Cooperation
ASEAN	Association of South East Asian Nations
EU	European Union
GDP	Gross Domestic Product
GMS	Grams
GFC	Global Financial Crisis
GFCF	Gross Fixed Capital Formation
ICSG	International Copper Study Group
IISI	International Iron and Steel Institute
IMF	International Monetary Fund
PNG	Papua New Guinea
PPP	Purchasing Power Parity
SOE	State Owned Enterprise
SX-EW	Solvent Extraction Electrowinning
US	United States of America
WBMS	World Bureau of Metal Statistics

P. Crowson (✉)
Centre for Energy, Petroleum and Mineral Law and Policy, University of Dundee, Dundee DD1 4HN, UK
e-mail: p.c.f.crowson@dundee.ac.uk

T. O'Callaghan, G. Graetz (eds.), *Mining in the Asia-Pacific*, The Political Economy of the Asia Pacific, DOI 10.1007/978-3-319-61395-6_13

Introduction

Copper is one of the first metals to be used by human beings; whether alone, or in alloys, such as bronze. Bronze has been alloyed with tin, zinc, nickel and manganese for centuries. It is one of the earliest alloyed metals, and one that was harder and more durable than iron alone. While there now are potential substitutes for copper, its resistance to corrosion, malleability, and conductivity ensure its continuing importance as a commodity. Indeed, it remains one of the essential building blocks of economic development. In terms of the tonnage used today, it ranks third behind iron and aluminium. Moreover, it is second only to iron in the value of its output.

This chapter examines the demand for copper in the Asia-Pacific Region. Like steel, most refined copper is processed into a range of semi-fabricated products before it is embodied in finished goods. Often it is alloyed with other metals, such as zinc, tin, and silver. Data on how copper is used only is available for semi-manufactured products, whether or not these have been made solely from refined copper or from copper alloys. Accordingly, the uses and production of and trade in, copper are analysed here, before a discussion of trends in the region's demand for copper. In the final section, I examine the future of the copper trade and the political risks that may weaken its commercial value.

The Uses of Copper

Table 13.1 shows the world's uses of copper, including remelted copper scrap, in 2012.[1] The Asia-Pacific Region accounted for three-fifths of global use, with the bulk of use in China.

A wide variety of electrical uses, mainly based on copper's properties as the best conductor of electricity, make up about three-quarters of the total use of copper, both globally and in the Asia-Pacific Region. Uses include: power generation, transmission and generation, building wire, electrical and electronic equipment, and electrical automotive components. In addition to its conductivity, copper's ductility and resistance to corrosion and creeping make it the preferred material for building wiring.

Building and construction is the largest market with some 29 per cent of total usage. Copper has faced growing competition in the plumbing sector from plastics and other materials, largely on cost grounds. However, losses in the sector have been more than offset by rising demand for electrical applications. Demand in

[1] Large quantities of scrap are generated during the fabrication and manufacturing processes. Most of this scrap is re-melted and recycled without further processing. Sometimes, it is added back into the smelting or refining stages. Copper-containing materials also are recovered and re-processed from obsolete products of all types. Secondary copper is a substantial contributor to annual supply, but is nowhere near sufficient to meet demand. That is satisfied only by new mine output, both from within the region and from global imports.

Table 13.1 Global Uses of copper in 2013

Building and construction	Plumbing	Water distribution, heating, gas, sprinkler	1612
	Building plant	Air-conditioning tube	180
	Architecture	Roofs, gutters, flashing, decoration, builders hardware	379
	Communications	Communications wiring in buildings	215
	Electric Power	Power distribution, earth ground, light, wire devices	6163
Infrastructure	Power Utility	Utility transmission and distribution network	3256
	Telecommunications	Telecommunications network	854
Industrial equipment	Electrical	Industrial transformers and motors	1657
	Non-electrical	Valves, fittings, instruments, plant equipment	2020
Transport	Automotive electrical	Harnesses, motors	2023
	Automotive non-electrical	Radiators and tubing	242
	Other vehicles	Railroad, shipping, marine	1353
Other equipment	Consumer and general products	Appliances, instruments, tools, other	2553
	Cooling equipment	Air-conditioning and refrigeration	2146
	Electronic Equipment	Industrial/commercial electronics and PCs	1018
Diverse	Other		3314
Total			28,958

Source: IWCC (2013)

power and telecommunications networks comprised 14 per cent of usage, and industrial equipment a further 13 per cent. Much of the demand in the transport sector, and in other equipment, also is directed toward infrastructure and fixed investment, rather than consumer markets. This characteristic of copper's commercial use, which is discussed in more detail later in the chapter, largely dictates the levels and the rates of change in demand.

Semi-Manufactured Copper Products

According to the International Copper Study Group (ICSG 2013a), approximately one fifth of the total annual use of copper consists directly of melted high-grade scrap. Most of this material is generated in the production of manufactured semis or in manufacturing processes. This direct scrap, which remains within the first-stage processing industries, or fabricators, arises in all countries that produce semi-manufactured copper products. Asian countries, principally China, Japan, and South Korea, collectively generate about 3.1 million tonnes World Bureau of Metal Statistics (WBMS 2013). As it is generated in the processing of refined copper, the subsequent tonnage fluctuates in step with usage, subject to lags between its generation and reuse, and to changes in processing technology and the final product composition of the semi-manufactured product.

Semi-manufactured copper products are made from direct scrap, as well as newly refined copper, with some of the latter obtained from the reprocessing of old scrap collected from obsolete copper-containing products. Semi-manufactures may contain just copper or copper alloys, such as brass (copper and zinc), bronze (aluminium or tin), or cupro-nickel (nickel). Through a wide variety of processing techniques, fabricators form wire, rod, tube, sheet, plate, strip, castings, powder and other shapes. These semi-fabricated products are the means by which estimates of total copper usage are established.

Apparent use best reflects changes in the usage of copper, subject to Table 13.1's incomplete coverage. China's rapid growth contrasts markedly with that of all other countries, especially in the post-2007 period. Its increased usage was more than sufficient to offset declines in Japan, South Korea, and Taiwan. The region swung from being a modest net importer to a small net exporter over the decade, which is due to changes in China's trade balance. Its production of semis rose faster than its apparent usage.

Usage of Refined Copper

The production of semis and other first use products is the basis for estimates of the usage of refined copper, the most widely used indicator of demand for copper. In that regard, it does not properly reflect the usage of copper *within* a country, but rather the usage *by* a country. Some used refined copper may be incorporated into exported semi-manufactures, whilst imports of semis are not taken into account in estimates of a country's copper usage. Thus, the usage of refined copper by Japan, China, and South Korea is partly in net exports of semis rather than domestic offtake, whereas domestic usage of copper containing semis is higher in Taiwan and the other Asia-Pacific economies than is implied by their usage of refined copper.

Accepting these statistical niceties, Table 13.2 compares the usage of refined copper in Asia-Pacific countries with global usage.

The Asia-Pacific Region has increased its share of global usage from nearly 40 per cent in 2002 to over 57 per cent in 2012. This growth mainly occurred in China, which is by far the largest user of refined copper and which also has exhibited the fastest rate of growth during the past decade. Indonesia also enjoyed continued growth over the period, but from a low base. China alone accounted for 43 per cent of global usage in 2012, compared with 18 per cent a decade earlier. In contrast to other countries, its growth of demand was the same in the second half of the decade as in the first. Most other countries, including Japan, South Korea and Taiwan expe-

Table 13.2 Usage of refined copper 2002–2012, Asia-Pacific and World

	'000 tonnes contained copper			Per cent per annum change		
	2002	2007	2012	2002–2007	2007–2012	2002–2012
China[a]	2775	4957	8845	12.3	12.3	12.3
Indonesia	125	200	239	9.9	3.6	6.7
Japan	1167	1252	985	1.4	−4.7	−1.7
South Korea	900	821	723	−1.8	−2.5	−2.2
Malaysia	170	184	189	1.6	0.5	1.1
Philippines	40	40	40	0	0.2	0.1
Taiwan	655	603	432	−1.6	−6.4	−4.1
Thailand	186	250	240	6.1	−0.8	2.6
Viet Nam	30	98	90	26.7	−1.7	11.6
TOTAL Asia-Pacific	6047	8405	11, 785	6.8	7.0	6.9
WORLD	15, 142	18, 196	20, 550	3.7	2.5	3.1
Rest of World	9095	9791	8765	1.5	−2.2	−0.4
Asia-Pacific as per cent of world	39.9	46.2	57.3			

Source: ICSG (2013b)
[a]Includes Hong Kong

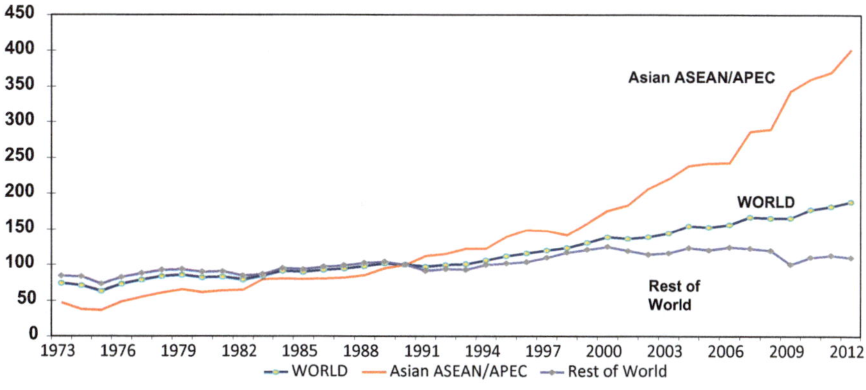

Fig. 13.1 Index numbers of the usage of refined copper. Source: ICSG (2013b) and private communication for pre-2003 data

rienced declining usage after 2007, or slow growth at best. In part, this sluggishness reflects the impact of the Global Financial Crisis (GFC).

The last decade's performance is set in a longer run perspective in Fig. 13.1, which compares index numbers, based on 1990 as 100, of the usage of refined copper in the Asia-Pacific Region with those of all other countries and of the world as a whole.

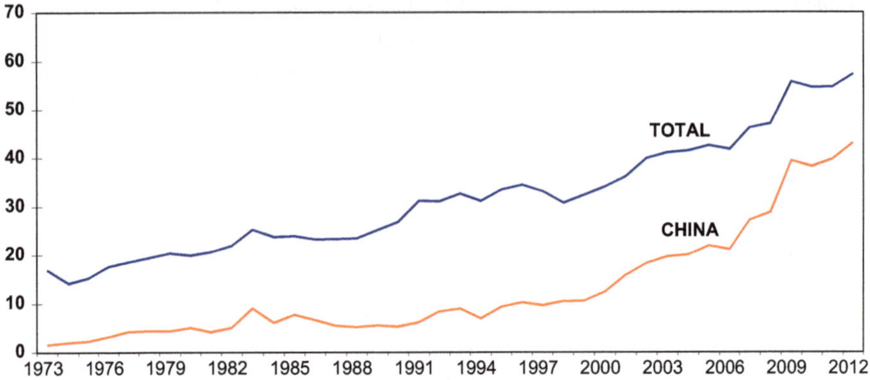

Fig. 13.2 Association of South East Asian Nations (ASEAN) and Asia-Pacific Economic Cooperation (APEC) per cent share of global usage of refined copper, 1973–2012. Source: ICSG (2013b) and private communication for pre-2003 data

Usage grew more rapidly in Asian countries than elsewhere for most of the period since 1973. It briefly levelled out in the mid-to-late 1980s with recessions in Japan and China, and dipped back in 1998–1999, due to the Asian Financial Crisis (AFC) and problems in Japan. Since then, the region's usage has roared ahead of that elsewhere. Usage in the rest of the world has been on a downward trend since the turn of the century. Figure 13.2 illustrates how the region's total share of global usage has changed since 1973. The separate series for China shows how the country has captured a growing share since the late 1990s.

Underlying Influences on the Usage of Refined Copper

One obvious reason for the rapid growth in China's usage of refined copper has been the strong rate of growth of its overall economy, as illustrated by the rates of change of real gross domestic product (GDP) in Table 13.3. The final line of the table shows the region's changing share of global GDP over the decade.

Equally, the decline in Japan and the sluggishness of South Korean and Taiwanese usage mirrors their overall economic performance. However, comparison of the rates of change shown in Tables 13.3 and 13.4 shows that usage of refined copper in Malaysia, the Philippines, and Thailand grew more slowly than their real GDP. Refined copper usage rose faster than GDP in China, Indonesia, and Vietnam in the decade as a whole. However, the two latter countries witnessed a marked adjustment in the relationship between the rates of change of copper usage and GDP in the second half of the decade.

The changing relationship between usage of refined copper and GDP from 1980 onward shows trends in copper intensity (measured as grams of copper per constant dollar of GDP) for the region's main copper using economies. Each country's GDP

Table 13.3 Annual average rates of change of real gross domestic product

	Per cent per annum changes[a]		
	2002–2007	2007–2012	2002–2012
China	11.6	9.3	10.4
Indonesia	5.5	5.9	5.7
Japan	1.8	−0.2	0.8
South Korea	4.3	2.9	3.6
Malaysia	5.9	4.3	5.1
Philippines	5.7	4.7	5.2
Taiwan	5.2	2.9	4.1
Thailand	5.6	2.8	4.2
Vietnam	7.4	5.8	6.6
World[b]	3.4	1.8	2.6
Asia-Pacific Region per cent share of world[c]	22.2	24.5	28

Sources: International Monetary Fund (IMF 2013; World Bank (2013)
[a]GDP at constant prices in national currencies
[b]GDP at constant 2005 international $ purchasing power parity rates
[c]Share, including Asian members of ASEAN/APEC, not shown separately at purchasing power parity (PPP) $ exchange rates

Table 13.4 Shares of gross fixed capital formation in GDP and copper intensities

	GFCF as per cent of GDP			Cu intensity 2011
	2002	2007	2011	(gms/$ of GDP)
China	36.3	39.1	45.5	0.643
Indonesia	19.4	24.9	32.0	0.174
Japan	22.9	22.6	20.7	0.219
South Korea	28.6	28.5	27.4	0.473
Malaysia	23.5	22.4	22.1	0.344
Philippines	20.6	19.9	19.1	0.095
Thailand	22.8	26.4	26.3	0.349
Vietnam	31.1	38.3	31.9	0.265
WORLD	20.3	21.7	19.4	0.238

Sources: ICSG (2013b); World Bank (2013); IMF (2013)

is measured in 2005 international dollar terms at purchasing power parity exchange rates. These give a better picture of the relative size of different countries' GDP than official exchange rates, because they adjust for differences in price levels.

The gradually declining trend in global intensity is shown for comparison with movements in the individual countries. The amount of refined copper used per dollar of GDP is below the global average in the Philippines and Indonesia, and has fallen rapidly under the average in Japan. Despite different experiences over the period, Thailand and Malaysia started and finished with similar levels of intensity. South Korea's intensity rose in the first two decades covered, but fell back in the last decade. Meanwhile, China's copper intensity has risen unevenly since the early

1990s, having previously fallen from a peak in 1983. One caveat about Chinese copper usage is that there sometimes have been large swings in unreported inventories, which may distort year-to-year comparisons. Estimates of usage for all countries are based on their production plus net imports and changes in reported stocks. For all countries, the earlier warning about differences between the usage of refined copper and total usage of copper needs to be recognised. Where copper semis are imported, usage of refined copper will be recorded in the supplying country. Conversely, countries that are net exporters of copper semis will appear to use more copper than in reality enters their domestic use.

The differing levels of the copper intensity of GDP between countries partly reflect differences in their per capita incomes, as suggested by the intensity of use hypothesis.[2] This posits an inverted U-shaped curve for the relationship between the intensity of use of materials and per capita income levels. When per capita incomes are low, demands for materials also are low. As an economy industrialises, its need for materials will expand faster than incomes to meet burgeoning needs for infrastructure and to supply developing manufacturing industries and construction. At much higher levels of income, the needs for infrastructure, machinery, and equipment increasingly will be satisfied and demand will shift toward less materials-intensive services. That will lead to intensity of use levelling out and then falling back. The precise nature of the relationship between intensity and incomes will vary between countries and over time, depending on prevailing economic structures and technologies.

As official exchange rates tend to understate the relative size of developing countries, in some cases substantially, the curves for intensity of use move closer together. Whilst the intensity of use hypothesis provides some rationale for changes in copper usage over time, it does not explain differences between countries with similar levels of per capita incomes. These partly result from differences in countries' industrial structures. Thus, a country like South Korea, in which manufacturing accounts for 31 per cent of total value added, has more copper using industries than economies like the Philippines, in which manufacturing makes up only 21 per cent (World Bank 2013). For comparison, manufacturing accounts for 30 per cent of Chinese and 19 per cent of Japanese total value added. Furthermore, South Korea's manufacturing industry is more biased towards the production of machinery and transport equipment, the most copper-intensive sectors, than the other economies. These sectors produced 47 per cent of South Korea's value added in manufacturing in 2008, compared with 37 per cent in Japan and 25 per cent in China, both in 2007 (World Bank 2013). Thus, machinery and equipment production account for almost 15 per cent (47 per cent of 31 per cent) of South Korea's GDP, compared with 7.5 per cent for China and seven per cent for Japan. Later data are not available for China, but they probably would show some increase, although not to South Korea's level.

Where countries largely produce the copper semis they use, like China, South Korea, and Japan, patterns of total expenditure also are important influences on cop-

[2] See Tilton (1990) for a discussion of the hypothesis. This originally is derived from the International Iron and Steel Institute (IISI 1972); Malenbaum (1978, 1975, 1973).

per usage. In particular, the share of GDP devoted to gross fixed capital formation (GFCF) is a major factor, as this is the most copper-intensive category of spending. It embraces expenditure on infrastructure, construction of all types, and investment in plant, machinery, and transport equipment. GFCF is lower than the often-quoted total investment, which includes spending on existing land and buildings and changes in inventories and work in progress. Table 13.4 compares the copper intensity of GDP in 2011, expressed in grams per current dollar at PPP exchange rates, with the share of GFCF in GDP in 2002, 2007, and 2011.

Future Trends in Usage

The conventional view is that the Asia-Pacific Region, spearheaded by China, will continue to take an expanding share of rising global demand for refined copper. The region's economic growth is expected to outstrip that of others in the foreseeable future. There still are unsatisfied needs for infrastructure and construction, and pent up demand for automobiles and consumer durables of all types. Even if the pace of growth slackens in China, due to rising labour costs, other countries in the region; for example, Laos and Cambodia, may take up the running as Chinese companies relocate some of their operations in search of lower labour costs. The tendency for manufacturing companies to move to lower cost areas helps explain some of the differential trends in copper usage shown earlier. China has captured market share and attracted copper-using manufacturing industries, not just from the advanced industrial economies, including Japan, but also from the Pacific Rim. However, China's rising relative wage costs will nullify its perceived locational advantages. Some American and European companies are moving their operations back closer to their home countries on cost, as well as logistical, grounds.

Sustained high rates of capital spending in China have created over-capacity in many sectors and have lowered rates of return. Chinese authorities are attempting the difficult task of changing the focus of total spending from investment and exports toward domestic consumption. Whilst the proportion of that consumption spent on automobiles and consumer durables will require copper, the tonnage involved probably will be lower than is incorporated in capital expenditure. Thus, the Chinese economy could become less copper intensive in the coming years. The impact of the changed focus on the demand for copper then will hinge not just on the overall rate of economic growth, but also on the pace of technical change. The declining intensity of copper use in advanced countries notably has not just reflected a diversion of the production of copper-intensive products to China and the Asia-Pacific Region, but also technical change and substitution.

This chapter is not the place for precise forecasts of copper demand. However, the conventional wisdom about continued rising Chinese and Asia-Pacific demand for refined copper probably is correct for the next few years. Nevertheless, there are risks and uncertainties that warrant caution. A plausible case could be made, if not for falling demand, then for much slower growth than is generally expected. The experiences of Japan, and more recently of South Korea and Taiwan, should be kept in mind.

One influence on demand that often is underrated is the behaviour of prices. In the short-run, demand may be unresponsive to movements in copper's relative price. It takes time to invest in appropriate equipment and change working practices in order to use substitutes. However, in the longer-term, demand responds much more to shifts in relative prices. These are governed by the interplay of demand and supply, and also by perceptions about future availability. The next sections therefore examine copper supply.

The Supply of Refined Copper

The production of refined copper metal involves several stages. There are two main routes. Some three-quarters of primary refined copper come from copper sulphide ores, which are mined and then processed usually at the mine site, to produce concentrates containing 25 to 45 per cent copper. These then are smelted in a furnace to produce copper matte (50 to 75 per cent copper plus iron, sulphur, and by-product metals), which is treated in an adjacent converter to produce blister copper and anodes, which contain 98.5 to 99 per cent copper. The anodes are processed further in an electrolytic refinery to produce copper cathodes assaying over 99.99 per cent copper. Typically, copper smelters and refineries are located away from the mines, and trade takes place in all the intermediate products leading to cathode production. In most, but not all, instances, smelting and refining are physically integrated on the same site.

The other route to refined copper begins with the leaching of oxidised ores with sulphuric acid. The leached copper is recovered by solvent extraction followed by electrolysis. This solvent extraction electrowinning process (SX-EW), which accounted for 22 per cent of the world's copper mine output in 2012, invariably takes place at the mine site. It produces refined copper metal of a comparable purity to that coming from the conventional pyro-metallurgical route. The share of SX-EW copper in total metal production has risen from negligible proportions in the 1980s to about 20 per cent of world production today. The main constraint on further increasing its share is the availability of suitable oxide ores. Figure 13.3 shows how the Asia-Pacific countries' usage of copper metal was supplied over the decade 2003–2012.

Mine output grew at an annual average rate of 12 per cent per annum, secondary output at 14.7 per cent, and net imports at 2.5 per cent. The secondary production shown in this figure was derived from both domestic sources and imports so that its rate of growth and the share of net imports are somewhat understated. That reservation aside, imports of ores, concentrate and metal accounted for an erratically declining share of usage over the decade, falling from 66 per cent in 2003 to 45 per cent in 2012. That disguises some large differences between individual countries' experiences. China's imports fell from 66 per cent of usage in 2003 to 48 per cent in 2006, but jumped to 66 per cent in 2009, before easing back to 62 per cent in 2012.

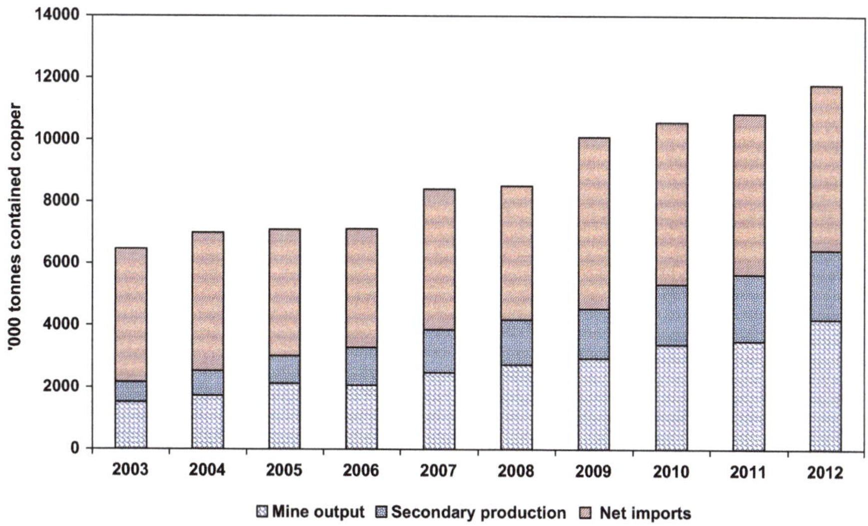

Fig. 13.3 Asia-Pacific sources of supply to meet usage of refined copper, 2003–2012. Note: Papua New Guinea (PNG) mine output is included. Source: ICSG (2013b)

Table 13.5 summarises how the main users of refined copper in the Asia-Pacific Region met their needs for metal in 2012.

China's total production of refined copper amounted to almost 66 per cent of its domestic usage, with nearly one third of that production based on secondary materials. These not only arose within China, but also were imported on a substantial scale from a wide range of countries. The European Union (EU) and the US together accounted for half of China's imports of 4.9 million tonnes. The country absorbs three-fifths of the global imports of copper and copper alloy scrap. Secondary production accounted for 18 per cent of Japan's and 16 per cent of South Korea's output of copper metal. Japan was a net exporter of copper bearing scrap, mainly to China, while South Korea was a net importer.

Malaysia, Taiwan, and Thailand rely entirely on imports of refined metal to satisfy their needs. A smelter/refinery briefly operated at Rayong in Thailand between 2004 and 2007, but it proved uneconomic and closed down. Taiwan once had a small local smelter/refinery, but that ceased production in 1990. China's primary production of refined metal is based on a mix of local mine output and imported materials. Neither Japan nor South Korea today mines copper to support their smelter/refineries, which now depend on domestic scrap and imported raw materials. Japan is a significant net exporter of metal, whereas South Korean production of refined copper is supplemented by net imports.

Japan's metal output originally was based on domestic ore production; however, its smelters and refineries turned to imports when local mines progressively became exhausted from the 1950s onward. The industry for many years was sheltered behind tariffs, which gave it a competitive advantage in purchasing concentrates.

Table 13.5 Usage and sources of refined copper in 2012 ('000 tonnes)

	Usage	Production		Imports	Exports
		Secondary	Primary		
China	8845	1885	3939	3402	274
Indonesia	239		207	125	69
Japan	985	245	1271	36	546
South Korea	723	96	493	302	170
Laos[a]			86		
Malaysia	189			210	
Myanmar[a]			19		
Philippines[b]	40		90	10	65
Singapore				38	22
Taiwan	433			434	1
Thailand	240			241	
Vietnam[c]	90		8		
Totals[d]	11,785	2227	6114	4796	1147

Source: ICSG (2013b)
Complete data on trade are not available for all countries
[a]Laos and Myanmar production was of SX-EW copper at mine sites
[b]In 2011, Philippines production was 164, 000 tonnes and exports 123, 000 tonnes
[c]Vietnamese imports were 107, 000 tonnes in 2011
[d]Imports and exports are shown gross before netting out intra-regional trade

Also, overseas copper mines were developed in the 1960s and early 1970s, with loan capital secured against Japanese purchase contracts. Japanese smelters and refineries kept abreast of technical change by concentrating on reducing their energy requirements and maximising the revenues they obtained from by-products. Many copper ores contain gold, silver, and a range of minor metals (such as selenium and tellurium) for which the mines often receive little or no payment. The sales revenue largely accrues to the copper smelters/refineries and supplements their incomes from their sales of copper. The Japanese industry has rationalised over the years, but six smelters continue producing. In most cases, refining is carried out on the same sites, but the individual smelters' anode output exceeds their capacity to manufacture refined metal. The excess is shipped to a separate refinery.

When South Korea's needs for copper expanded in the 1970s and 1980s, it followed Japan's model of domestic metal production rather than reliance on imports of metal. A relatively small smelter/refinery has operated for decades. Most South Korean metal is produced at the Onsan smelter/refinery, which commenced operations in the mid-1970s and since has been greatly expanded.

Laotian and Burmese production consists of electrowon copper from mines using the SX-EW process. Both Indonesia and the Philippines have local mine production and their smelters/refineries were constructed in response to governmental pressures. In the Philippines, the smelter on Leyte, which now is owned by GlencoreXstrata, commenced operation in 1983 with government financial support in order to process locally produced copper concentrates. At that time, the

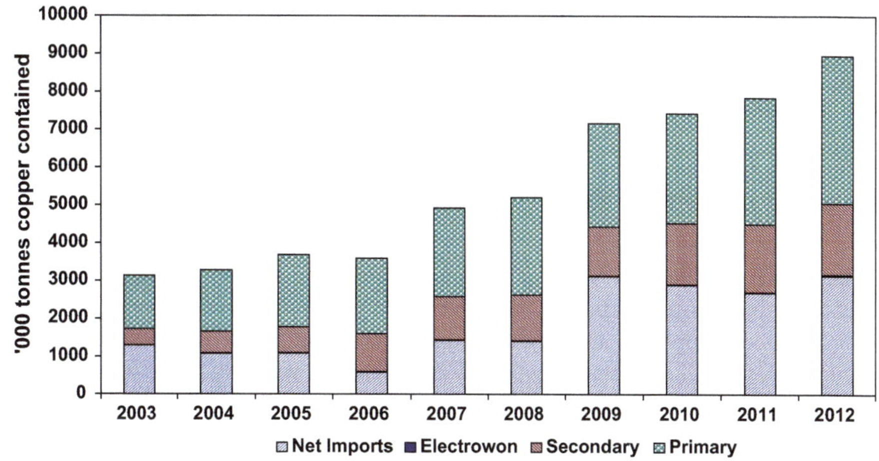

Fig. 13.4 Sources of Chinese supplies of refined copper metal, 2003–2012. Source: ICSG (2013b)

Philippines' mine output was about 0.3 million tonnes of contained copper, and mines were induced to supply part of their output to the smelter. However, the local mines' ore reserves gradually were exhausted and not replaced by new facilities. As a consequence, the smelter has had to source its materials from overseas in competition with larger smelters in China, Japan, and South Korea. Its output fell back sharply in 2012 after its facilities were damaged by Typhoon Haiyan. Indonesia's smelter/refinery at Gresik, East Java, which began operating in 1999, is majority owned by Japanese smelting companies led by Mitsubishi. Freeport McMoRan Copper and Gold Inc., whose Indonesian copper mine output greatly exceeds the smelter's capacity, owns 25 per cent. The smelter's capacity currently is being expanded. Vietnam's insignificant production of metal is based on local mines.

China is the region's largest importer of refined metal to supplement its domestic output. The latter is based on a mixture of domestic mine production, including a modest quantity of electrowon metal from an SX-EW mine, and imported concentrates. Figure 13.4 shows the relative contributions of production and net imports to Chinese supplies of copper metal over the past decade.

The share of primary production, including electrowon metal, barely has changed over the decade (45 per cent in 2003 and 44 per cent in 2012), but the output of secondary metal has gained at the expense of imports of metal. As noted earlier, a considerable proportion of that secondary metal is derived from imported materials. Exports of refined metal have fluctuated according to the balance between domestic production and local needs. They rose sharply in 2012 to 274 kilotons.

Nearly all the copper metal exported by Asia-Pacific countries went to other countries in the region, and such intra-regional exports made up one-fifth of the region's total imports. Non-Asian APEC members supplied 49 per cent: Chile (37 per cent), Australia (8 per cent), the European Union (10 per cent), India (6 per cent), and roughly four per cent each from Kazakhstan, Zambia, and The Democratic Republic of Congo. Viewed from the originating countries, the region took nearly all the exports of copper metal from Australia and India and substantial shares of the shipments of Chile, the US, Peru, Brazil, Kazakhstan, and the African exporting countries.

Sources of Copper Raw Materials

The Asia-Pacific Region is a substantial net importer of primary copper raw materials to support its production of refined metal. This is shown in Table 13.6, which shows mine production and trade in concentrates and blister copper for selected countries in 2012. Imports of secondary raw materials were discussed above.

Except for China and South Korea, the region has negligible trade in blister and anode copper. Both China and South Korea import small quantities from a range of sources. Chile is the main supplier to South Korea (70 per cent of the total) and a major supplier to China (32 per cent of the total). China is the largest mine producer of copper in the region, and has become the world's second largest producer after Chile. Nonetheless, its mines only can supply 35 per cent of the primary copper used in China's refineries, and only 17 per cent of the country's total usage of refined copper, a lower share than in 2003. Mine output has grown rapidly, but by no means rapidly enough. Figure 13.5 shows how China's mine output has changed in comparison with that of the region's other countries.

China's growing output contrasts with that of the other countries, most notably, Indonesia and PNG. New mines have been developed in Laos and production has been revived in the Philippines from its 2004–2005 nadir. It remains a shadow of the

Table 13.6 Asia-Pacific production and trade in primary copper raw materials in 2012

	Mine production		Concentrates trade		Blister/anode trade		Total primary raw materials
	Concentrates	SX-EW	Imports	Exports	Imports	Exports	
China	1450.0	40.0	2191.7	0.2	522.6		4204.1
South Korea	–	–	522.6	–	62.8	–	585.5
Japan			1432.2		4.0	0.2	1436.0
Indonesia	399.5	–	11.7	359.5	–	–	51.7
Laos	63.3	86.3	–	–	–	–	149.6
Myanmar		19.0	–	–	–	–	19.0
PNG	125.3	–	–	125.3	–	–	0
Philippines	65.7		100.9	67.4	0.3	1.2	98.3
Vietnam	11.0	–	–	–	–	–	11.0
Thailand	–	–	0.5	–	0.5	–	1.0
Malaysia	–	–	–	–	0.5	–	0.5
Taiwan	–	–	–	–		1.4	−1.4
Singapore	–	–	–	–	0.7	0.8	−0.1

Source: ICSG (2013b)

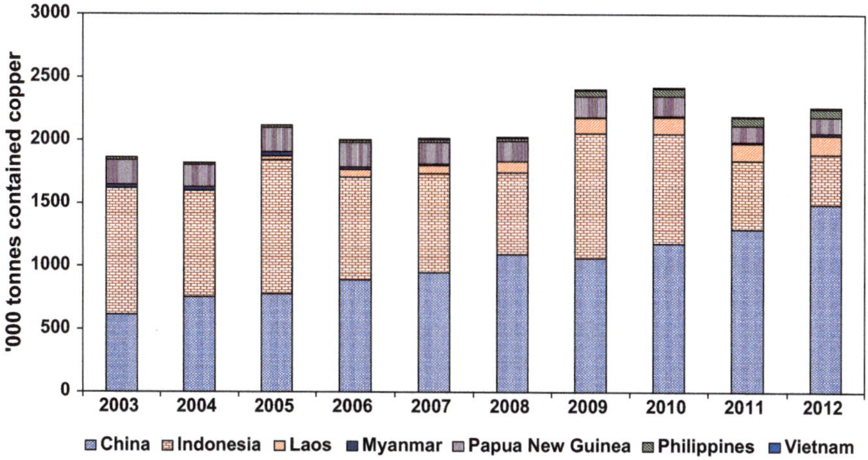

Fig. 13.5 Mine output of copper in the Asia-Pacific Region, 2003–2012. Source: ICSG (2013b)

peak of just over 0.3 million tonnes reached in 1980–1981. In PNG, the sole producer, Ok Tedi, has suffered both from a combination of reduced ore throughput and declining ore grades. Between 2003 and 2012, its ore output fell 20 per cent, but its production of contained copper dropped by 36 per cent (IntierraRMG 2013). Since 2008, its 22 per cent fall in output solely has been the result of lower grades. Indonesia's decline in production has been even more pronounced. It has two operating companies. The Batu Hijau copper-gold mine, operated by Newmont with Japanese partners, opened in 1999 and increased its production to 325 thousand tonnes in 2005, but produced only 71 thousand tonnes in 2012. A combination of labour unrest, equipment damage, and production difficulties, coupled with lower grades, reduced output at Indonesia's other copper mining operations at the Grasberg mine. The mine's peak production in 2005, and again in 2009, was roughly 0.75 million tonnes, but it was barely half that level in 2012. Even if ore grades did not recover at either operation, Indonesia's mine output of copper would be around 45 per cent above its 2012 level, when the mines operated to capacity.

The region accounts for just over two-thirds of global imports of copper concentrates and for similar shares of the exports of the major concentrate shippers. Chile is the most important supplier, with one-third of the region's imports, followed by Peru with 17 per cent. In total, the non-Asian members of APEC supply 73 per cent of their Asian partners' requirements. China is the largest importer with 52 per cent of the Asia-Pacific total, followed by Japan with 34 per cent. The Japanese and South Korean smelters have long-established links with mines in British Columbia, Indonesia, PNG, and Chile, and in many instances, either having provided loan finance when their mines were started, or owned a proportion of their equity. Japan's imports peaked in 2007 and only were 16 per cent higher in 2012 than in 2003. South Korea's imports rose by nearly one-third over the period. China's imports, in contrast, were 2.5 times higher in 2012 than in 2003. Consequently, its smelters have had to establish new sources of supply in competition with the longer established importers of concentrates, and they will have to continue doing so to meet their rising future needs.

Constraints on Copper Mining

In that regard, China and the other Asia-Pacific importers of copper are competing in a global market both for concentrates and refined metal. Shortages or surpluses in one region are rapidly transmitted throughout the world. Since the mid-2000s, global supply has been well below the level implied by estimated copper mining capacity and its historical average rate of utilisation.[3]

Admittedly, mine capacity is a somewhat nebulous concept. As mines extract and process ore, their capacity is properly expressed in tonnes of ore rather than of contained metal. The latter will vary not just with the number of hours the mine and processing plant operate, but also with the copper content (head grade) of the ore being processed and the proportion of the contained copper that actually is recovered (the recovery rate). Estimates of the capacity of any copper mine necessarily are based on assumptions about these factors. The main reason for the decline in capacity utilisation appears to have been a marked fall in head grades at a number of major mines (Crowson 2012, 2009).

Although the drop in head grades may have been temporary in some cases, it appears to be more permanent at many major mines. It reflects the exhaustion of their higher-grade ore and a fall in the copper content of their annual ore production to the average grade of copper in their ore reserves. Over the past decade, the ratio of head grades at copper mines to their reserve grades has fallen substantially. In 2000, the head grade equalled or fell short of ore grades in just under 32 per cent of mines, and in 14 per cent of mines head grades were at least double the ore reserve grades. By 2008, the mines whose head grades were the same as, or less than, their ore grades had risen to 50 per cent of the total (Crowson 2012). Comparable data for 2012 are that 58 per cent of mines had head grades equal to or less than their ore reserve grades, and in a further 21 per cent, head grades were under 1.25 times their reserve grades. Only four per cent had head grades that were more than double their reserve grades (IntierraRMG 2013). Put differently, the average ratio of head to ore grades, weighted by each mine's production of copper in the relevant year, fell from 1.61 in 2000 to 1.43 in 2006, 1.31 in 2010 and 1.24 in 2012. Future trends in copper head grades are unclear, but the ore grades of new mines entering production or planned are in many instances greater than the average of existing mines. Also, it is important to remember that a large proportion of copper deposits contain other pay-

[3] The data published by the ICSG (2013b; 2008; 2007) show that mine capacity utilisation averaged well over 90 per cent up to 2004, but dropped to a low point of 81.1 per cent in 2011 rising to 81.9 per cent in 2012.

able products besides copper. The profitability of mines greatly can be affected, for good or ill, by the revenues earned from by- or co-products. Falling copper grades alone do not necessarily imply reduced viability.

The impact of falling ore grades on capacity utilisation was exacerbated by a series of natural disasters, by over-stretched suppliers of equipment and spares being unable to meet demand, and by an upsurge in labour unrest and disruption. This unrest is primarily a response to a surge in copper prices and corporate profitability, after an extended period of weak prices and low profitability. In a buoyant copper market, the workers' bargaining power increases, and they seize the opportunity to press for improved wages and conditions.

Some of the factors holding down mine utilisation, such as shortages of spares and equipment, have eased, with a rate of 85.2 per cent achieved in August 2013, the highest since early 2008 (ICSG 2013c). However, effective capacity, after allowing for changes in head grades, is much lower than the published statistics suggest. Some existing mines may have scope to produce more, but additional mine capacity is required to satisfy rising demand for copper. Some large mines are nearing the end of their expected lives and there is always a need for new capacity to offset closures and reductions in grade.

There is no shortage of known ore deposits awaiting development. In November 2013, IntierraRMG listed 50 new copper mines or expansions under construction, 91 at the feasibility stage, 66 at the prefeasibility stage, and 226 conceptual projects (IntierraRMG 2013). All these projects are of varying sizes and likelihood. They are additional to mines like Oyu Tolgoi in Mongolia, which started producing in 2013, but which has not yet reached full operating capacity. Typically, new mines take several years to reach their designed capacity.

In February 2013, the International Copper Study Group (2013d) forecast that global mine capacity would rise from 20.4 million tonnes in 2012 to 27.7 million tonnes by 2016. However, as Fig. 13.6 demonstrates, there has been a tendency for previous forecasts of capacity to be overly optimistic.

2010 forecasts for capacity in 2012 was inflated by 6.7 per cent, the equivalent of almost 1.4 million tonnes. The forecasts made in February 2013, which were for a 36 per cent rise in mine capacity between 2012 and 2016, were overtaken by subsequent developments. Concern about rising capital and operating costs, difficulties in securing sufficient finance, and changed corporate priorities have led to the postponement or cancellation of some large projects (for example, BHP Billiton's Olympic Dam expansion in Australia, Anglo-American's withdrawal from the Pebble project in Alaska, and GlencoreXstrata's cessation of work at Tampakan in the Philippines). This does not mean that copper mine capacity will stagnate, nor that the delayed projects will never eventually produce copper, but merely that the timing, and possibly the scope of mine expansions and start-ups, will change.

By no means all potential investors in new copper mine capacity are subject to the same pressures from financial institutions as the major multinational companies. Chinese State-Owned Enterprises (SOE's), for example, are involved in a number of overseas projects, as well as in domestic mine developments. They are less con-

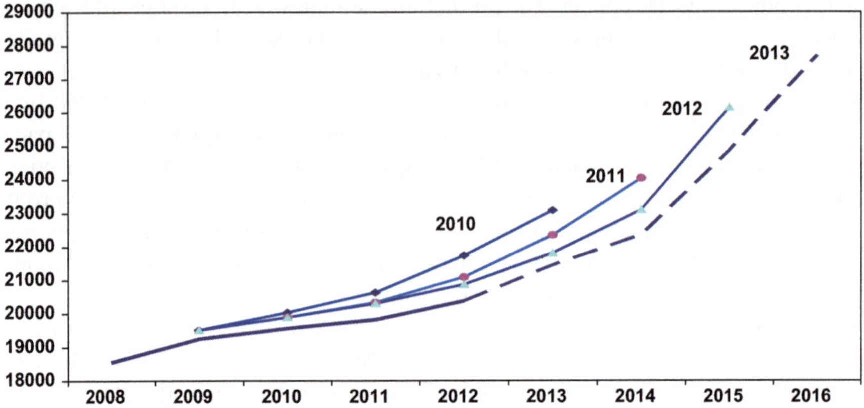

Fig. 13.6 ICSG forecasts of global copper mine capacity, 2010–2013 ('000 tonnes contained copper). Source: ICSG (2013d)

cerned about political risks than the large multinationals (Feher in this volume). Those who claim, for example, that only very limited opportunities remain for developing large new open pit mines largely are ignoring the activities of Chinese companies in areas that private sector companies might regard as unacceptably risky.

The past decade's surge in copper prices and profitability has failed to flow through to tax receipts in host countries on anything like the same scale as predicted, and has led to a widespread reassessment of fiscal charges by host countries. Many have imposed new taxes or raised existing rates, but this was after a period when fiscal terms were largely in the investors' favour (Johnson in this volume). In most jurisdictions with sizeable copper production potential, after-tax rates of return are acceptable for all but marginal projects.

Recent rises in capital costs are forcing some reassessment by all companies of the desirability of opening new mines at their eventual expected throughput rate, rather than building up slowly to that rate through incremental investments over a much longer period. Not only does such an approach make it easier to raise the necessary funds, but also it means that adverse environmental impacts more easily can be accommodated.

Conclusion

Various risks and opportunities have been identified in the preceding sections of this chapter. Whatever happens to copper prices, the Asia-Pacific Region will drive the performance of the global copper market for the foreseeable future. However, history cautions against the projection of recent trends into the indefinite future. The future always springs surprises, giving rise to new risks and opportunities for users and producers alike.

On the demand side, the region likely will experience continued growth in demand for copper as its economic development proceeds, but at a slower pace than in the past decade. China probably will remain the main focus of growth, but its rising labour costs could divert copper-using industries to other countries, both within and outside the region. Also, the rebalancing of the Chinese economy from its over-dependence on capital spending and exports adversely could affect its copper requirements, without other countries taking up any slack. These macro-economic risks to the growth of regional demand may be exacerbated by continued technological change in copper's end-use markets and by the commercialisation of substitute products and processes in major uses.

Domestic mine production and locally generated supplies of secondary materials are insufficient to meet demand in the region's main copper-using countries. The consequent dependence on imports of raw materials, intermediate products, and refined metal provides healthy opportunities for global copper producers, especially in Latin America, Australia, and China's Asian neighbours, but creates significant risks for the importing countries. There are increasing pressures for downstream processing in countries such as Indonesia, which exports ores and concentrates. To the extent that such pressures become embodied in domestic processing obligations, copper smelters in China, Japan and South Korea will face mounting difficulties in obtaining raw materials. The need for domestic processing is but one aspect of mounting resource nationalism over the past decade. Rising royalty rates and fiscal burdens in most mineral producing countries, and calls for increased state participation are other manifestations. In addition, host communities are becoming more assertive in their attitudes to mine development, regardless of national policies. To the extent that host communities raise barriers to investment these varying and cumulative political pressures may impede the timely development of the additional capacity that may be required to meet rising demand. That, in turn, increases the risks of price volatility.

The political and social risks to supply are additional to the normal technical and economic risks associated with mining. Falling average ore grades and depleting mines mean that continuing heavy capital spending is required merely to sustain mine production, let alone meet rising demand. Many of the known undeveloped copper ore deposits face logistical, environmental, and social impediments to their speedy development. This is nothing new as the development of new mines in virgin areas has always been risky, but the traditional mining investors' appetite for risk appears to have weakened. This creates opportunities for new entrants that are not circumscribed by the preconceptions of existing major producers or by the belief that large-scale is the only route to successful mine development. It also provides scope for innovative mining and processing techniques that both can reduce costs and minimise energy and water requirements.

References

Crowson, P. (2012). Some observations on copper yields and ore grades. *Resources Policy*, (37(1), 59-72.

Crowson, P. (2009). Copper production from the late 1990s - A paradox? *Copper Studies*, December 2009. CRU International Limited, London.

International Copper Study Group. (2013a). The world copper factbook 2013. Lisbon.

International Copper Study Group. (2013b). ICSG 2013 statistical yearbook. Lisbon.

International Copper Study Group. (2013c). Copper: preliminary data for August 2013. Lisbon.

International Copper Study Group. (2013d). Press releases: on successive issues of the Directory of Copper Mines and Plants. Lisbon.

International Copper Study Group. (2008). Press release: advertising the 2007 statistical yearbook. Lisbon.

International Copper Study Group. (2007). Press release: advertising the 2006 statistical yearbook. Lisbon.

International Iron and Steel Institute. (1972). Projection 85: World Steel Demand. International Iron and Steel Institute, Brussels.

International Monetary Fund. (2013). World economic outlook database. Washington. Accessed 14 June 2014.

IntierraRMG. (2013). Raw materials database. Accessed 9 December 2013.

International Wrought Copper Council. (2013). Copper's end uses. London. http://www.copper-council.org/. Accessed 9 December 2013.

Malenbaum, W. (1978). *World demand for raw materials in 1985 and 2000*. New York: McGraw-Hill.

Malenbaum, W. (1975). Law of demand for minerals. Proceedings of the Council of Economics, 104th Annual Meeting of the American Institute of Mining, Metallurgical and Petroleum Engineers, pp.146-155. New York.

Malenbaum, W. (1973). Material requirements in the United States and abroad in the year 2000. Research Project prepared for the National Commission on Materials Policy. Philadelphia: The University of Pennsylvania.

Tilton, J. (1990). World metal demand, trends and prospects. *Resources for the Future*. Washington.

World Bureau of Metal Statistics. (2013). World metal statistics yearbook 2013. London.

World Bank. (2013). World development indicators. http://data.worldbank.org/data-catalog/world-development-indicators. Accessed 9 December 2013.

Chapter 14
Coal Seam Gas

Cristelle Maurin

Abstract This chapter focuses on the development of the coal seam gas industry in Australia. The rapid development of unconventional energy resources since the mid-2000s has left the community divided with regard to the benefits and risks involved, and has risen to prominence in the Australian policy debate. In 2015, the Australian Government adopted its Domestic Gas Strategy, which emphasises the development of onshore gas resources as a national priority and recognises the need to ensure the responsible development of coal seam, shale and tight gas resources for the benefit of all Australians. "Introduction" section of this chapter discusses the significance of coal seam gas resources in the Australian context. "The Growth of the Coal Seam Gas Industry in Australia" section considers the issues that have arisen in relation to the growth of the coal seam gas industry and its impacts on communities. "Coal Seam Gas, Water, Air Pollution and Communities" section reflects on the regulation of coal seam gas activities. The conclusion addresses questions about the future of the industry.

Abbreviations

CO_2	Carbon Dioxide
COAG	Council of Australian Governments
CSG	Coal Seam Gas
EIA	Energy Information Administration
IESC	Independent Expert Scientific Committee on Coal Seam Gas and Large Coal Mining Development
PJ	Petajoules
SCER	Standing Council on Energy and Resources
US	United States of America

C. Maurin (✉)
Office - DVC (Research and Innovation), Research and Innovation Division, University of Newcastle, Callaghan, NSW 2308, Australia
e-mail: Cristelle.Maurin@Newcastle.edu.au

T. O'Callaghan, G. Graetz (eds.), *Mining in the Asia-Pacific*, The Political Economy of the Asia Pacific, DOI 10.1007/978-3-319-61395-6_14

Introduction

The development of unconventional gas resources has become a key agenda item for governments in resource-rich jurisdictions due to their energy potential and economic dividends. While the emissions output of burning natural gas still is subject to debate (Howarth et al. 2011; World Watch Institute 2011; Day et al. 2012), the global shift toward cleaner sources of energy has spurred the development of onshore gas fields, including from unconventional sources. It is assumed that increased production of natural gas will help phase out coal and, therefore, help reduce global carbon emissions.[1] This assumption rests on the fact that natural gas combustion produces less carbon dioxide (CO_2) emissions than other fossil fuels. In the United States (US), the substantial increase in shale gas production has led to the country's transition toward energy self-sufficiency, with natural gas overtaking coal as the top source of electric power generation in 2015.[2]

With the advancement of new technologies, natural gas can now be commercially developed from source rocks, such as shale formations or coal seams. The anticipation of increased global energy demand, and the decrease in costs of extraction, has spurred investment in the sector. The growth of the unconventional gas industry has been exponential worldwide, notably in North America and Australia where it has become a significant part of the gas industry. The 'unconventional gas revolution' that has occurred in the USA has led other countries to follow the same path, as unconventional gas resources are found in various regions around the world.

Australia's substantial unconventional gas resources are located in several basins across the country, and while there are few prospects yet for turning Australia's shale resources into a profitable industry, coal seam gas (CSG) has become an integral part of Australia's natural gas industry (EIA 2015a).[3]

This chapter focuses on the development of the coal seam gas industry in Australia. The rapid development of unconventional energy resources since the mid-2000s has left the community divided with regard to the benefits and risks involved, and has risen to prominence in the Australian policy debate. In 2015, the Australian Government adopted its Domestic Gas Strategy, which emphasises the development of onshore gas resources as a national priority and recognises the need to ensure the responsible devel-

[1] Ongoing scientific research will determine the level of emissions over the life-cycle of unconventional gas projects and whether the development of unconventional gas represents a positive scenario from a climate perspective.

[2] According to the US Energy Information Administration (EIA), energy imports in the US declined from 30% of total energy consumption in 2005 to 13% in 2013 (EIA 2015). About 31% of electric power generation in April 2015 came from natural gas and 30% from coal.

[3] Australia's shale basins—437 trillion cubic feet (tcf)—are found in: the Perth and Canning Basins in Western Australia; Georgina and Amadeus in the Northern Territory; the Cooper Basin in South Australia and Queensland; and the Otway Basin in Victoria and South Australia (EIA 2015a). A decision by the Queensland Government in 2000 that required 13% of all power supplied to the State electricity grid to be generated by gas by 2005 was the first trigger in the growth of the industry. The development of a gas export industry in Queensland has accelerated the development of coal seam gas reserves.

opment of coal seam, shale and tight gas resources for the benefit of Australians. The "Introduction" section of this Chapter discusses the significance of coal seam gas resources in the Australian context. The "The Growth of the Coal Seam Gas Industry in Australia" section considers the issues that have arisen in relation to the growth of the coal seam gas industry and its impacts on communities. The "Coal Seam Gas, Water, Air Pollution and Communities" section reflects on the regulation of coal seam gas activities. The conclusion critically adresses questions about the future of the industry.

The Growth of the Coal Seam Gas Industry in Australia

Coal seam gas is naturally occurring methane gas (CH_4) in coal seams.[4] Referred to as unconventional gas, CSG is, like conventional natural gas, a type of petroleum derived from the compressed remains of organic matter over geological times and principally is comprised of methane.[5] It is the rock formation and techniques required for its extraction that have led to its nomination as 'unconventional gas'. Conventional gas migrates from a source rock into a reservoir, where it is trapped under an impermeable cap-rock; here, it can be extracted with traditional vertical drilling techniques. In contrast, the coal seam is the source rock from which the gas originates. The gas is held in the coal seam by water pressure and, therefore, its extraction involves the removal of the associated water to depressurise the coal seams in order for the gas contained in the coal pores to be released and flow to the surface. In some circumstances, hydraulic fracturing can be used to stimulate the release of the gas. Also known as 'fracking', hydraulic fracturing is a process that is used to create fractures in a geological formation using a fluid composed of sand, water, and chemicals.[6] Injected at high pressure into the formation, the fracking fluid opens up cracks and creates a path for water to flow back to the surface. For the production of shale gas, hydraulic fracturing is necessary for the recovery of the gas from the highly permeable rock formation. However, in the case of coal seam gas, only certain types of geology—where the natural permeability of the coal seam is insufficient—would require the application of hydraulic fracturing to stimulate the gas recovery. Of the 1844 coal seam gas wells that were drilled in Australia over 15 months during 2012 and 2013, only 6% were hydraulically fractured.

Exploration for CSG began in 1976 in Queensland's Bowen Basin, where two exploratory wells were drilled by the company, Houston Oil and Minerals of Australia. However, it was only in the late 1990s that the commercial development of CSG started in Queensland and New South Wales.

[4] Coal seam gas also is called coal seam methane (CSM) and coal bed methane (CBM).

[5] During the earliest stage of coalification (the process that turns plants into coal), biogenic methane is generated as a by-product of microbial action, which is similar to the mechanism that generates methane in landfills.

[6] The composition of fluids used in hydraulic fracturing activities varies depending on site requirements. Benzene, toluene, ethyl-benzene, and xylenes (BTEX chemicals) are banned from use in hydraulic fracturing in New South Wales and Queensland.

Ninety two percent of Australia's 33 tcf of CSG resources are found in the Surat (69%) and Bowen (23%) Basins, which are located along the east coast of Australia in Queensland. Other reserves, the equivalent of 2904 petajoules (PJ), are found in several basins across New South Wales (Gunnedah, Clarence-Moreton, Gloucester, and Sydney Basins).[7] Following developments in North America in which the extraction of coal bed methane became central to the booming gas industry, gas companies operating on the east coast of Australia realised the potential in developing the methane contained in Australia's coal seams. In 1996, Australia's first CSG was produced at the Dawson Valley project in central Queensland, while CSG production in New South Wales started in 2001 in the Sydney Basin. Australia's annual production of CSG has increased from 1 PJ in 1996 to 240 PJ in 2010–2011, with 97% produced in Queensland (234 PJ/0.2 tcf) from the Bowen (121 PJ/0.1 tcf) and Surat (113 PJ/0.1 tcf) Basins, while only 6 PJ was produced from the Sydney Basin in New South Wales. In 2011, CSG activity in Queensland comprised 600 production and exploration wells in the Bowen, Galilee, and Surat Basins, with more than 3500 land access agreements signed by CSG operators. The number of CSG wells reached 1600 in 2013–2014; this number is expected to reach 30,000 in the decades to come (Government of Queensland, Department of Natural Resources and Mines 2015).

In the decade since the start of its commercial development, CSG reached about 10% of Australia's total gas production. Several factors explain the growth of the industry. The initial driver was a decision made in 2000 by the Queensland Government requiring that 13% of power supplied to the State electricity grid had to be generated by gas by 2005. In an effort to substitute gas for coal in electricity generation, this requirement subsequently was increased to 15% by 2010 and 18% by 2020 (Australian Government, Geoscience Australia n.d.).

From the mid-2000s, another factor has contributed to the growth of the CSG industry in Australia. The increase in gas prices on world markets has been a major trigger for the development of the country's gas infrastructure. Gas companies with tenures in Queensland and New South Wales have sought to benefit from increasing global demand for gas and the resultant higher prices. As a consequence, over AUD$60 billion has been invested in the development of export facilities in Queensland alone. Three gas projects with a combined capacity of ~25 Mtpa have been developed to connect the Australian Eastern gas region to world markets (Sims 2015), and technology has been developed to convert CSG into liquefied natural gas (LNG).[8] While the opening of the

[7] Australia's identified CSG reserves—economically demonstrated resources—were estimated to be 33 tcf (35,905 PJ) in January 2012. According to the Australian Bureau of Resources and Energy Economics (BREE), economically demonstrated resources means the quantity of the resource that is judged to be economically extractable under current market conditions with current technology. Geoscience Australia estimated total proved plus probable commercial reserves of natural gas at 132 tcf (99 tcf of conventional natural gas and 33 tcf of coal CSG) in 2012. Most of the conventional gas resources (about 92%) are located in the North West Shelf in the offshore Carnarvon, Browse, and Bonaparte Basins (EIA 2015a).

[8] Queensland Curtis LNG (QCLNG)—operated by BG Group in a venture with China National Offshore Oil Corporation (CNOOC) and Tokyo Gas—will have a capacity of 8.5 Mtpa from two trains, which commenced operations in December 2014 and July 2015 respectively. Gladstone

Australian Eastern gas region to global demand has been a major factor in the growth of CSG projects, it also has altered the dynamics of the domestic gas market.

Australia's gas market comprises three regions, which are divided on the basis of the gas basins and pipelines that supply them. Of the three regions, only the *Eastern Region was unconnected to overseas markets until recently.*[9] Relying on several conventional and unconventional gas basins that contain around one third of Australia's gas reserves, the Eastern Region is an interconnected gas grid that links the eastern and southern states (South Australia, New South Wales, Victoria, and Queensland). The supply of gas at low costs within the region has benefited domestic consumers for decades. With the nascent gas export industry in Queensland, and completion of LNG projects, this gas market is undergoing major structural change.[10] The connection to overseas markets exposes domestic consumers to international competition with an increase in the gas price, as well as to the issue of the availability of gas supplies (Australian Competition and Consumer Commission 2015). Despite a significant increase in CSG production in the Eastern Region, the reduced availability of gas for the domestic market has triggered concerns about domestic energy costs. In New South Wales, which has been highly reliant on gas imports from other states, availability of gas supplies has become an issue of political significance. With an estimated 511 billion cubic metres, the State's CSG reserves could provide energy independence (New South Wales Government, Legislative Council, General Purpose Standing Committee No. 5 2012). However, the development of CSG in New South Wales has been slow due to a growing anti-CSG movement that has emerged in opposition to the industry.

Coal Seam Gas, Water, Air Pollution and Communities

Despite the prospect of significant economic benefits for the Australian community, the development of CSG has been met with mixed feelings in resource communities in Queensland and in New South Wales.[11]

LNG (GLNG)—operated by Santos in partnership with Petronas, Total, and Korea Gas Corporation (KOGAS)—will have a capacity of 7.8 Mtpa from two trains, and produced first gas in third quarter 2015. Australia Pacific LNG (APLNG)—operated by Origin Energy (upstream) and ConocoPhillips (downstream) in a venture with Sinopec—will have a capacity of 9 Mtpa from two trains, with first gas produced in final quarter 2015.

[9] Australia's other two gas regions—the Western Region, with over 50% of the country's gas reserves, and the Northern Region have focused on exports due to their connection to international markets.

[10] In the absence of a national wholesale market for natural gas in Australia, the resource is traded bilaterally via long-term contracts between producers and purchasers—energy retailers and large industrial gas users—related through pipelines that connect gas fields to consumers. However, supply and demand had been in balance and steady for several decades, based on long-term contracts that were periodically renewed.

[11] According to the New South Wales Minerals Council (2012), in 2012, mining companies contrib-

In October 2015, the Australian Government Department of Industry, Innovation and Science released its *Review of the Socioeconomic Impacts of Coal Seam Gas in Queensland* (Australian Government, Office of the Chief Economist 2015). The report highlights the main issues that have arisen in the CSG regions in Queensland, and emphasises the 'strong need for sustainable coexistence between the gas industry, local landholders and communities'. The expansion of the CSG industry has precipitated concerns about its impacts on communities along the east coast of Australia. Concerns about the environmental—and social—impacts of the industry particularly have been acute given the proximity of projects to agricultural land. The Surat and Bowen Basins historically have been agricultural areas—cropping and livestock (sheep and beef production)—and are relatively densely populated. Although some Queensland communities already were acquainted with extractive projects, the sudden surge of activities associated with the development of CSG projects has affected a large number of communities across the state. Compared to other onshore extractive projects, CSG projects are spatially dispersed due to the characteristics of coal seam geology and the wide distribution of the resource; CSG projects require a large number of production wells to recover the gas. The dispersed footprint of gas projects has exacerbated concerns about potential surface and groundwater pollution, and the broader environmental footprint. Existing literature on the topic has identified several potential issues related to the production of CSG that may trigger environmental impacts. These potential impacts have been a major source of concern among rural communities. The anti-CSG movement on Australia's east coast in part has grown because of the perception that stakeholders' concerns inadequately were being addressed by policy-makers and regulators. This is particularly the case with the extraction of water during the mining process.

Extracting natural gas from coal seams requires the removal of associated water to reduce the pressure that holds the gas in the coal seam formation. This water, also called 'produced water', commonly is a brine with a high salt content; it also may contain other chemicals that either are naturally occuring or are present as a result of hydraulic fracturing (New South Wales Government, Chief Scientist and Engineer 2013). Although not all CSG wells require hydraulic fracturing, the method has been used in certain circumstances to increase well productivity. As the process involves the injection of chemicals, concerns have arisen about the risks of underground water contamination from the fracking fluids, and their disposal once they are pumped to the surface. The amount of produced water can vary for each well. In Queensland, it is estimated that water production averages about 20,000 litres per well per day (CSIRO 2012). Given the number of CSG wells in a field, a single field can withdraw millions of litres of water each day.

---

uted $9.3 billion ($2.6 billion in wages and salaries and $6.7 billion in the purchase of goods and services and community contributions). This generated another $17.3 billion in indirect economic activity. The Hunter Region was the recipient of the largest proportion of direct expenditure from the companies surveyed ($4.6 billion), followed by Sydney ($1.8 billion), the Illawarra ($956.6 million), and the Central West ($858.4 million).

In Australia, access to groundwater is crucial given the centrality of water resources for agriculture. How the extraction of groundwater will affect the water table is one of the most frequently raised questions by communities in the vicinity of CSG projects. The lack of clarity—and scientific evidence—regarding the long-term impacts of the removal of produced water on groundwater systems, and ecosystems more broadly, has raised fears about the impacts on other water users. The disposal of produced water is another significant issue raised by rural communities, as the reuse of water is limited without treatment. Historically, untreated CSG water was disposed in evaporation ponds with high risks of leakage of saline waters into soils, aquifers, and rivers (CSIRO 2012). Under current legislation, water treatment usually is required. The treatment process raises another question about the disposal of salt residues from the produced water, which need to be managed to avoid contaminating soil and surface water (Council of Australian Governments, Standing Committee on Energy and Resources 2013). Another way of dealing with produced water is to reinject it into underground formations, which is common practice in the US. However, this method also may presents risks, particularly with regard to the link between increased seismic activity and underground water injection (Davies et al. 2013; Gibson and Sandiford 2013; Vaneckova and Bambrick 2014).

Another cause of community concern relates to air emissions associated with CSG developments, particularly, the potential impact on the environment and public health.[12] Given methane is a more powerful greenhouse gas than CO_2, its potential for global warming is significant—approximately 21 times that of carbon dioxide. Methane emissions and other atmospheric pollutants arising from CSG production increasingly have been the subject of scrutiny by the scientific community and the public. The amount of fugitive emissions across the life-cycle of CSG projects continues to be a subject of debate—as well as the accuracy with which fugitive emissions are measured. Accordingly, scientific data on the greenhouse gas impact of unconventional gas developments (New South Wales Government, Chief Scientist and Engineer 2013), as well as the potential health impacts, are the subject of controversy. Although there are limited data suggesting some increased health risk, the possible impact on human health has received considerable attention in Australia and overseas (Vaneckova and Bambrick 2014). While scientific inquiries into the CSG industry have highlighted some of the risks to environmental health due to exposure to CSG operations (New South Wales Government, Chief Scientist and Engineer 2014), uncertainty about environmental health impacts—notably about the cumulative impacts of exposure to methane and other chemicals on human health and climate change—has fuelled angst in resource communities. On the east coast of Australia, video footage of methane-leaking bores being set alight have had a significant impact on communities' perceptions of CSG developments. There additionally remains a high level of scientific uncertainty about the empirical risks and impacts associated with CSG production, particularly regarding long-term and cumulative impacts. The peak scientific body in Australia, the CSIRO (2012), sug-

[12]The US documentary, *Gasland*, has had a profound impact on communities' perceptions of the unconventional gas industry.

gests that groundwater impacts may not become evident immediately and that it may take decades before the consequences of CSG developments fully are understood.

Poor well integrity is a major cause of hydraulic connectivity between otherwise isolated aquifers with different water qualities, which potentially can lead to the contamination of subsurface water and the migration of methane and/or other pollutants into surrounding aquifers, wells and bores, and the surface. While the integrity of unconventional gas wells therefore is essential to minimising the possibility of gas or other fluids leaking from a well and interfering with groundwater resources, the lack of scientific knowledge about underground water systems and connectivity remains a cause of concern (Council of Australian Government's Standing Committee on Energy and Resources 2013).

Regulating the Coal Seam Gas Industry

In Australia, the states and territories have primary responsibility for regulating mining and petroleum activities, including setting licence conditions for mineral and petroleum exploration and development, assessing environmental impacts, and monitoring and enforcing industry compliance, as well as collecting royalties. The Commonwealth's jurisdiction applies only in limited circumstances, for example, for matters of national environmental significance such as the *Environment Protection and Biodiversity Conservation Act 1999 [Cth.]*. As the industry expanded in the early-2000s, the existing regulatory frameworks for petroleum activities were applied to CSG projects. A diversity of laws and regulation still govern CSG activity across the country. However, critics have pointed to the inherent limitations of existing frameworks to deal with the scale, impacts, and new risks associated with CSG projects. While the industry would point toward an excessively burdensome and costly regulatory regime for a nascent industry, grassroots and environmental organisations denounce the lack of effective regulatory capacity to manage the new risks posed by CSG development (Environmental Defender's Office NSW 2011). This ongoing debate over CSG reflects a highly polarised environment with, on the one hand, the view that extractive development is central to the Australia's economic growth—and should be supported with more streamlined regulatory governance and, on the other hand, a strongly held belief that stronger regulation is needed to ensure priority is given to ecologically sustainable development. The rapid expansion of the industry from the mid-2000s also coincidences with a shift in the resource governance landscape. The spread of factual and scientific information on the environmental and socio-economic impacts of the mining industry through social media has transformed the public debate about resource development, connecting grassroots movements across the globe and placing local communities as important governance actors. The emergence of new norms and actors has challenged the traditional governance structures of the mining industry , and increased demands from communities for more socially responsible and ecologically sustainable practices.

In this context, the public, and the extractive communities, have sought reassurance and evidence that CSG development can occur without negative environment impacts (Lloyd et al. 2013). Social and political pressure has stopped the development of unconventional gas in some countries. Bans and moratoria are in place in several jurisdictions for shale or CSG development, including in Australia (Victoria), France, and the US (New York). However, other jurisdictions generally have proceeded with a more pragmatic approach involving scientific inquiries and changes to their regulatory frameworks to enhance mechanisms for environmental monitoring and compliance.

In Australia, while the regulation of resource development falls under state and territory jurisidictions, the Australian government showed interest in supporting states and territories with *developing a governance framework for unconventional gas development as early as 2012.* The Council of Australian Governments' (COAG) National Partnership Agreement on Coal Seam Gas and Large Coal Mining Developments was established in March 2012 between the Commonwealth, New South Wales, South Australian, Victorian, Queensland and Northern Territory Governments with the objective of strengthening the regulation of CSG and large coal mining developments by improving scientific knowledge and ensuring that future government decisions on development proposals are informed by scientific advice. In November 2012, the Australian Government established an Independent Expert Scientific Committee on Coal Seam Gas and Large Coal Mining Development (IESC) under the *Environment Protection and Biodiversity Conservation Act 1999* (Cth) to provide advice to Australian governments on the impact the industries may have on water resources. One of the principal tasks of the IESC is to address gaps in knowledge. Accordingly, the Committee has been given the responsibility to carry out 'bio-regional assessments' in priority regions to gather environmental data about the relationships between ground and surface water. Further responding to concerns about the impacts of the industries on water resources, the Australian Parliament passed an amendment to the Environment Protection and Biodiversity Conservation Act in June 2013, which made water resources a matter of national environmental significance in relation to CSG and large coal mining developments (Australian Government, Department of Environment 2014). This means that a separate process from state regulation, requiring Commonwealth assessment and approval, can be triggered for CSG projects if they have a 'significant impact' on water resources. The Council of Australian Government's Standing Council on Energy and Resources (2013) simultaneously released the *National Harmonised Regulatory Framework for Natural Gas from Coal Seams*, which establishes eighteen leading practices to mitigate potential impacts in relation to four key areas associated with CSG developments: well integrity, water management and monitoring, hydraulic fracturing, and the use of chemicals. Based on the premise that CSG is 'an important component of eastern Australia's domestic gas supply' and that the industry will deliver export income to Australia's economy, the Framework provides guidance to regulators about developing more effective regulatory regimes across all Australian jurisdictions for the management of CSG projects.

In Australia, the Crown owns the mineral and petroleum resources. State and territory governments, on behalf of the Crown, may grant licences to companies to explore for and extract resources. Given CSG deposits predominantly are located under agricultural land, the issue of coexistence with other land users has presented challenges. Project proponents are required to negotiate agreements with individual landholders to access their land and provide compensation for the disruption to, and impacts on, their land. The land access system in Australia does not provide landowners with a legal right to refuse access to their land. This legal situation has exacerbated concerns among farmers, including: not being able to refuse access to companies; the impacts on their assets and lifestyle; the extraction activities; and land productivity. The surface footprint of CSG infrastructure also has been the subject of controversy given the large footprint of CSG operations and the scale of activities related to well drilling during the development phase (CSIRO 2012). Despite governments and industry claiming that agriculture and CSG can coexist, current practice has not fully convinced landowners and the broader community that this is the case. Opponents of the industry claim CSG projects are developing faster than the science, and regulatory frameworks provide insufficient time for scientific evidence to demonstrate the long-term impacts on underground water systems. Although the extraction of CSG generally does not require hydraulic fracturing, the characteristics of the geology may require further stimulation to increase gas flow and, therefore, fracking may be used for some CSG projects.

Given the proximity of CSG projects to urban and agricultural areas, the issue of coexistence between different land use(r)s has become critical in New South Wales and Queensland. The *Review of the socioeconomic impacts of coal seam gas in Queensland* (Australian Government, Department of Industry, Inovation and Science, Office of the Chief Economist 2015) notes that as economic and social changes occur due to CSG activities, there is a role for governments to play in ensuring mechanisms are in place to support coexistence, for example, with regard to land access and the payment of compensation to landholders, the protection of agricultural land, or sharing benefits among communities.

The *National Harmonised Regulatory Framework for Natural Gas from Coal Seams*, recognises that Australia cannot reap the benefits from CSG activity if the industry's social licence is not established and maintained. This is based on the view that the CSG industry and agriculture can coexist if the regulatory settings it details are implemented. The Framework recommends that 'regulatory and legislative settings should be underpinned by the principle of co-existence' where 'a shared commitment exists between the resources industry, other land users, local communities and governments to multiple, merit-based and sequential land use that provides certainty for industry and improved community confidence in land use decision-making' (Australian Governments, Standing Council on Energy and Resources 2013).

In Queensland and New South Wales, the challenges that have accompanied CSG development have triggered reforms of the regulatory and policy frameworks—and the development of new institutions—in an attempt to better address issues of coexistence.

The Government of New South Wales established an 18-month scientific inquiry into the risks and impacts of the CSG industry led by the State's Chief Scientist and Engineer. The report of the Chief Scientist concluded that the risks associated with CSG development effectively could be managed under appropriate regulation, and recommended the government commit to adopting a vigilant, transparent, and effective regulatory and monitoring system to ensure highest standards of compliance and performance (New South Wales Government, Chief Scientist and Engineer 2014). Following the inquiry, the Government of New South Wales released its *Gas Plan*, which committed the government to enhancing transparency and control in the allocation of titles for resource exploration with the community interest as its prime objective (NSW Government n.d.). The regulatory reforms that implement the *Gas Plan* will establish a new framework for a fair, transparent, and balanced approach to land use (New South Wales Government, Department of Industry, Resources and Energy n.d.), with the aim to ensure that the development of the State's CSG resources benefits the community. The establishment of a *Strategic Release Framework* for the granting of various mining and petroleum prospecting titles introduces new processes for the designation of certain areas of the State for resource exploration.[13] Seeking to bring the issue of coexistence to the forefront, the *Strategic Release Framework* is based on the preliminary assessments of resources, as well as the economic, social, and environmental issues within a region. It intends to provide more clarity and transparency in the allocation of coal and petroleum prospecting titles, enhanced control for the government, and preliminary consultations with communities for the allocation of land for resource exploration. A Gas Community Benefits Fund is being established that will require contributions from titleholders to the communities in which they operate; an approach that provides for the financial benefits from gas exploration and production to be shared with landowners and communities.

Much of the academic debate on the impacts and sustainability of the mining industry seems to have focused on the responsibility of project proponents to enhance their environmental and social performance to meet communities' expectations. However, a recent trend of the regulatory reforms—as exemplified in New South Wales—sees issues of coexistence between the proponents of extractive developments and communities placed under the regulatory realm, with specific provisions being developed to address land access, community relations, and the share of the benefits. This is a positive development.

Conclusion

The development of natural gas has been touted as a key element in the transition toward a cleaner energy future. With the advancement of technologies and recovery techniques, reserves and supplies of gas are increasing globally, with the potential recovery of gas from unconventional fields. However, this positive scenario is tainted by a number of issues, in particular, the growing gap between communities and the proponents of developments. Although natural gas is the lowest emission fossil fuel,

[13] For further information, see: New South Wales Government, *Mining and Petroleum Legislation Amendment (Grant of Coal and Petroleum Prospecting Titles) Bill 2015* (NSW).

the merits of natural gas in relation to the goals of climate change mitigation continues to be debated, most notably in the case of unconventional gas. It also has been suggested that the current focus on natural gas development might impede progress toward the ultimate goal of large-scale deployment of renewable energies.

Given the uncertainties around CO_2 emissions and the long-term environmental impacts of unconventional gas development, the new industry has proven highly controversial. This is best exemplified in the Australian context, where an anti-CSG movement has grown in reaction to the rapid expansion of the onshore gas industry. Although CSG features as a priority for Australian governments that foresee the potential economic benefits from developing the resources, acknowledgement of the risks has slowed public acceptance and opened a debate on the adequacy of the governance frameworks for resource development.

The recent legislative amendments in Queensland and New South Wales characterise a reactive approach by governments to the risks caused by CSG development. They also reflect careful thinking from governments aware that community trust has become central to the success of extractive projects. Although the concept of the 'social licence to operate' has become accepted in the CSG industry, the recent regulatory reforms in resource-rich jurisdictions also sees issues relating to stakeholder relations, land access, and community well-being being placed within the regulatory realm.

At a time when the contemporary debate in Australia around resource development, and particularly unconventional gas, becomes so polarised, new governance models need to be considered to involve stakeholders in defining priorities for energy use, and the associated policies of land use and energy resource development. An enhanced dialogue with the community is required for a better understanding of the potential roles of natural gas in enhancing energy diversity, economic prosperity, and climate change mitigation—as well as the risks involved. Much of the current discourse has focused on either natural gas or renewable energy as discrete components of the world's energy systems. While focusing on the competitive impacts of one over the other, the policy—and public—debate has missed the opportunity to concentrate on the potential synergies between the two energy sources to create synergies based on their complementarities. A coordinated, strategic, and holistic approach between relevant institutions and levels of government should be developed to enable consideration and integration of multiple perspectives and objectives, and to ensure that social and environmental values are placed at the core of energy and resource-development policies in order to balance environmental, social, and economic outcomes for land use and energy resource development.

References

Australian Government, Department of the Environment. (1999). The environmental protection and biodiversity conservation act, 1999. http://environment.gov.au/epbc/. Accessed 12 July 2015.

Australian Competition and Consumer Commission. (2015). East Coast gas inquiry: issues paper. https://www.accc.gov.au/system/files/Eastpercent20Coastpercent20Gasper cent20Inquiryper-cent 20-percent20Issuespercent20paper.pdf. Accessed 12 July 2015.

Australian Government, Department of Environment. (2014). Hydraulic fracturing ('fraccing') techniques, including reporting requirements and governance arrangements. http://www.resourcesandenergy.nsw.gov.au/landholders-and-community/coal-seam-gas/facts-maps-links/independent-research-on-csg. Accessed 20 July 2015

Australian Government, Department of Industry, Innovation and Science. (2015). Review of the socioeconomic impacts of coal seam gas in Queensland. http://www.indutry.gov.au/Office-of-the-Chief-Economist/Publications/Documents/coal-seam-gas/Socioeconomic-impacts-of-coal-seam-gas-in-Queensland.pdf. Accessed 28 September 2015.

Australian Government, Geosciences Australia. (n.d.). Coal factsheet. Australian atlas of minerals resources, mines and process centres. http://www.australianminesatlas.gov.au/education/fact_sheets/coal.html. Accessed 28 September 2015.

Council of Australian Governments, Standing Council on Energy and Resources. (2013). The national harmonised regulatory framework for natural gas from coal seams 2013. http://www.scer.gov.au/sites/prod.energycouncil/files/publications/documents/National-Harmonised-Regulatory-Framework-for-Natural-Gas-from-Coal-Seams_1.pdf. Accessed 16 March 2015.

CSIRO. (2012). Coal seam gas extraction and co-produced water. http://www.iesc.environment.gov.au/system/files/resources/3b6a1b5a-dde5-4c24-9b16-5520321c2765/files/fact-sheet-coal-seam-gas-extraction-co-produced-water.pdf. Accessed 4 July 2015.

Davies, R., Foulger, G., Bindley, A., Styles, P. (2013). Induced seismicity and hydraulic fracturing for the recovery of hydrocarbons. Marine and Petroleum Geology, 45, 171-185.

Day, S., Connell, L., Etheridge, D., et al. (2012). Greenhouse gas emissions from coal seam gas production in Australia. CSIRO, October.

Energy Information Administration. (2015a). EIA's annual energy outlook 2015. http://www.eia.gov/forecasts/aeo/section_energyprod.cfm. Accessed 5 December 2015.

Energy Information Administration. (2015b). EIA's annual energy outlook 2015: Australia. http://www.eia.gov/beta/international/analysis.cfm?iso=AUS. Accessed 5 December 2015.

Energy Information Administration. (n.d.). Australia. https://www.eia.gov/beta/international/analysis_includes/countries_long/Australia/australia.pdf. Accessed 5 December 2015.

Environmental Defenders Office, NSW. (2011). Ticking the box: flaws in the environmental assessment of coal seam gas exploration activities. http://d3n8a8pro7vhmx.cloudfront.net/edonsw/pages/291/attachments/original/1380666573/ticking_the_box.pdf?1380666573. Accessed 5 April 2014.

Gibson, G., Sandiford, M., (2013). Seismicity & induced earthquakes. http://www.chiefscientist.nsw.gov.au/__data/assets/pdf_file/0017/31616/Seismicity-and-induced-earthquakes_Gibson-and-Sandiford.pdf. Accessed 6 March 2015.

Howarth, R., Santoro, R., Ingraffea, A. (2011). Methane and the greenhouse-gas footprint of natural gas from shale formations. http://www.acsf.cornell.edu/Assets/ACSF/docs/attachments/Howarth-EtAl-2011.pdf. Accessed 31 September 2014.

Lloyd, D., Luke, H., Boyd, W. (2013). Community perspectives of natural resource extraction: coal-seam gas mining and social identity in Eastern Australia. Coolabah, No. 10.

New South Wales Government. (2015). The Mining and Petroleum Legislation Amendment (Grant of Coal and Petroleum Prospecting Titles) Bill. https://www.parliament.nsw.gov.au/bills/DBAssets/bills/SecondReadSpeechLA/675/2Rpercent20Miningpercent20andpercent20cognatespercent20(5).pdf. Accessed 14 December 2015.

New South Wales Government, Chief Scientist and Engineer. (2014). Final report: September. http://www.chiefscientist.nsw.gov.au/reports/coal-seam-gas-review/final-report-september-2014. Accessed 12 October 2015.

New South Wales Government, Chief Scientist and Engineer. (2013). Initial report on the independent review of coal seam gas activities in NSW. http://www.chiefscientist.nsw.gov.au/__data/assets/pdf_file/0016/31246/130730_1046_CSE-CSG-July-report.pdf. Accessed 12 October 2015.

New South Wales Government Legislative Council. (2011). General Purpose Standing Committee No. 5. Inquiry into coal seam gas. http://www.nswic.org.au/pdf//Memberspercent20only/2011/NLper cent20-per cent20CSGper cent20Transcript.pdf. Accessed 5 March 2014.

New South Wales Government. (n.d.). Mining exploration and mining. http://www.resourcesandenergy.nsw.gov.au/miners-and-explorers. Accessed 15 March 2015.

New South Wales Mineral Council. (2012). NSW mining 2012: a snapshot. http://www.nswmining.com.au/NSWMining/media/NSW-Mining/Publications/NSW-Mining-2012-A-Snapshot-FINAL.pdf. Accessed 12 October 2015.

Queensland Government, Department of Natural Resources and Mines. (2015). Queensland's petroleum and coal seam gas. https://www.business.qld.gov.au/invest/investing-queenslands-industries/mining/resources-potential/petroleum-gas. Accessed 3 January 2016.

Sims, R. (2015). Keynote address: inquiry into Australia's East Coast gas market. https://www.accc.gov.au/speech/keynote-address-inquiry-into-australiapercentE2percent80percent99s-east-coast-gas-market. Accessed 12 October 2015.

Vaneckova, P., Bambrick, H. (2014). Approaches to baseline studies of human health in relation to industries with potential environmental impact. http://www.chiefscientist.nsw.gov.au/__data/assets/pdf_file/0005/56894/140903_Human-CSG_completed_report.pdf. Accessed 15 June 2015

World Watch Institute. (2011). Comparing life-cycle greenhouse gas emissions from natural gas and coal. https://www.db.com/cr/en/docs/Natural_Gas_LCA_Update_082511. Accessed 26 February 2014.

Part III
The Major Issues

Chapter 15
Mining and Sustainable Development

Glen Corder

Abstract This aim of this chapter is to introduce the reader to the key elements of sustainable development. To achieve this aim, the chapter is subdivided into the following sections. Section one offers a short history of sustainable development in the mining industry. In the following section, an overview of sustainable development principles and frameworks are examined. The following three sections examine the relationship between a social licence to operate and sustainable development, and issues surrounding implementation and measurement. In the penultimate section, a case study of the Philippines is covered. The paper concludes that sustainable development is becoming increasingly central to the future of the mining industry in the region.

Abbreviations

4QA	Four Quadrant Analysis
ASM	Artisanal and small-scale Mining
GMI	Global Mining Initiative
GRI	Global Reporting Initiative
ICMM	International Council on Mining and Metals
IFC	International Finance Corporation
MMSD	Mining, Minerals and Sustainable Development
MMSS	Mining and Metals Sector Supplement
PNG	Papua New Guinea
SOHO	Sustainability Opportunities and Hazards Overview
SPeAR	Sustainable Project Appraisal Routine
UN	United Nations

G. Corder (✉)
Centre for Social Responsibility in Mining, University of Queensland,
St Lucia, QLD 4072, Australia
e-mail: g.corder@uq.edu.au

T. O'Callaghan, G. Graetz (eds.), *Mining in the Asia-Pacific*, The Political Economy of the Asia Pacific, DOI 10.1007/978-3-319-61395-6_15

Introduction

The concept of sustainable development has emerged in the mining industry over the past three decades. Although sustainable development can cover a wide range of definitions and interpretations, it is viewed here as responsible development and growth, rather than one that refers to sustaining the industry itself. While I employ the term *sustainable development* in this chapter, the term *sustainability* is also used within the mining industry. It is also commonplace to see it used in other sectors as well, especially in the forestry and chemical industries. The difference between the two terms is debatable, but here 'sustainability' is regarded as the end goal of 'sustainable development' (Dunphy et al. 2000). Based on this distinction, the contemporary mining industry should focus on the sustainable development of the industry, as its contribution to the end goal is the sustainability of the planet's resources.

Since the industry began grappling with the concept of sustainable development, its perception has evolved from a broadly philosophical approach, promoted by senior corporate managers, to relevant initiatives at an operational level. Moreover, the motive for embracing sustainable development has changed and matured with time: from a philanthropic approach (it is the right thing to do) to a sound business strategy (shared value between industry and its stakeholders), although different companies and different parts of the industry are at various stages on this journey.

The most common definition of sustainable development is the Brundtland definition: 'Sustainable development is development that meets the needs of the present without compromising the ability of future generations to meet their own needs' (United Nations 1987). This high-level definition of sustainable development is helpful from a global—and even a corporate—perspective. Yet for those working within the industry the definition generally is considered too abstract and removed from their day-to-day activities to be useful. Other definitions of sustainable development are similar in intent and do not effectively bridge this gap. Consequently, the mining industry over several years has attempted to develop frameworks and tools that will help deliver outcomes that support the aims of sustainable development and which can be measured with various indicators.

Approaches for generating sound and long-lasting sustainable development credentials are dependent on the region(s) in which the mining operations are situated. For instance, mining operations that are situated in dry, sparse, and remote regions have to deal with different sustainability issues compared with operations that are situated in wet, bio-diverse regions, surrounded by local communities. This contextual element means that solutions to address sustainable development issues need to be customised from site-to-site, requiring various degrees of innovation rather than a 'one size fits all' approach. Balancing these differing aspects for site-to-site, region-to-region, and country-to-country contexts requires multi-disciplinary and holistic approaches, which draw together people with technical, environmental, social, and financial expertise.

This aim of this chapter is to introduce the reader to the key elements of sustainable development. To achieve this aim, the chapter is subdivided into the following sections: (1) a short history of sustainable development in the mining industry, (2) an overview of sustainable development principles and frameworks, (3) the relationship between a social licence to operate and sustainable development, (4) implementation of sustainable development, (5) measuring sustainable development, (6) a case study of mining activity in the Philippines, and, (7) the increasing importance of sustainable development for the mining industry.

Short History of Sustainable Development in the Mining Industry

While the ideals of sustainable development have been broadly articulated, promoted and applied by various sectors of society for at least half a century, the terms *sustainable development* and *sustainability* only really have gained significant traction with industrial organisations over the last quarter of a century. Seminal documents include:

- The Club of Rome's *Limits to Growth* (Meadows and The Club of Rome 1972),
- *Our Common Future* (often referred to as the Brundtland Report),
- International Union for the Conservation of Nature, and
- The United Nations World Charter for Nature.

These, and other documents, laid the foundations for a broader acceptance of sustainable development across civil society, government, academia, and industry. In particular, *Our Common Future* was a critical trigger for the 1992 Earth Summit, formally known as the United Nations Conference on Environment and Development, held in Rio de Janeiro, Brazil. This conference produced the *Rio Declaration on the Environment and Development* (UNEP 1992) and *Agenda 21* (United Nations 1992), both of which aimed at fostering and guiding sustainable development across the globe (Dresner 2008).

These initiatives, and their respective outcomes, were instrumental in the mining industry's embrace of sustainable development principles. A milestone in heightening the acceptance of sustainable development in the mining industry was the Global Mining Initiative (GMI). The GMI was borne from a recognition by several mining industry Chief Executives that the industry needed to rethink the way in which it argued its case, and that sustainable development was an appropriate vehicle to achieve this (Littlewood 2000). The GMI led to a major study of mining and sustainability: *Mining, Minerals and Sustainable Development* (MMSD) (Littlewood 2000). Over forty companies and organisations contributed to the GMI and MMSD, with the aim of identifying the main challenges and possible strategies to foster a more sustainable future for the industry (Azapagic 2004). The MMSD offered an independent review of how the mining industry performed in relation to broad-ranging sustainable development issues and produced a final report called *Breaking New Ground*. This drew on the MMSD's two-year process of consultation and research. At the time of publication in 2002, it described the industry's activities through the concept of sustainable development, and offered a detailed plan for immediate action on the environment (IIED 2002).

A key development catalysed by the GMI and the MMSD project was the creation of the International Council on Mining and Metals (ICMM), which in fact was a transformation of the International Council on Metals and the Environment through the broadening of the antecedent organisation's mandate. At the conclusion of the GMI, a major global conference, *Resourcing the Future*, was held in Toronto in 2002, and the ICMM member companies signed the *Toronto Declaration* committing the ICMM to continue the work started by the MMSD project and to engage in constructive dialogue with key stakeholders (ICMM 2015c).

Since the early part of this century, there has been a wide range of organisations assisting and/or working with the industry in operating and growing in a responsible and sustainable manner. In 2016, the ICMM has 23 mining and metals companies and 34 national and regional mining and commodity associations focused on addressing core sustainable development challenges, and is now an important institution for sustainable development in the mining industry (ICMM 2015a). Other national and industry bodies (for example, the Minerals Council of Australia and the World Gold Council), non-government organisations (for example, Oxfam), and research institutions (for example, the Sustainable Minerals Institute at The University of Queensland) are helping to deliver sustainable development outcomes across the mining industry.

Sustainable Development Principles and Frameworks

Sustainable development principles and frameworks have provided the catalyst for driving better overall industry performance. An important step for the ICMM in addressing the aforementioned challenges at the conclusion of the GMI was the adoption of ten guiding principles, which would become the first element of the ICMM's Sustainable Development Framework (Table 15.1).

The ten principles, and each principle's supporting statements, form the cornerstone of the ICMM's Sustainable Development Framework. The principles are well accepted by mining companies, and often corporate-level policies are aligned with the principles.

While the ICMM's work on sustainable development has been the flagship of the mining industry, other sets of principles, frameworks and standards have influenced the sustainable development performance of the industry. These include: *The Equator Principles*, *the United Nations (UN) Global Compact*, the International Finance Corporation's (IFC) *Performance Standards*, *the Dow Jones Sustainability World Index*, and the *Global Reporting Initiative* (GRI). A brief summary of each of these is presented in Table 15.2

Table 15.1 The ICMM's 10 guiding principles

1	Implement and maintain ethical business practices and sound systems of corporate governance
2	Integrate sustainable development considerations within the corporate decision-making process
3	Uphold fundamental human rights and respect cultures, customs and values in dealings with employees and others who are affected by our activities
4	Implement risk management strategies based on valid data and sound science
5	Seek continual improvement of our health and safety performance
6	Seek continual improvement of our environmental performance
7	Contribute to conservation of biodiversity and integrated approaches to land use planning
8	Facilitate and encourage responsible product design, use, reuse, recycling and disposal of our products
9	Contribute to the social, economic and institutional development of the communities in which we operate
10	Implement effective and transparent engagement, communication and independently verified reporting arrangements with our stakeholders

Source: ICMM (2016)

Table 15.2 Relevant sustainability principles, frameworks, and standards

The Equator Principles	Risk management framework adopted by financial institutions for determining, assessing, and managing environmental and social risk in projects
The UN Global Compact	A strategic policy initiative for businesses committed to aligning their operations and strategies with ten universally accepted principles in the areas of human rights, labour, environment, and anti-corruption
IFC Performance Standards	Defines IFC clients' responsibilities for managing their environmental and social risks
The Dow Jones Sustainability World Index	Tracks the stock performance of the world's leading companies in terms of economic, environmental, and social criteria
The Global Reporting Initiative	A framework providing metrics and methods for measuring and reporting sustainability-related impacts and performance within organisations

There are also examples of sustainable frameworks that, although applicable to the mining industry across the world, have been developed by regional organisations. For example, the Minerals Council of Australia developed the framework, *Enduring Value*, which builds on the Australian Minerals Industry Code for Environmental Management, to provide implementation guidance on the ICMM's ten Principles and their application at the operational level (MCA 2005). Another example, which emerged from North American regional activity associated with the MMSD, is the *Seven Questions to Sustainability* framework. The aim of this framework is to provide a guide for the assessment of whether or not the net contribution of a project to sustainability is positive over the long-term (International Institute for Sustainable Development et al. 2002).

Table 15.3 Features of the Five Capitals

Natural Capital	Refers to the natural resources (matter and energy) and processes that produce and deliver goods and services. They include renewable and non-renewable resources, sinks that absorb, neutralise or recycle wastes, and processes, such as climate regulation, which maintain life. Natural capital forms the foundation for all other capitals
Human Capital	Consists of people's health, knowledge, skills, motivation and capacity For relationships. These facets are required for productive work, and the creation of a better quality of life. Human capital can be fostered through improving opportunities for learning, creativity, stimulation and enhanced health
Social Capital	Concerns the institutions that help societies maintain and develop human capital in partnership with others. It includes such institutions as families, communities, businesses, trade unions, schools, and voluntary organisations. A critical component of social capital is the development of trust
Manufactured Capital	Comprises material goods, or fixed assets, that contribute to the production process or the provision of services, rather than being part of the output itself. It includes, for example, tools, machinery, buildings and infrastructure
Financial Capital	Plays a critical role in our economy, enabling the other types of capital to be owned and traded, for example, through shares, bonds or banknotes. Financial capital is the traditional primary measure—the 'single bottom line' of business performance and success

More recently, Moran and Kunz (2014) developed a hierarchical framework to assess the progress of mining, minerals and energy supply and demand networks toward sustainable development. The framework begins at the global level and ends with the unit level. The novelty of this framework lies in the phrase 'operating sustainably', rather than simply 'sustainable mining', as 'it is more inclusive that mining *per se* and addresses more of the value chain and commodity use and re-use' (Moran and Kunz 2014).

A more general sustainable development framework that has gained a growing degree of traction in the mining industry is the Five Capitals Model. It provides a helpful basis for conceptualising sustainability/sustainable development and uses the concept of capitals—natural, human, social, manufactured and financial—to encompass the dimensions of sustainability (Forum for the Future n.d.). An important understanding is that financial capital is a function of the other four capitals and has no intrinsic value on its own, but rather is a mechanism for trade between the other capitals. Descriptions of each capital are provided in Table 15.3.

The Five Capitals model can be extended to include a sixth capital, Intellectual Capital, which comprises organisational, knowledge-based intangibles, including intellectual property—patents, copyrights, software, rights, and licences (Corder et al. 2014)—with this model having now been adopted by some mining companies (Exxaro.com 2015).

Social Licence to Operate

While sustainable development principles and frameworks typically are focussed on the developers and operators of mining projects, the concept of the 'Social Licence to Operate' is associated with stakeholders' acceptance of, or acquiescence in, mining activities. The term has become entrenched in the language of the mining industry, even though as a concept it can have a range of interpretations, which often are dependent on one's perspective. From a sustainability perspective, it is a highly relevant concept, as it reflects the relationship and influence that critical stakeholders have with mining operations, projects, and companies. While the term, 'licence to operate', means, in general, acquiring the necessary approvals from the relevant regulatory authorities and is clearly defined in one or more formal outputs, social licence to operate is less tangible. However, it does encompass local, regional, and national interests associated with mining developments, and if stakeholders believe that these interests are compromised, mining operations can lose their social licence (Corder et al. 2014). An operation could consider to have a social licence to operate when it achieves ongoing acceptance or approval from the local community and other stakeholders who can affect its profitability; without this social licence, it is difficult for a mine to operate effectively or profitably.

The connections between the concepts of social licence to operate and sustainable development are palpable. Most of the sustainable development principles presented earlier relate to the benefits and impacts on affected stakeholders who, directly or indirectly, influence social licence. In fact, personnel working for the mining industry either have viewed sustainable development as interchangeable with social licence to operate, or as an umbrella under which social licence to operate is described as the product of social sustainable development (Lacey et al. 2012).

The growing importance of the concept of the social licence to operate has been well documented by scholars studying the social aspects of the mining industry (Owen and Kemp 2013; Prno 2013; Prno and Slocombe 2012; Solomon et al. 2008). The social dimensions of the mining industry now are becoming critical to business success, yet in many ways these aspects are the most challenging in terms of the business concept of sustainable development (Solomon et al. 2008). The traditional approaches to extractive developments no longer suffice for local communities and affected stakeholders. They are now demanding a greater share of benefits and increased involvement in decision-making, with these demands aligning closely with the sustainable development principles and policies espoused by the mining industry.

In a similar vein, Prno and Slocombe (2012) concluded that it is widely recognised that mining companies need to gain a social licence to operate from local communities in order to avoid potentially costly conflict and exposure to social risks. In a later article, Prno (2013) conducted a comparative case study of four international mining operations: Red Dog Mine in Alaska; Minto Mine in the Yukon, Canada; the proposed Tambogrande Mine in Peru; and the Ok Tedi Mine in Papua New Guinea (PNG)—in an effort to identify the key determinants of social licence to operate outcomes in the mining industry. Five lessons for earning a social licence to operate emerged from this analysis: (1) context is key, (2) a social licence to operate is built on relationships, (3) sustainability is a dominant concern for communities, (4) local benefits provision and public participation play a crucial role, and, (5) adaptability is needed to confront complexity. Again these lessons align closely to core principles of sustainable development discussed earlier in this chapter.

Owen and Kemp (2013) claim that, while the social licence to operate as a concept has highlighted in a positive fashion the profile of social issues within the industry, it has not been able to articulate a collaborative developmental agenda or a pathway forward for restoring the lost confidence of affected communities, stakeholders, and pressure groups. They argue that the industry needs a less defensive and more constructive approach to stakeholder engagement and collaboration and, as a first step, should reconcile its internal risk-orientation with the expectations of external stakeholders. This exemplifies the potential for a breakdown in communication between the industry and affected stakeholders, due to the inability collectively to adopt a common language for developing a shared agenda.

These findings show that social licence to operate is inextricably linked to sustainable development. While the concept of the social licence to operate is not easily defined and is even harder to acquire, it is not possible to imagine that an individual mining operation, a new mining project, or a mining company will be able to gain and maintain a social licence to operate without meeting the ideals of sustainable development. In addition, a mining operation, a new project, or a mining company needs a social licence to operate to ensure a long-lasting and successful business to reduce the likelihood of project delays and/or production stoppages, which quickly and seriously can affect revenue and, accordingly, profits. What this means is that setting and delivering on a sound and thorough sustainable development agenda is critical for business success in the mining industry today. Key to this success is the idea that sustainable development is based on mutual benefit and robust relationships between the industry and affected stakeholders, as opposed to a solely philanthropic approach, typified by one-off financial payments for institutions, such as hospitals, schools, and public amenities and services.

Implementing Sustainable Development

An important challenge for the mining industry concerns how to implement systematically and rigorously initiatives that satisfy the core aims of sustainable development and, in so doing, achieve a social licence to operate. Since the turn of the century, and in some cases earlier, a range of approaches have been developed by industry that aim to incorporate into, and/or assess, sustainability aspects of the project design process and management systems. Several engineering companies and consultancy firms have developed their own tools and methodologies for assessing or incorporating sustainability into their projects (Table 15.4).

Table 15.4 A sample of sustainability tools utilised by industry

Tool name	Description
Sustainable Project Appraisal Routine (SPeAR®)—ARUP	SPeAR®, originally developed for the built-environment and infrastructure project business sectors, but which now has been applied more widely, appraises projects based on key themes (e.g., transport, biodiversity, culture, employment, and skills) and utilises a traffic light system graphically to indicate performance in each area (ARUP 2013)
GoldSET—Golder Associates	GoldSET is a set of web-based tools to evaluate alternatives, or to monitor on-going projects based on geospatial information management, forecasted project performance, and utilising multi-criteria analysis (Golder Associates 2014)
Four-quadrant analysis (4QA), sustainability opportunities and hazards overview (SOHO), FutureWatch™—Hatch Associates	A suite of tools, utilising a range of workshop programs, for identifying opportunities and hazards, as well as assessing sustainability performance using contemporary thinking around sustainability (Medveçka and Bangerter 2007)
EcoNomics™—WorleyParsons	EcoNomics™ comprises three components—sustainable decisions, sustainable project delivery, and sustainable operations—as well as a carbon management service, Carbon EcoNomics™, and is aimed at enhancing risk management and improving sustainability performance across the asset life-cycle (WorleyParsons 2014)

Characteristics of the above approaches include the comparison of different sustainability impacts through schematic diagrams, sustainability opportunities and threats analyses, and economic valuation of sustainability impacts. In addition, individual mining companies also have produced their own tools to develop better sustainability outcomes.[1] Moreover, the ICMM has produced a range toolkits to assist with the implementation of good sustainability practice, covering community development, materials stewardship, partnerships and mine closure planning (ICMM, 2015b). For example, the Community Development Toolkit comprises a set of 20 tools providing guidance on the community development process, from exploration, through to closure and maintenance (ICMM 2012).

Furthermore, the Australian government has supported the development of the *Leading Practice Sustainable Development Program for the Mining Industry*, which aims to promote sustainable development and industry self-regulation through the adoption of leading practice principles. The outputs of the program, which are in the public domain, comprise 15 handbooks relating to sustainable development topics, such as biodiversity management, community engagement and development, mine closure and maintenance, stewardship, water management) in the mining industry, as well as a companion publication to the handbooks called the

[1] For example, Anglo American's Socio-Economic Assessment Toolbox (SEAT) (Anglo American 2014) and the company's Mine closure Toolbox (Anglo American 2013).

Social Responsibility in the Mining and Metals Sector in Developing Countries (Australian Government 2015).

Although there is a range of tools and methodologies that attempt to incorporate sustainability concepts at the design and operational level of mining projects, McLellan et al. (2009) have argued that there is no consistent, integrated approach to support the mining industry in incorporating a greater level of sustainability into the design process. As a result, many of the aforementioned approaches have not gained the necessary traction with the project management systems that drive the development of a new mining projects through the project phases (concept, prefeasibility, feasibility, etc.), nor have they been consistently applied on an industry-wide basis for improving the overall sustainability performance of an mining operation on a year-by-year basis. Since McLellan et al.'s (2009) article, there has been some progress in this area, for example, the *Initiative for Responsible Mining Assurance's Standard for Responsible Mining* (IRMA 2015).

In Australia, the *Cooperative Research Centre for Sustainable Resource Processing* developed the SUSOP (Sustainability Risks and Opportunity) framework through a collaborative research project with the aim of producing a sustainable development standard for project and operational engineering (Corder et al. 2012a). There were several reasons for its development, including the absence of an industry standard for sustainable design of industrial processing plants, government initiatives not focusing on a whole-systems approach, and the lack of sustainability objectives to guide project design and operational activities. While the SUSOP framework has been well documented elsewhere (2013; Corder et al. 2013, 2012a, b; Corder and Green 2011, 2012). A brief summary is provided here.

The SUSOP framework utilises the Five Capitals Model to facilitate a contribution to sustainability by the industrial facilities being studied, designed, built, or operated. It uses a multi-disciplinary study team of technical, environmental, social, and management practitioners, and comprises three major elements:

- Sustainability opportunities and risks identification,
- Preparation of action plans for conducting a detailed evaluation of the shortlisted or high-priority opportunities and risks, and
- Decision support for providing assistance with decision-making at the end of project phases.

The SUSOP Knowledge Base supports the framework and the main outputs are recorded in a sustainability register, which works in a similar manner to a conventional risk register, and includes sustainable development balance sheets to show schematically the positive and negative impacts across the Five Capitals Model.

In essence, all of the above approaches have an overriding aim of assisting the initiation, design, and delivery of enhanced sustainability outcomes in the mining industry. Even though there are numerous examples throughout the industry of good sustainable development initiatives, examining and evaluating potential initiatives has not become routine engineering practice in the same uncompromising way that safety is treated in the industry. Accordingly, many initiatives that have implemented and achieved sound sustainable development credentials have resulted in productive outcomes.

Measuring Sustainable Development

An important aspect of implementing sustainability in the mining industry, particularly for technically oriented personnel, is the ability to measure sustainability performance, both quantitatively and qualitatively. This can be for assessing the sustainability benefits of different project options, or for determining sustainability improvements of an operation on a regular (annual) basis. Sets of indicators and examples are presented below.

Shortly after the release of the outcomes of the MMSD project, Azapagic (2004) proposed a framework for sustainability indicators as a tool for performance assessment and improvement, specifically for use by the mining industry. The framework recognised that that the indicators, which covered economic, environmental, social, and integrated aspects, could be used both internally, for identification of problem areas, and externally, for sustainability reporting and stakeholder engagement. The framework was compatible with the Global Reporting Initiative (GRI) indicators, with the inclusion of several sector-specific indicators to reflect the industry's characteristics, such as closure and rehabilitation, mineral resources, resettlement, and fly-in and fly-out (Azapagic 2004).

This work was the forerunner for the development of the Mining and Metals Sector Supplement (MMSS). The MMSS was developed through a multi-stakeholder working group and is now used by organisations in the mining industry to cover key aspects of sustainability performance which are not sufficiently covered by GRI guidelines (ICMM 2015d). The current version, *Mining and Metals Sector Disclosures*, includes specific indicators for aspects such as materials, biodiversity, effluents and waste, labour/management relations, local communities, artisanal and small-scale mining, resettlement and closure planning (GRI 2013). The full set of indicators covers the following sustainability themes: economic, environmental, labour practices and decent work, society, and product responsibility (GRI 2012).

Mining, Sustainable Development and The Philippines: A Short Case Study

Mining has been an important part of the economy of the Asia-Pacific Region. Countries such as China, Indonesia, Malaysia, The Philippines and Papua New Guinea each have had a long history of mining. More recently, countries including Laos, Myanmar (Burma), Cambodia and Vietnam have seen the growth of their mining industries, or the potential for growth therein.

While there have been obvious traditional economic benefits from mining developments, including jobs, business activity, taxes and royalties, there also have been some negative impacts from mining, which adversely have affected the development of new and future mines. For example, the government of The Philippines estimates the country's mineral wealth is $1 trillion (comprising the second-largest gold deposits after South Africa, one of the largest copper deposits in the world, and rich nickel, chromite, and zinc deposits) (Greenlees 2008). However, even though foreign ownership of mines is permitted in the country, the level of foreign investment is low as major mining companies have been cautious in developing new projects. A crucial reason for this is that mining suffers from a significant domestic image problem as a result of past serious environmental damage, which has antagonised local communities and powerful interest groups, including the Catholic Church (Greenlees 2008).

The most significant environmental incident was the Marcopper copper mine disaster in 1996, on the island of Marinduque. A fracture in the drainage tunnel of a large pit containing old mine tailings led to a discharge of toxic mine waste into the Makulapnit-Boac river system and caused flash floods in areas along the river. While the owners (including the Canadian company, Placer Dome) paid $70 million in compensation to affected villagers, the event led not only to a strong backlash against the mining industry by communities and environmentalists, but was followed by a legal challenge to the 1995 mining law that allowed foreign interests to wholly own a mining operation. This effectively stopped new foreign investment, and was not resolved until 2005, when the country's Supreme Court upheld the constitutional validity of the law (Greenlees 2008).

The Marcopper disaster illustrates the critical influence that such an event can have on future mining development in a locality, region, or at a national level, even when there is huge mineral wealth in geological terms. Although there have been few major mining projects since the mid-1990s (the Tampakan Copper-Gold Project in Mindanao (SMI Inc. 2015) being the main exception), a substantial artisanal and small-scale mining (ASM) industry has emerged in recent times, with many of the operations being performed illegally, and without the usual safety and environmental practices required by legal operations (Scholz in this volume). At the gold-rich Compestela Valley of the Philippines, 36 miners were killed on 5 January 2013 after heavy rainfall triggered a landslide, even though the government had previously ordered the miners off the land due to safety concerns. However, the miners had returned once they recognised that there was no enforcement of the order (GBR 2013). The lack of resources for enforcement of government orders, coupled with the economic circumstances of miners, produces a complex set of issues that in the longer-term is unsustainable.

While other countries in the region have had similar concerns or issues with mining development (for example, Papua New Guinea as a result of riverine tailings disposal at Ok Tedi, or the unrest that precipitated the closure of the Bougainville Copper Mine in 1989, and similar issues at the Freeport mine in Indonesia's West Papua province), the issues in the Philippines touch on most, if not all, key sustainable development areas. Even with its large mineral wealth, it is the broad range of sustainable development issues that has over recent times prevented the Philippines from successfully developing the country's mineral resources for national economic and social benefit. This is supported by the Fraser Institute's annual Survey of Mining Companies 2015, in which the Philippines ranked 71 out of 110 jurisdictions on the Investment Attractiveness Index, slipping from a ranking of 61 out of 112 in the equivalent survey in 2013 (Fraser Institute 2014).

Comparing the above mining issues in the Philippines with the sustainable development frameworks and principles mentioned earlier in this chapter, it is possible to identify a direct connection with each of the principles. Furthermore, it is feasible to state that if mining development in the Philippines were framed by these principles, the likelihood of detrimental events occurring would be reduced considerably.

The practices employed by the mining industry in developing new projects, the operation of existing mine sites, and during mine closure have improved over the last generation. This is a result of the heightened awareness of sustainable development principles, frameworks, and toolkits. Improved regulatory frameworks, aimed at better safeguarding the environment and affected host communities and stakeholders from unwanted negative impacts, has also been beneficial. Mining companies now have to balance the benefits to a mining development that transcends regulatory compliance requirements beyond what is required by law with the value of the deposit to the company, as well as balancing support for community development programs that allow for a more harmonious relationship with local communities, with the benefits in being seen as a good corporate citizen. This added value is becoming more critical, given the increasing influence that stakeholders and the wider civil society have on the mining industry.

To date, new mining operations have been routinely designed, built and run in a similar manner to existing operations. Tried and tested solutions that are perceived to have lower technical and financial risks continue to be chosen over more innovative initiatives that have better alignment with sustainable development principles. Commonly, such principles are used to ensure that the project or operation is at best compliant once all major decisions are made, leaving little scope for innovative initiatives that could improve environmental and social outcomes. Brewer (2007) argues this effect is exacerbated by business operational pressures:

> Lacking time, however, usually means that choices fall into one or a limited number of types: incrementalism, standard operating procedures, vacillation and indecision, and doing nothing at all. Creativity in any case is seldom sought or celebrated.

Key to driving sustainable development in the industry is to have proper consideration, analysis and assessment of innovative initiatives in the project development systems, so that new mining projects will not continue to imitate existing mining operations. Innovative initiatives that deliver benefits will help build and maintain a strengthened social licence to operate. A common misconception is that good sustainable development outcomes will come at a cost to the project compared to the corresponding conventional approach. In fact, sound initiatives, such as a smaller residue facility with water recycling, or a wetland compared with a reverse osmosis plant for effluent treatment, are not more expensive in terms of capital or operating costs, but may be perceived to be more risky and therefore are not considered in the usual engineering analysis. More commonly, project development processes routinely do not investigate options that could deliver strong sustainability benefits and satisfy technical and financial requirements, the opposite being more common practice where sustainability benefits are considered too late in the development process. The potential for new and existing mining projects to engage local businesses and thereby build up robust relationships with local stakeholders is growing within the industry, and forward thinking leaders recognise that such initiatives strengthen their reputation and their social licence to operate. Creating a robust and transparent relationship with stakeholders also can help with misperceptions regarding potential environmental impacts (even if these impacts are not grounded by scientific evidence), which can create risk to the business (for example, the perceived impact of mining discharges on fish populations). Finally, technology has an increasing and vital part to play in managing social and environmental risks, as alternative technical solutions that may increase costs could also help to reduce the overall risk profile of an operation or project, thereby warranting additional investment.

Conclusion

In this chapter, I have presented an overview of sustainable development practice in the mining industry. The complexities facing the industry worldwide, and in the Asia-Pacific Region in particular, need to be underpinned by a sound and workable sustainable development agenda that aims to deliver benefits to all affected stakeholders, including project proponents. In so doing, extractive operations will acquire and maintain a stronger social licence to operate. However, to deliver on a sustainable development agenda requires holistic and systematic approaches that must be context and environment specific. To date, these have been lacking across the industry. Through the use of more sophisticated systematic and holistic approaches—some of which have been surveyed in this chapter—enhanced sustainable development outcomes, such as water and energy savings, integrated rehabilitation plans, enhanced capacity building in local communities, greater skills development, shared infrastructure, and the reduction of business risk, can be better incorporated into the design and operation of mining projects to deliver benefits to project proponents and critical stakeholders.

References

Anglo American. (2014). Socio-economic assessment toolbox (SEAT). http://www.angloamerican.com/~/media/Files/A/Anglo-American-Plc/docs/seat-toolbox-v3.pdf. Accessed 25 March 2104.

Anglo American. (2013). Mine closure toolbox. Anglo American. http://www.angloamerican.com/~/media/Files/A/Anglo-American-Plc/pdf/development/toolbox/toolbox-main-brochure-lr.PDF. Accessed 11 March 2015.

ARUP. (2013). SPeAR® (Sustainable project appraisal routine). http://www.arup.com/projects/spear. Accessed 22 March 2014.

Australian Government, Department of Industry, Innovation and Science. (2015). Leading practice sustainable development. http://www.industry.gov.au/resource/Programs/LPSD/Pages/default.aspx. Accessed 23 March 2015.

Azapagic, A. (2004). Developing a framework for sustainable development indicators for the mining and minerals industry. *Journal of Cleaner Production*, 12(6), 639-662.

Brewer, G. (2007). Inventing the future: scenarios, imagination, mastery and control. Sustainability Science, 2(2), 159-177.

Corder, G., Currey, N., Becker, G. (2013). Sustainability, made easy; a business improvement case study. In D. Pollard and G. Dunlop (Eds.), MetPlant 2013: Metallurgical Plant Design and Operating Strategies. Proceedings, pp. (259-271.

Corder, G. (2013). Developing better projects through the early identification of sustainability opportunities and risks. http://www.conference.net.au/chemeca2013/papers/26958.pdf. Accessed 22 March 2015.

Corder, G., Green, S. (2012). Using a sustainability assessment framework to achieve enhanced legacy outcomes. Life of Mine (311-317). Brisbane: Australia.

Corder, G., Green, S. (2011). Integrating sustainability principles into mineral processing plant design (264-279). MetPlant 2011 Metallurgical Plant Design and Operating Strategies – World's Best Practice, 8-9 August 2011 Perth. AusIMM.

Corder, G., Keith, A., Dyer, L. (2014). A capitals based approach – leading innovation in planning for life-of-mine sustainability. Life of Mine (499-511). Brisbane: Australia.

Corder, G., McLellan, B. Green, S. (2012a). Delivering solutions for resource conservation and recycling into project management systems through SUSOP®. *Minerals Engineering*, 29, 47-57.

Corder, G., McLellan, B., Green, S., et al. (2012b). Engineering-in sustainability through the application of SUSOP®. *Chemical Engineering Research and Design*, 90(1), 98-109.

Dow Jones (2013). Dow Jones sustainability indexes. Available: http://www.sustainability-indices.com. Accessed 11 January 2014.

Dresner, S. (2008). *The principles of sustainability.* London: Earthscan.

Dunphy, D., Benveniste, J., Griffiths, A., Sutton, P. (2000). *Sustainability: the corporate challenge of the 21st century*. Sydney: Allen & Unwin.

Exxaro Resources. (2015). Exxaro awarded Bronze Class medal in sustainability investment specialist. http://www.exxaro.com/index.php/exxaro-awarded-bronze-class-medal-in-sustainability-investment-specialist-robecosams-2015-sustainability-yearbook/. Accessed 21 February 2015.

Forum for the Future (n.d.). The Five Capitals Model – a framework for sustainability. http://www.forumforthefuture.org/sites/default/files/project/downloads/five-capitals-model.pdf. Accessed 9 August 2016.

Fraser Institute. (2015). Annual Survey of Mining Companies. https://www.fraserinstitute.org/studies/annual-survey-of-mining-companies-2015. Accessed 20 March 2015.

Global Business Reports. (2013). Mining in the Philippines: revisiting the rim. A report by GBR for E&MJ. http://gbreports.com/wp-content/uploads/2014/09/Philippines_Mining2013.pdf. Accessed 15 April 2013.

Golder Associates. (2014). GoldSET: sustainability. https://golder.goldset.com/portal/default.aspx. Accessed 22 March 2014.

Greenlees, D., (2008). Miners shun mineral wealth of the Philippines. *The New York Times*, 14 May. http://www.nytimes.com/2008/05/14/business/worldbusiness/14iht-mine.1.12876764.html?_r=0. Accessed 30 November 2008.

Global Reporting Initiative. (2012). The global reporting standards. http://www.globalreporting.org. Accessed 13 July 2013.

Global Reporting Initiative. (2013). Sector disclosures: mining and metals. https://www.globalreporting.org/resourcelibrary/GRI-G4-Mining-and-Metals-Sector-Disclosures.pdf. Accessed 12 March 2015.

International Council on Mining and Metals. (2016). Sustainable development framework. http://www.icmm.com/our-work/sustainable-development-framework. Accessed 24 February 2011.

International Council on Mining and Metals. (2015a). Our history http://www.icmm.com/about-us/our-history. Accessed 13 February 2015.

International Council on Mining and Metals. (2015b). About us. http://www.icmm.com/about-us/about-us. (Accessed 13 February 2015).

International Council on Mining and Metals. (2015c). Public reporting http://www.icmm.com/our-work/sustainable-development-framework/public-reporting. Accessed 12 March 2015.

International Council on Mining and Metals. (2012). Community development toolkit. http://www.icmm.com/community-development-toolkit. Accessed 1 February 2013.

International Finance Corporation. (2012). Environmental and social performance standards and guidance. http://www.ifc.org/wps/wcm/connect/topics_ext_content/ifc_external_corporate_site/ifc+sustainability/our+approach/risk+management/performance+standards/environmental+and+social+performance+standards+and+guidance+notes - 2012. Accessed 13 February 2015.

International Institute for Environment and Development. (2002). Breaking new ground: mining, minerals and sustainable development. http://www.iied.org/mmsd-final-report. Accessed 25 June 2011.

International Institute For Sustainable Development. (2002). Seven questions to sustainability: how to assess the contribution of mining and minerals activities. http://www.iisd.org/pdf/2002/mmsd_sevenquestions.pdf. Accessed 13 June 2014.

Initiative for Responsible Mining Assurance. (2014). Standard for responsible mining draft v1.0. http://www.responsiblemining.net/images/uploads/IRMA_Standard_Draft_v1.0percent2807-14percent29.pdf. Accessed 23 March 2015.

Lacey, J., Parsons, R., Moffat, K. (2012). Exploring the concept of a Social Licence to Operate in the Australian minerals industry: results from interviews with industry representatives. *EP125553*. CSIRO. http://scholar.google.com.au/citations?view_op=view_citation&hl=en&user=EbqNGkQAAAAJ&citation_for_view=EbqNGkQAAAAJ:u5HHmVD_uO8C. Accessed 30 October 2012.

Littlewood, G. (2000). The Global Mining Initiative. http://www.icmm.com/document/104. Accessed 13 August 2014.

Minerals Council of Australia. (2005). Enduring value: an Australian minerals industry framework for sustainable development. http://www.minerals.org.au/__data/assets/pdf_file/0006/19833/EV_SummaryBooklet_June2005.pdf. Accessed 24 February 2011.

McLellan, B., Corder, G., Giurco, D., Green, S. (2009). Incorporating sustainable development in the design of mineral processing operations - review and analysis of current approaches. *Journal of Cleaner Production*, 17(16), 1414-1425.

Meadows, D., Randers, J., Meadows, D., Behrens, W. (1972). *The limits to growth; a report for the Club of Rome's project on the predicament of mankind.* New York: Universe Books.

Medveçka, J., Bangerter, P. (2007). Engineering SD into industry: unlocking institutional barriers. 6th International Copper/Cobre Conference. Toronto, Canada: Canadian Institute of Mining, Metallurgy and Petroleum. https://www.researchgate.net/publication/285006283_Engineering_sustainable_development_into_industry_Unlocking_institutional_barriers. Accessed 17 March 2009.

Moran, C., Kunz, N. (2014). Sustainability as it pertains to minerals and energy supply and demand: a new interpretative perspective for assessing progress. *Journal of Cleaner Production*, 84, 16-26.

Owen, J. Kemp, D. (2013). Social licence and mining: A critical perspective. *Resources Policy*, 38(1), 29-35.

Prno, J. (2013). An analysis of factors leading to the establishment of a social licence to operate in the mining industry. *Resources Policy*, 38(4), 577-590.

Prno, J. and Scott Slocombe, D. (2012). Exploring the origins of 'social license to operate' in the mining sector: perspectives from governance and sustainability theories. *Resources Policy*, 37(3), 346-357.

Sustainable Minerals Institute. (2015). Welcome to Sagittarius Mines, Inc.'s Tampakan project. http://www.smi.com.ph/EN/Pages/Home.aspx. Accessed 20 March 2015.

Solomon, F., Katz, E., Lovel, R. (2008). Social dimensions of mining: research, policy and practice challenges for the minerals industry in Australia. *Resources Policy*, 33(3), 142-149.

The Equator Principles Association. (2011). Equator Principles. http://www.equator-principles.com. Accessed 12 January 2014.

United Nations Environment Programme. (1992). Rio declaration on environment and development. http://www.unep.org/Documents.Multilingual/Default.asp?documentid=78&articleid=1163. Accessed 13 February 2015.

United Nations. (2013). Overview of the UN Global Compact. https://http://www.unglobalcompact.org/AboutTheGC/index.html. Accessed 13 February.

United Nations. (1992). Agenda 21. https://sustainabledevelopment.un.org/content/documents/Agenda21.pdf. Accessed 13 February 2015.

United Nations. (1987). Report of the world commission on environment and development: our common future. General Assembly Resolution 42/187, 11 December 1987. http://www.un-documents.net/wced-ocf.htm. Accessed 5 April 2015.

WorleyParsons. (2014). EcoNomics. http://www.worleyparsons.com/aboutus/economics/Pages/default.aspx. Accessed 22 March 2014.

Chapter 16
Artisanal and Small-Scale Mining

Tamsin Scholz

Abstract This chapter argues the significant issues affecting artisanal and small-scale mining in the Asia-Pacific Region are caused by the profit-driven nature of the industry, which encourages poor practices to lower operational costs. Three key issues—environmental damage, health and safety and security—seriously impact on the industry, and will not be resolved without better legalisation of the industry. Environmental damage caused by poor mining practices is a key concern, and is not only an environmental issue. It also has a significant impact on the health of miners and their local communities, which is further compounded by poor safety standards and a lack of knowledge and skills. The establishment of artisanal and small-scale mining camps also has lead to increased violence as a result of community breakdown, and security is further threatened by conflicting relationships with large-scale mining corporations and state governments. The hazardous nature of the industry limits its potential, and without the introduction of standardised regulations, it will continue to be affected negatively by these issues.

Abbreviations

ASM	Artisanal and Small-scale Mining
CASM	Community and Small-Scale Mining
LSM	Large-scale Mining
PNG	Papua New Guinea
WHO	World Health Organization

T. Scholz (✉)
Centre for International Risk, School of Communication, International Studies and Languages, University of South Australia, Adelaide, SA 5001, Australia
e-mail: schts001@gmail.com

T. O'Callaghan, G. Graetz (eds.), *Mining in the Asia-Pacific*, The Political Economy of the Asia Pacific, DOI 10.1007/978-3-319-61395-6_16

Introduction

Artisanal and small-scale mining (ASM) activity plays an important role in the economy of developing countries, directly and indirectly supporting an estimated 100 million people globally (International Council on Mining and Metals 2009). The term, ASM, refers in a general manner to subsistence mining, where workers are independent from any large-scale mining company. The lack of equipment required, coupled with the lack of environmental and health regulations, make the industry particularly viable for the poor in many countries. However, though it is an economically attractive activity for local populations, it is affected by issues, including environmental damage, threats to health and safety, child labour, gender equity, and security threats. In this chapter, I explore three significant issues—environmental damage, health and safety, and security threats, focusing on ASM in the Asia-Pacific Region. I draw on examples from Indonesia, the Philippines, Myanmar, and Papua New Guinea.

Mercury use in artisanal gold mining is the principal issue facing the industry, and both Indonesia and the Philippines have experienced extensive environmental damage due to its use. Indonesia also has undergone extensive deforestation, as a result of artisanal miners, damaging the environment and affecting fauna in the country. Without first legalising ASM, environmental regulations cannot and will not be imposed effectively, and environmental damage likely will continue unabated. Health and safety is the other key issue that is experienced by artisanal miners, as well as their local communities. Mercury again plays a role here, affecting both the artisanal miners who use it and local populations who ingest it through local food sources after it has contaminated the environment. Artisanal miners also are more prone to accidents than those employed in conventional mining projects, Indeed, landslides, rock falls and machinery accidents are commonplace. This is the case in all the countries examined here. Finally, the personal safety of artisanal miners is also threatened by conflicts with governments, large-scale mining corporations, and by the fragmentation of communities that often occurs when ASM mining camps are established. Assault, drug use, prostitution, threatening behaviour, gun violence, and murder often increase as a result of an influx of artisanal miners. However, to explore these issues, we must first describe artisanal and small-scale mining, and examine the industry's impact on local populations.

Describing ASM

ASM is a broad ranging term with no clear or single definition. Too specific a definition fails to encompass differences in the industry, while a general definition does not adequately explain the nature of the activity (Hruschka and Echavarria 2011). Therefore, instead of defining the term, I describe the activity in some detail.

ASM refers to a range of different mining activities characterised by high-level labour intensity, low-level capital input and limited technological capabilities. Hilson (2002) argues a small-scale mining operation is one with:

- Intense labour activity with fewer than 15 workers per operation,
- Remote and isolated location—situated well out of the reach of urban centres,
- Rudimentary techniques and low technological awareness,
- Low degree of mechanization and a chronic shortage of capital, and,
- Low levels of environmental, health and safety awareness.

ASM is not limited to any single mineral or geographical area, but occurs throughout the world. The activity particularly is common in rural and remote areas of developing countries where the local population is poor and uneducated. People engaged in ASM activities also are likely to be unskilled, and have no formal training in the mining sector. For these reasons, ASM is associated with health and safety risks (Hilson 2002). The small profit margins and lack of alternative employment make cheaper, dangerous techniques more economically viable, and therefore there is little incentive for artisanal miners to change work practices. For example, the use of mercury as an amalgam in gold mining is common around the world, because it is cheap and readily available (Zolnikov 2012). Mercury is used despite the health risks. It is also used despite the risks its use poses to the environment. The lack of proper work practices, encouraged by the miners' socio-economic status, can result in severe environmental damage, such as soil erosion, acid mine drainage, habitat destruction and the afore-mentioned mercury contamination (Hilson 2002). ASM activity often makes use of the remnants of large-scale mining (LSM) operations, such as reprocessing abandoned tailings, and extracting remaining minerals from abandoned mines (Hruschka and Echavarria 2011). This can be hazardous, particularly with regard to health concerns and environmental impacts.

ASM also is often perceived as, and associated with, illegal mining (Seccatore et al. 2014). The actual status of ASM differs between countries, but the issue of legality has received more attention in recent years. According to the International Council on Mining and Metals (2009), many countries do not recognise ASM as a legitimate industry and, as such, the ASM workers are often considered to be mining illegally. In other cases, ASM is recognised as a legitimate industry practice, with mining permits issued by the state. However, it is frequently the case that the rural poor are unable to register their claims. This is due to the cost of registration. This can bring ASM into conflict with large-scale mining operations, which have been granted legal leases. It means effectively that ASM workers are trespassing on legitimate claims. In some cases, these miners have been shot and killed. Nevertheless, despite the many hazards that are associated with the activity, ASM brings a number of benefits to local communities, as the industry is an important source of income. In this way, the activity is important to community development.

A Positive Influence on Local Economies

While accurate data is difficult to find, it is estimated that as many as 100 million people benefit directly and indirectly from ASM, with almost all of these living in developing countries (International Council on Mining and Metals 2009). The number of practicing miners is estimated to be approximately 13 million (International Institute for Environment and Development 2002). In the Asia-Pacific Region, Indonesia has around 109,000 people engaged in the practice, 185,000 in the Philippines, and between 50,000 and 100,000 in Papua New Guinea (PNG) (Javia and Siop 2010). Without doubt the country with the most miners is China, where there are thought to be some 250,000 mines that employ some three million people. There are also approximately four million women engaged in the activity, as well as an unknown number of children.

ASM is an important industry in the small, rural communities, because of the economic benefits associated with the activity (International Institute for Environment and Development 2002). It increases capital in the local community, and provides a diverse source of income in regions dominated by subsistence agriculture. When the industries supporting the community are so few, and opportunities limited, self-employment through ASM is an attractive prospect. ASM therefore can provide employment for people with few other opportunities. This includes employment for workers rendered unemployed by the closure of large-scale mines and, in many cases, for women. Large-scale mine employees often are retrenched *en masse* when mine profits decrease and cutbacks are made. This means there are often many people unemployed at the one time, and other work opportunities in their field are limited. ASM provides an income for these workers, in an industry in which they have some knowledge and skill. Women make up an estimated one-third of the ASM workforce, and are involved in different aspects of the industry around the world. This may allow them to have a level of independence and an income, not available to them through traditional work streams, such as agriculture. Nevertheless, woman continue to face gender barriers in the industry.

There are a variety of issues regarding women's participation and role in artisanal mining, the industry itself offers great potential to empower women through their involvement. They can gain a level of independence unavailable to them in more traditional sectors. The mining industry more broadly has the potential to erode social constructs that limit the behaviour of women and the role they can play in society which in turn positively affects the economic and social development of under-developed countries (Hinton 2007). In artisanal mining, where most participants are from low socio-economic backgrounds, the empowerment of women through the industry has the potential to not only benefit women but also lift their communities out of poverty.

Though the industry provides the opportunity for female empowerment they still face a range of gender barriers preventing not only their full participation as members of the workforce, but also threatening their health and safety. Legalisation and regulation of artisanal mining must occur to allow women to make the most of the opportunities the industry can provide, but beyond that there must be serious consideration of the role of women as stakeholders in the sector. Only when their specific needs, which differ from those of men, are factored into any attempts to legalise and monitor the industry will women be able to reap the benefits associated with it—and benefit their wider communities in the process.

Women play a variety of different roles in artisanal mining, depending on the region and society in which they operate. They can be found scavenging through tailings, processing ore and waste rock, or even extracting ore itself (Hilson 2002). They can also be found working as prostitutes in mining camps, operating in incredibly insecure environments prone to violence (Aljazeera 2014). Domestic violence and sexual assault in artisanal mining camps is a common occurrence, and while this reflects broader societal attitudes it is the insecurity of these illegal camps that increases the risk of harm to women. Cases of increased domestic and sexual violence have been well documented in Papua New Guinea in particular (Human Rights Watch 2010, p. 38). As well as the threat of physical and sexual violence, women are particularly affected by pollutants in tailings, as they (along with children) are more likely to be engaged in this kind of activity (Hilson 2002, p. 49). While these are just a few examples, it is clear the risks posed to women in the industry are great.

As ASM is viable in areas with little infrastructure, it also plays an important role in alleviating poverty in these regions. ASM therefore can provide a level of financial security and independence that is not available in other industries. Although there are obvious health and safety issues, as well as environmental impacts, the relatively high returns for ASM make it an appealing industry, and it plays an integral role in lifting small communities out of poverty. In comparison with the wages people commonly earn in these areas, for example from agriculture, the income from ASM is relatively high (International Council on Mining and Metals 2009). Consequently, the impacts on health and wellbeing are often disregarded.

This industry also has wider benefits for developing countries: ASM can reduce rural-urban migration, increase the demands for locally produced goods, and allow countries to exploit mineral deposits that are not feasible for large-scale mining operations. As a relatively low cost industry for a developing country, the regulation and legalisation of ASM could result in increased state revenues through taxation, as well as reduce poverty in the region. In many cases, then, the benefits, may be certainly worth the risk, not only for the local miners, but also for developing countries as a whole. However, the issues that threaten the industry must be managed for ASM to be viable and sustainable over the long-term. Environmental damage resulting from ASM, is one of the foremost concerns, and has significance globally as well.

The Negative Environmental Impact Caused by Unsafe Mining Practices

The environmental impacts of ASM are of great concern, as they can, and do, affect large land areas. Deforestation, mercury contamination, erosion, acid rock drainage, landscape destruction, the disposal of tailings, and damage to river health, particularly from effluent and siltation are some of the most pressing issues (International Institute for Environment and Development 2002). Of these, mercury contamination is the most pressing challenge. As stated above, mercury is commonly used as an amalgam in artisanal gold mining. However, it is also a pollutant, and can stay present in the environment between 6 months up to 2 years (Li et al. 2009). While mercury contamination occurs globally, in the Asia-Pacific Region it is most common in the Philippines and Indonesia.

On Buru Island, in the Mollucas, for example, gold was discovered in 2011. This led to an influx of aspiring artisanal miners, who for several years remained unregulated (Male et al. 2013). While mining licences since have been introduced. This has established a level of control over population increases. However, the environmental impact of mining methods has not been subject to regulation. On Buru, mercury is used in the trommel method as a way of extracting the gold. This requires large volumes of water.[1] Fine sediments and mercury are wash into the Wamsait River, and from there into Kayeli Bay, which supports a population of roughly 50,000 people (Male et al. 2013). The study found that sites where trommels have been used by miners will likely be unusable as a consequence of legacy issues. A high concentration of mercury was found in trommel waste and sediments, showing the scale of impact this practice has on the local environment. Male et al. also list food contamination and the 'long-term transport of mercury throughout the ecosystem' as other areas of concern in the future. This is just one example of mercury contamination in Indonesia. There are many other studies which focus on different regions in the country, where ASM has led to high levels of contamination. The North of Sulawesi is one such region. However, the practice is not limited to Indonesia.

In the Philippines, the island of Mindanao is rich in gold and, accordingly, has hosted artisanal and small-scale gold mining activities for decades (Appleton et al. 1999). In the 1980s, the use of mercury in mining activity in the headwaters of the Mamunga River, in eastern Mindanao, resulted in high levels of mercury found in fish in the Agusan River, which joins Mamunga. The study by Appleton et al. found mercury concentration in the air exceeded the World Health Organization's (WHO) exposure limit in 65% of gold commercialisation shops. It also found extremely high mercury concentration levels in water samples taken along the Mamunga, with samples 14 km downstream of the mining area, still two times higher than they should be (Appleton et al. 1999). This demonstrates the far-reaching impact mercury can have on the environment, extending far beyond the immediate mining site. A follow up study by (Appleton et al. 2006) found that irrigation of rice paddy crops using water from the Mamunga River had reduced yields from 6000to 3000 kg/ha because of fine suspended silt in the water. Significant numbers of water buffalo have also died from cyanide poisoning. The effects of mercury on the environment, therefore, not only is far-reaching, but also long-lasting. As the damage often occurs in remote areas where the population is poor, it cannot easily be remedied. Indeed,

[1] A trommel is a screened cylinder that separates fine materials from larger material. In this case, it separates the larger material from the finer material, which contains the gold. This fine material then runs into a sluice box.

the number of deaths associated with mercury contamination in these areas may never be known.

Mercury contamination is not the only environmental issue caused by ASM, however. These two examples of gold mining in the Philippines and Indonesia, additionally, show the damage caused to rivers by siltation and the disposal of tailings and effluents into waterways. There also are environmental impacts as a consequence of land degradation, deforestation, and soil erosion. Again, with specific reference to Indonesia, an estimated 1,000,000 hectares of land has been destroyed by local artisanal coal miners in the Kalimantan region, with around 10,000 hectares destroyed yearly (Aspinall 2001). Artisanal miners dig pits and tunnels, which can cause soil erosion and contribute to landslides. As well as causing environmental damage, this type of unregulated mining also affects the health and safety of local residents and animals. Though it is illegal, the prosecution of miners is rare.

Environmental regulation of artisanal and small-scale mining is difficult to implement because of the often illegal nature of the industry. While the activity remains unregulated, and the potential profit from ASM remains high enough to ignore environmental externalities. Nevertheless, there is hope for the future in countries where ASM has been legalised. In PNG, ASM is legal and regulated by the Mines Department (Burke 2006). This has allowed an education program to be developed, which focuses on informing local miners about the hazards of mercury use. The program has expanded to include safe mining training. Legalising ASM is a key step. However, without more countries following PNG's example, there likely will continue to be serious environmental damage from ASM.

While this problem may not be as large in the Asia-Pacific Region, as it is in Africa and parts of South America, still it is significant. After all, ASM, not only affects the environment, but also the lives of the individuals working in and around the industry.

Health and Safety Concerns Among Artisanal Miners and Small-Scale Miners and Their Local Communities

There are two aspects of health and safety affecting the artisanal and small-scale mining industry. First, the personal safety of the miners, and the health and safety of the people affected by negative environmental impacts. In both cases, mercury contamination plays a long-term role, affecting the health of miners and local communities. There also are concerns about more immediate safety issues, including landslides, rock falls, mining shaft collapses, and machinery accidents. According to Hinton (2007), fatality rates in ASM are up to 90 times higher than those at large-scale mines in industrialised countries. Non-fatal accidents are six to seven times higher, despite underreporting. In the artisanal coal mining industry in China alone, around 6000 fatalities are reported every year. This demonstrates the enormous difference between small and large-scale mining. Artisanal miners also are often exposed to social health problems, such as the prevalence of sexually transmitted diseases, as well as alcoholism and substance abuse. These factors combine to make the ASM industry hazardous for those involved, some of whom are ignorant of the level of risk associated with the activity.

First, the lack of occupational health and safety regulations in the industry allow artisanal miners to increase their incomes by minimising the costs associated with mining. For example, protective equipment such as safety glasses, helmets, dust masks and ear protectors, often are foregone, either because the miners are unable to afford such equipment, or they are unwilling to reduce their incomes. Explosions, machinery accidents, electrocution, shaft collapses, poor ventilation, and dust and noise pollution are common problems, due to a lack of regard for safety measures in the industry (Hinton 2007). The illegal nature of the industry also has an effect on the ability of artisanal miners to be safe in the workplace. In November 2015, a landslide at a jade mine in Myanmar killed an estimated 120 people (Fuller 2015). The deceased were illegal artisanal miners who earned a living by sorting through the waste piles left by large mining companies. The miners lived in a camp on the waste pile in shacks made from rubbish and tarps, and were sleeping when the landslide occurred (Fuller 2015). The exact number of people missing is unknown, as there are no formal records of who was at the camp that night. This disaster is just one of many that occur at ASM sites, where a lack of knowledge and regulation leads to environmental disasters that cost lives. The workers at the mine site in Myanmar were well aware that the work was dangerous. However, the potential income made the risks acceptable ones.

It is clear ASM is hazardous work, and the immediate health and safety impacts of it are not the only areas of concern. While landslides, rock falls and workplace accidents are significant risks, the use of chemicals also has a dangerous impact on health. Accordingly, an artisanal miner's lack of knowledge and skill does not only affect their ability to manage environmental disasters such as landslides, it also affects their ability to manage water pollution, and adequately to protect themselves against inhaling dangerous chemicals.

Human exposure to mercury vapour can cause damage to the respiratory and neurological systems (Castilhos et al. 2016), affecting the health of miners, and those who live on, or near, ASM sites. However, it is miners most who most commonly are exposed. Miners working in the artisanal gold mining industry, repeatedly are found to have higher levels of mercury in their urine, hair and blood. A study carried out at gold mining sites in North Sulawesi and Central Kalimantan in Indonesia also had similar results. They found participants from exposed areas showed significantly higher levels of mercury in their urine, hair and blood, while those who worked as amalgam-smelters were particularly affected (Bose-O'Reilly et al. 2010). Among participants in exposed groups, the standard symptoms of mercury intoxication were widespread and it was calculated that 55% of amalgam smelters in North Sulawesi were diagnosed with chronic mercury intoxication. Twenty-four per cent of mineral processors were also similarly diagnosed. In Mindanao in the Philippines, several incidences of mercury poisoning among miners were recorded in the 1980s, including one incident resulting in 11 injuries and one fatality (Appleton et al. 1999). Fish also are affected by mercury contamination in waterways, which in turn can affect the local population, as fish are a principal food source. When entering waterways, Mercury can be transformed into methylmercury, which has much higher levels of toxicity (Castilhos et al. 2015). Mercury is often present in fish in artisanal gold mining areas, and its ingestion can lead to the nervous system being affected irreversibly by neurotoxins and teratogens. In the Asia-Pacific Region, the presence of methylmercury in the environment has also been found to pose a threat to food safety on Buru Island in Indonesia, affecting the wider local population, as well as artisanal miners. Male et al. (2013) argue this not only poses a threat to those who consume contaminated food sources, but also to babies through transfer through breast milk and to the foetus through the placental barrier. In Mindanao, following large-scale mercury contamination from artisanal gold mining, 38% of the local population were classified as mercury intoxicated, due to eating contaminated rice and fish.

Threats to the Security of Miners in the ASM Industry

In this section, I explore two issues that affect the security of those involved in ASM. The first focuses on the security of artisanal miners and their communities, when faced with increased violence, drug use, sexual assault and theft, that result from the presence of the industry. The second addresses the safety of artisanal miners when coming into contact with large-scale mining companies, armed militia, and state authorities. Both of these issues affect the security of those involved in the industry, but to different degrees. For example, the competition for resources can result in outbreaks of violence between miners, and when combined with drug use, can have a devastating impact on the families of mine workers. Conversely, artisanal miners may find their personal security threatened by state authorities, the security teams of large-scale mining companies, or armed militia.

The International Council on Mining and Metals (2009) lists security threats as one of the principal issues affecting the ASM industry. Prostitution and theft are common challenges when large mining camps are established, and when dealing with high-value minerals, it is possible for groups within the ASM community to use force to protect their livelihoods. In PNG, the rate of violent crimes increased substantially in and around the Porgera gold mine, where illegal ASM occurs alongside the larger mine (Human Rights Watch 2010). The establishment of the mine led to an influx of people from other areas of PNG, seeking an income from artisanal gold mining (Bose-O'Reilly 2009). The rates of murder, assault, violence against women and children, and sexual assault reportedly increased over the last decade. Alcoholism also has become a widespread problem in the area, exacerbating the existing security issues (Human Rights Watch 2010). The Porgera example shows the establishment of ASM camps and the influx of people seeking work can precipitate social breakdown on a community-wide scale. It not only affects the artisanal workers themselves, who are likely to be the subject of violence, but also the surrounding community, which suffers from increased domestic violence. Drug use, prostitution and public violence also are rife within the artisanal jade mining community in Myanmar (Aljazeera 2014). The associated social impacts that emerge when these unstable influx communities develop are one of the principal causes of insecurity for artisanal miners and local communities. The environment in which these workers operate not only is affected by such domestic insecurity, but also by conflict between artisanal miners and other groups.

There is a great deal of literature about the conflicts that occur between artisanal miners, large-scale mining operations, and state authorities, though most of this literature focuses on mining in Africa. In Sierra Leone, artisanal diamond miners have come into conflict with large-scale mining companies over land rights, with artisanal miners citing this as one of the main reasons for clashes between the groups (Wilson 2013). The state was accused by local miners and communities of ignoring current ASM licences in the region in favour of leases for large-scale operations, souring relations between artisanal miners and both the state and the mining companies operating in the country. The conflicts between them range from simple disagreements to violence. In Zimbabwe, a large community of artisanal miners developed after the discovery of diamonds in Chiadzwa in 2006. The miners violently were driven out from the region by police and the military in 2008 and 2009, under orders from the Zimbabwean government to stop the 'illegal activity' (Nyamunda and Mukwambo 2012).

In the Asia-Pacific Region, PNG and Myanmar are of significant concern, with the *New York Times* reporting that artisanal miners in Myanmar's secretive jade industry unofficially are taxed by officials (Fuller 2015). In 2014, the jade trade was estimated to be $31 billion, almost half the national gross domestic product. Most of the income is said to be controlled by military elites, drug lords and crony companies (Regencia 2015). This control fuels the conflict that is ongoing in the Kachin Region, where there is a significant deposit of jade. An episode on 101 East, an Aljazeera television program, demonstrates the industry also is affected by a well-developed drug trade, with a high percentage of miners addicted to heroin and methamphetamines (Aljazeera 2014). While information on the region is scarce, the nature of the artisanal jade industry, its control by military elites, the existing drug trade, and violent conflict in the region lead one to conclude that violence and intimidation will continue to occur in the country. The security of artisanal miners in Myanmar is particularly precarious in the precious gem trade.

In PNG, artisanal miners working on the fringes of large-scale operations often have suffered violence at the hands of security guards employed by these operations (Bose-O'Reilly et al. 2009). For example, Canadian mining company Barrick Gold Corporation has employed security guards for this purpose (Dashwood 2012). It is, therefore, not surprising that the company has been accused of human rights violations. Deaths, assaults, shootings, and sexual violence are alleged to have occurred in the region. Indeed, the company has had to pay compensation to 137 women raped by security guards, (McVeigh 2015). Human Rights Watch

(2010) has also found that artisanal miners are not innocent, however, as they engage in organised armed raids on the mine to steal ore from the pit, seriously endangering the safety of the security guards and themselves. The clashes between the local population and mine employees in Porgera, with violence on both sides, demonstrate the tensions that can occur when large-scale mining and ASM clash. This contributes to community breakdown, which, as previously stated, is one of the consequences of an influx of workers seeking employment through the industry.

Conclusion

The range of issues involved in artisanal and small-scale mining, of which only three have been explored here, significantly impact, not only the miners, but also the environment, and the surrounding community. Whilst the industry's economic viability is the reason people in developing countries continue to seek out gold and other minerals for mining, the consequences of poor practices are particularly damaging. Siltation, the disposal of tailings and effluents into waterways, deforestation, soil erosion, and land degradation are significant environmental impacts. In Indonesia, thousands of hectares of land have been destroyed in the process of artisanal coal mining. Mercury use in gold mining, which has occurred particularly in Indonesia and the Philippines, also has been a key factor in environmental damage, through water pollution and the contamination of waste. The use of mercury has also led to high concentrations of the chemical in their bodies, and artisanal gold miners are regularly found to have chronic mercury intoxication. Local communities also suffer with the presence of methylmercury in the environment, threatening food safety. Mercury pollution in the water can lead to contamination of fish in the region, affecting not only the environment, but also the health of local communities, who may depend on fish as a principal food source.

Mercury is not the only health hazard, with a lack of safety equipment and knowledge of what constitutes a safe work environment. This contributes to the high accident and mortality rates in the industry. Machinery accidents, shaft collapses, dust, noise pollution, explosions, electrocution, and poor ventilation are all common, and put the health and safety of artisanal miners at risk. Finally, the security of ASM workers is also threatened by conflicts between miners, large-scale mining operations, and state governments. The influx of miners to mineral-rich areas often leads to community breakdown, as evidenced by examples in PNG. This also has an impact on the safety of artisanal miners and their communities. The response by LSM and state authorities has been lacklustre at best. While these issues, have serious, wide-reaching impacts, the benefits of artisanal and small-scale mining help to reduce poverty in rural and remote areas. However, without appropriate regulatory governance of the industry, the challenges associated with ASM are likely to continue.

References

Aljazeera. (2014). 101 East – Myanmar's jade curse. *YouTube*, 2 December. https://www.youtube.com/watch?v=O7FOAWq-kK8. Accessed 10 December 2015.

Appleton, J., Weeks, J., Calvez, J., Beinhoff, C. (2006). Impacts of mercury contaminated mining waste on soil quality, crops, bivalves, and fish in the Naboc River area, Mindanao, Philippines. *Science of the Total Environment*, 354(1/2), 198-211.

Appleton, J., Williams, T., Breward, N., Apostol, A., Miguel, J., Miranda, C. (1999). Mercury contamination associated with artisanal gold mining on the island of Mindanao, the Philippines. *Science of the Total Environment*, 228(2/3), 95-109.

Aspinall, C. (2001). Small-scale mining in Indonesia. Institute for Environment and Development & World Business Council for Sustainable Development. http://pubs.iied.org/pdfs/G00725.pdf. Accessed 4 December 2015.

Bose-O'Reilly, S., Drasch, G., Beinhoff, C. et al. (2010), Health assessment of artisanal gold miners in Indonesia. *Science of the Total Environment*, 408(4), 713-725.

Burke, G. (2006). Opportunities for environmental management in the mining sector in Asia'. *Journal of Environment & Development*, 15(2), 224-235.

Castilhos, Z, Rodrigues-Filho, S, Cesar, R, Rodrigues, A., et al. (2015). Human exposure and risk assessment associated with mercury contamination in artisanal gold mining areas in the Brazilian Amazon. *Environmental Science and Pollution Research*, 22(15), 11255-11264.

Dashwood, H. (2012). *The rise of global corporate social responsibility: mining and the spread of global norms*. Cambridge: Cambridge University Press.

Fuller, T. (2015). Myanmar continues search after deadly landslide at jade mine. *New York Times*, 23 November http://www.nytimes.com/2015/11/24/world/asia/myanmar-landslide-jade-mine.html?_r=0. Accessed 28 November 2015.

Hilson, G. (2002). Small-scale mining and its socio-economic impact in developing countries. *Natural Resources Forum*, 26(1), 3-13.

Hinton, J. (2007). Communities and small-scale mining: An integrated review for development planning. Communities and Small-Scale Mining Initiativ. http://www.eisourcebook.org/cms/Juneper cent202013/CASM,per cent20anper cent20Integratedper cent20Reviewper cent20forper cent20Developmentper cent20Planning.pdf. Accessed 8 December 2015.

Hruschka, F., Echavarria, C. (2011). Rock solid chances for responsible ASM. Alliance for Responsible Mining. http://www.responsiblemines.org/images/sampledata/publicaciones/Rockper cent20Solidper cent20Chancesper cent20forper cent20responsibleper cent20artisanalper cent20mining/059_RSC_FINAL_web_low.pdf?phpMyAdmin=cde87b62947d46938306c1d6ab7a0420. Accessed 28 November 2015.

Human Rights Watch. (2010). Gold's costly dividend: human rights impacts of Papua New Guinea's Porgera gold mine. https://www.hrw.org/sites/default/files/reports/png0211webwcover.pdf. Accessed 11 December 2015.

International Council on Mining and Metals. (2009). Working together: how large-scale mining can engage with artisanal small-scale miners. http://www.icmm.com/document/789. Accessed 27 November 2015.

International Institute for Environment and Development. (2002). Breaking new ground: mining, minerals and sustainable development. http://pubs.iied.org/pdfs/9084IIED.pdf. Accessed 1 December 2015.

Javia, I., Siop, P. (2010). Paper on challenges and achievements on small scale mining and gender: Papua New Guinea. http://www.un.org/esa/dsd/resources/res_pdfs/csd-18/10may/pm-SIDSday/presentation_javia.pdf. Accessed 28 July 2016.

Li, P, Feng, X, Qiu, G, et al. (2009). Atmospheric mercury emission from artisanal mercury mining in Guizhou Province, Southwestern China. *Atmospheric Environment*, 43(14), 2247-2251.

Male, Y T, Reichelt-Brushett, A J, Pocock, M., et al. (2013). Recent mercury contamination from artisanal gold mining on Buru Island, Indonesia – Potential future risks to environmental health and food safety. *Marine Pollution Bulletin*, 77(1/2), 428-433.

McVeigh, K. (2015). Canada mining firm compensates Papua New Guinea women after alleged rapes. *The Guardian*, 4 April. http://www.theguardian.com/world/2015/apr/03/canada-barrick-gold-mining-compensates-papua-new-guinea-women-rape. Accessed 11 December 2015,

Nyamunda, T., Mukwambo, P. (2012). The state and the bloody diamond rush in Chiadzwa: unpacking the contesting interests in the development of illicit mining and trading, c. 2006-2009. *Journal of Southern African Studies*, 38(1), 145-166.

Regencia, T. (2015). Myanmar accident highlights dangers facing jade miners. *Aljazeera*, 23 November. http://www.aljazeera.com/news/2015/11/tragedy-highlights-myanmar-jade-mining-practices-151123045546779.html. Accessed 10 December 2015.

Seccatore, J, Veiga, M, Origliasso, C, Marin, et al. (2014). An estimation of the artisanal small-scale production of gold in the world, *Science of the Total Environment*, Volume 496(1), 662-667.

Wilson, S. (2013). Company-Community conflicts over diamond resources in Kono District, Sierra Leone. *Society & Natural Resources: An International Journal*, 26(3), 254-269.

Zolnikov, T. (2012). Limitations in small artisanal gold mining addressed by educational components paired with alternative mining methods. *Science of the Total Environment*, 419(1), 1-6.

Chapter 17
Mining, Corporate Social Responsibility, and Corporate Reputation

Terry O'Callaghan and Belinda Spagnoletti

Abstract We examine the corporate social responsibility (CSR) experience of the mining industry and the limitations of CSR activities. We consider the dominant ideologies that underpin these activities and then juxtapose those ideologies with academic and practitioner critiques of CSR. This is followed by a short discussion of Royal Dutch Shell's long-standing commitment to CSR, which has resulted in marginal improvement to its corporate reputation at best. We then provide an analysis of the recent publications of the International Council on Mining and Metals (ICMM) on development partnerships. We argue that this body's shift in emphasis from CSR to development partnerships signifies a move away from the CSR discourse of the mining industry toward an approach centred on genuine community development. In the final section, we consider the potential strengths and limitations of the development partnerships approach. To that end, we adapt Arnstein's (1969) 'Ladder of Citizen Participation' to the contemporary mining industry. In so doing, we attempt to illustrate the complexities associated with participation in the development paradigm. While the ICMM appears to be tracking in a more altruistic direction through its promotion of development partnerships, we believe that a greater focus on the complexities associated with participations may add value to companies engaged in the mining industry in their future pursuits to achieve positive community development outcomes.

Abbreviations

BOP	Bottom of the Pyramid
CoW	Contract of Work
CSR	Corporate Social Responsibility

T. O'Callaghan (✉)
Centre for International Risk, School of Communication, International Studies and Languages, The University of South Australia, Adelaide, SA 5001, Australia
e-mail: Terry.OCallaghan@unisa.edu.au

B. Spagnoletti
Nossal Institute for Global Health, University of Melbourne, Melbourne, Australia
e-mail: bsspagnoletti@students.umelb.edu.au

T. O'Callaghan, G. Graetz (eds.), *Mining in the Asia-Pacific*, The Political Economy of the Asia Pacific, DOI 10.1007/978-3-319-61395-6_17

ICMM	International Council on Mining and Metals
LTO	Licence to Operate
UK	United Kingdom
US	United States

Introduction

Companies engaged in mining are subject to political risk. These risks include, among other events, corruption, civil unrest, regulatory uncertainty, and political instability. Some companies have had to manage these risks for the life of their operations. Freeport McMoRan's experience in the Indonesian province of Papua is noteworthy in this regard. Since the first Contract of Work (CoW) was signed in 1969, the company has endured ongoing negative media coverage, due to its alleged involvement in human rights abuses and the environmental impacts of its operations. The company also has had to manage the consequences of numerous employee and activist fatalities around its mine site (Leith 2002; Taylor and Supriatna 2013). Freeport's ongoing exposure to political risk is not an anomaly in the mining industry. Indeed, mining and energy companies regularly experience political risk. It is the price that mining companies must pay to develop their deposits.

Political risk can arise at any point during the life of an extractive project. For example, when a company begins conversations with governments and other stakeholders about a project, opposition from landowners, interference from non-government organisations, and bureaucratic hurdles can stall, or even thwart, the granting of an exploration licence. It is not unheard of for regulatory approvals to take years to be granted (O'Callaghan 2010). The approval process can be slow and costly for extractive companies, as the experience of Sagittarius Mines in The Philippines shows. The company first sought a licence to operate (LTO) the Tampakan mine deposit in 1996; it still is not operational and has recently been mothballed.

The other political risk factor affecting the mining industry is the continued threat of asset expropriation. For example, in 2007, Venezuela nationalised a number of foreign oil and gas projects in the country (Ellsworth 2007). Russia and Bolivia also have expropriated foreign oil and gas assets. A number of Latin American countries have adjusted their taxation regimes to weaken the revenue streams of extractive companies, and Asian countries appear to be moving in the same direction. In China, the export of rare earth elements has been restricted (Ting in this volume). Indonesia's 2009 regulatory framework required that mining companies process their ores in-country, in a move that widely was viewed as a step toward resource nationalism (Johnson in this volume). In Mongolia, the national government has sought to gain a greater share of the revenues from Rio Tinto's Oyu Tolgoi mine. For its part, Australia attempted to introduce a new tax regime for its mining industry in 2010; arguably, with limited success.

According to Jacobsen (2010), 'the international mining industry—and, indeed, global business in general—has suffered political risk in virtually all corners of the developing and emerging world.' Consequently, dealing with political risk has become integral to the mining industry approach to managing their existing operations and generating new business. Political risk management, therefore, almost is as critical to the viability of an extractive project as technical and feasibility issues.

Extractive companies rely on a suite of risk management strategies to manage political risk. These include retaining legal advisors, engaging locally based consultants to provide on-the-ground intelligence, establishing joint venture partnerships, and purchasing political risk insurance policies (Spagnoletti and O'Callaghan 2011). At the softer end of the strategies to mitigate political risk are commitments to corporate social responsibility (CSR). A company's CSR portfolio may involve adhering to various international codes of conduct, committing to support sustainable development, treating employees fairly and equitably, and making a positive difference to the communities in which they operate. As such, CSR is a strategy used by mining companies to signal to their stakeholders that they understand their ethical responsibilities to their host communities; that they are good corporate citizens.

Our rationale for considering the future value of CSR tools and processes to the mining industry is three-fold. First, CSR increasingly is the subject of critique in the academic literature, with scholars from diverse ideological backgrounds concerned about the value and legitimacy of the concept. Second, in practice, CSR does not appear to assist mining companies to: achieve better relations with host communities—including indigenous communities; improve the realisation of their commercial goals; or shield them from political risk. Finally, there is some evidence that the mining industry-specific peak organisation, the International Council on Mining and Metals (ICMM), tacitly acknowledges the limitations of CSR. Indeed, research publications by the organisation seem to ignore CSR in favour of a 'development partnerships' approach (ICMM 2011, 2012). We interpret these factors as an important step away from a shallow CSR agenda, toward a more tangible development role for mining companies. This agenda is one that purports to: add value to communities; grow the reputation of companies engaged in mining; and, importantly, enhance the reputation of the industry as a whole. This is not to say that mining companies are becoming development agencies; rather, that positive development outcomes increasingly are being linked to commercial success.

In this chapter, we examine the CSR experience of mining companies and the limitations of their CSR activities. We consider the dominant ideologies that underpin these activities and then juxtapose these with academic and practitioner critiques of CSR. This is followed by a short discussion of Royal Dutch Shell's long-standing commitment to CSR, which, at best, has resulted in marginal improvement to its corporate reputation. We then provide an analysis of the recent publications of the ICMM on development partnerships. We argue that the organisation's shift in emphasis from CSR to development partnerships signifies a move away from the CSR discourse of the industry toward a more ideological approach centred on genuine community development. In the final section, we consider the potential strengths and limitations of the development partnerships approach. To that end, we adapt Arnstein's (1969) 'Ladder of Citizen Participation' as it relates to the contemporary mining operations. In so doing, we attempt to illustrate the complexities associated with participation in the development paradigm. While the ICMM appears to be tracking in a more altruistic direction through its promotion of development partnerships, we believe that a greater focus on the complexities associated with participation may add value to companies engaged in the industry in their future pursuits to achieve positive community development outcomes.

Why Do Companies Adopt CSR Policies?

CSR is the voluntary integration of social and environmental concepts into a mining company's business strategy (Graetz 2015). It is an approach to business that attempts to place stakeholders at the centre of business conduct. The key assumption underpinning CSR is that, in acting as a good corporate citizen, will have a positive impact on a company's operational performance (Zadek 2007). CSR also has a self-regulatory function for companies that implement its core ideals. It helps them to manage their social and environmental responsibilities. Once a company accepts the need to adopt CSR policies, it compels them to adjust their behaviour to show to stakeholders that they are, indeed, good corporate citizens. The CSR narrative now is pervasive across the mining industry and, in essence, is regarded as its ethical face (Fombrun 1996; Parsons et al. 2014).

Porter and Kramer (2006) contend that there are four justifications that motivate companies to practise CSR. The first is that companies have a moral obligation to the communities in which they operate. Practising CSR signifies that a company understands that acting ethically is essential to successful business outcomes. The second justification is that companies need to embrace sustainability so that future generations are able to meet their own needs. Embracing CSR enables a company to demonstrate to stakeholders that it is committed to sustainable development and, particularly, intergenerational equity. The third justification is the need for companies to gain a 'social licence to operate'. Commitment to a CSR agenda implies that a company seeks to reinforce its legitimacy by gaining the acceptance of its stakeholders. The final justification for CSR is its perceived ability to strengthen corporate reputations. Companies increasingly are aware that a strong corporate reputation is a driver of corporate success (O'Callaghan 2016; Knox and Maklan 2004).

We suspect that most companies use all of these justifications when considering their societal relationships. However, in this chapter, we are concerned with the relationship between CSR and corporate reputation, primarily as a means to reduce the incidence of political risk events. The link between these two concepts is well established in the academic literature, which emphasises the role of CSR in the development and maintenance of a strong corporate reputation (Larkin 2003; Doorley and Garcia 2007). Nevertheless, we question the effectiveness of CSR as a means to enhance the corporate reputations of extractive companies. These companies may benefit to some degree from practising CSR. For individual companies, it allows for comparison against competitors. CSR also provides a set of broad guidelines through which indicators of corporate social performance may be aligned with community expectations. CSR, of course, has value for extractive companies, but we question its long-term benefit to the mining industry. Indeed, CSR appears to be a corporate strategy in flux. Not much has changed since White's observations in 2005:

> CSR is at a crossroads ... Will external events and company choices relegate CSR to a passing fad, leading to its fading from corporate and public agendas? Or will CSR reach full fruition as it becomes aligned, integrated and fully institutionalized in company strategy and operations? Or, alternatively, is something more transformational on the horizon as CSR morphs into a deeper change mode, becoming a force for altering corporate purpose at the most fundamental and systemic levels?

In our view, CSR is something of a promissory note, but it is difficult for companies to achieve in practice. Despite the best intentions and corporate implementation of social programs and policies, a tailings dam can collapse without warning, precipitating significant social, environmental, business, and political risks and impacts. Environmental activist groups can arrive at a mining operation unannounced, forcing the suspension of operations, or bringing unwanted negative media coverage and reputational damage. Moreover, the political and/or regulatory landscape governing mining developments of a jurisdiction can change unexpectedly. CSR cannot prevent these, or most other *political* risks from hindering the viability of a new or existing mining operation. Consequently, we do not believe that CSR offers a sound foundation for building strong corporate reputations in the industries that so inherently are prone to political risk. A different approach therefore is needed. To this end, the ICMM's development partnerships approach, which we discuss below, may be able to bridge this gap, and is a sign of a changing discourse within the mining industry.

Why Is CSR So Challenging for the Mining Industry?

Over a decade ago, and with specific regard to the mining industry, Hutchins et al. (2005) explored whether CSR could become a unifying discourse. Their conclusion was ambivalent:

> It remains an open question whether the emerging discourse of corporate social responsibility can lead to truly mutually acceptable resolutions of inherent conflicts with local stakeholders, or is simply a new means to promote a corporate agenda.

This ambivalence is noteworthy, for the study highlighted the inherent weaknesses in the reliance of mining companies on CSR to meet various corporate and stakeholder expectations. We share this ambivalence, but for broader reasons: different ideological perspectives appear to converge on the potential value of CSR strategies to the mining industry. We accept both the reservations of the left and the right in this regard.

Neoliberal scholars argue that CSR is an assault on capitalism. Henderson (2001, 2009) has mounted a strong challenge to the efficacy of CSR. For him, acceptance of CSR by business is a form of 'appeasement' to the demands of anti-business lobby groups. While acknowledging the pervasiveness of the concept, he argues that CSR is burdensome, raises costs, impairs corporate performance, reduces competition, and 'will make the world poorer and more over-regulated' (Henderson 2009). He further suggests that CSR has a negative impact on the global economy, diverting scarce resources away from productive investment, toward social and environmental programs. For Henderson, none of these activities generate wealth; rather, they squander it. Another business advocate, Griffin (2008), asserts that his audit of corporate CSR policies reveals that the concept is 'implemented very inconsistently', and is 'unwieldy' and 'ill thought through' in many organisations. He argues that CSR often is interpreted as a means to correct 'corporate irresponsibility'. For him, CSR is a negative, reactive response by companies. Griffin prefers the notion of corporate citizenship, because, in his view, it has more positive connotations:

> The language of 'corporate citizenship' is much more helpful. It is far more positive and describes more accurately the desired positioning of companies in wider society. Good companies should behave like good citizens.[1]

Porter and Kramer (2006) also have concerns about CSR. They argue that the justification for CSR fails to offer adequate guidance for the difficult decisions companies must make in practice. The CSR discourse relies on an inherent tension between business and society. They assert that this is a misunderstanding of the relationship between business and society. The relationship is interdependent, rather than based on incompatible interests. When viewed as being incompatible, CSR can lead companies to make short-term commitments that end up being little more than public relations exercises.

In seeking to satisfy external stakeholders, companies can cede control of their CSR agendas to outside interests. Porter and Kramer (2006) therefore contend that the test of good CSR is not 'whether a cause is worthy but whether it presents an opportunity to create shared value—that is, a meaningful benefit for society that is also valuable to business.' Much of what passes for CSR in extractive companies appears to fail this test.

At the other end of the ideological spectrum are those who argue that the problem with CSR is that its adoption is voluntary. CSR commitments are not legally binding; it is left to the discretion of companies as to how they implement, monitor, and evaluate their commitments. This perspective asserts that companies use CSR to camouflage their poor social and environmental performance. The commonly stated resolution to this challenge is to strengthen government regulation of large, often multinational, corporations. As Bakan (2005) argues, 'Government regulation should be reconceived, and relegitimated, as the principal means for bringing corporations under democratic control and ensuring that they respect the interests of

[1] Contrary to this view, Windsor (2001) argues that '[t]he corporate citizenship notion conflates citizen (which a firm cannot be) and person (which a firm can be), but only as a legal fiction.'

citizens, communities and the environment.' In a similar vein, Moody (2007) is critical of the claim that mining activity is sustainable. As he puts it, we need to 'separate out obfuscation from reality, primarily from the perspective of those most affected by both big and small mining' (Moody 2007). He goes on to argue that the idea that long-term extraction can be sustainable is little more than a 'transparent oxymoron' (Moody 2007). Others have claimed that the CSR commitments of mining companies are nothing more than an industry-wide form of greenwash. Whitmore (2006) claims that '[a]ttempts by the mining industry to greenwash itself as a new, improved, sustainable industry simply will not wash, as even some corporate mining cheerleaders have pointed out.'[2] Accompanying the claim of greenwash is a high degree of cynicism regarding the CSR claims that mining companies make.

While CSR initiatives in the mining industry have grown exponentially in the last decade, there is a dearth of evidence about the success of these initiatives. It is difficult to ascertain the degree to which a 'development dividend' flows from CSR programs to benefit local stakeholders. As Blowfield and Frynas (2005) argue, 'We know very little about the impact of CSR initiatives in developing countries, and what we do know raises questions about both the efficacy of CSR approaches and the tangible benefits for the poor and marginalized.' With such varied claims about the value and merits of CSR strategies and initiatives, it is easy to understand how the concept has become a contentious issue for the mining industry. It begs the question as to whether CSR really is the worthwhile concept that many extractive companies believe it to be.

The Intersection of CSR, Corporate Reputation, and Political Risk

According to Larkin (2003), 'Reputation in a corporate context is based on perceptions of the characteristics, performance and behaviour of a company.'[3] Our view is that, while these all are important elements, it is corporate behaviour that is the key to developing a strong corporate reputation. This is the reason why most companies now are active consumers of the CSR narrative. It is not that they are ethical entities by design, but that they perceive commercial benefits from being seen to act ethically.

There are a number of reasons why positive corporate behaviour now is critical to business success. The cost of behaving 'badly' can be exceedingly expensive. For example, the cost to BP following the devastating 2010 oil spill in the Gulf of Mexico at its Deepwater Horizon oil rig is estimated at up to US$80 billion. The

[2] The chairman of Rio Tinto acknowledged the depth of the problem in the late 1990s. 'There is a perception held, often tacitly, by a growing number of people that the global mining industry is incompatible with sustainable development' (Burton 2002).

[3] For other definitions of corporate reputation, see: Fombrun (1996); Peters (1999); Jackson (2004); Doorley and Garcia (2006).

emissions software crisis that engulfed Volkswagen in 2015 is likely to cost that company billions of dollars as well.

There has been a shift in public sentiment in recent years, which has compelled companies to think beyond their narrow business interests. While stakeholders expect companies to be profitable, they also wish them to be socially and environmentally responsible. Companies also have realised that there are real financial and non-financial benefits from acting ethically. Benefits may include increased value in the goodwill companies earn on sale, the ability to source better employees, improved workplace morale, internal efficiency gains, a larger consumer base, and a reduction in political and other risks. Finally, there is growing understanding that good corporate behaviour leads to a strong corporate reputation (Brammer and Pavelin 2006; Grayson and Hodges 2004).

However, CSR remains a difficult balancing act. It tacitly binds companies to a set of ethical commitments, which are difficult to achieve practically. A fall in the commodity price, a downgrade from a ratings agency, or confrontation with a local community can undermine a company's good intentions. Corporate social responsibility is, in all likelihood then, a victim of change in commercial and economic realities. Royal Dutch Shell's long-standing commitment to CSR and its ongoing reputational woes are a case in point. In the 1990s, the company had a public confrontation with Greenpeace over the dumping of the Brent Spar oil storage facility. The company also was accused of being complicit in the execution of the Ogoni Nine, as well as of precipitating widespread environmental damage around its oil fields in Nigeria (Ekatah et al. 2011). In response, Shell embarked on a strong CSR program, particularly focused on sustainable development, with the aim being to improve its reputation. Despite the company's efforts to commit to a seemingly robust CSR agenda, Shell has been unable to avoid criticism. The company has been found to have engaged in cartel behaviour in Europe on more than one occasion; fined both in the United States (US) and the United Kingdom (UK) for overstating its oil reserves; and has been fined for numerous health and safety breaches. In addition, Shell's environmental record continually has been questioned (Sluyterman 2010). The UK's Advertising Standards Authority has reprimanded the company for claiming its oil tar sands operation in Canada promotes a sustainable energy future. Most recently, the company's Kulluk drilling rig ran aground off the coast of Alaska. A US Coast Guard report that followed concluded that Shell's actions in the lead up to the incident were reckless and ill-timed.

If Shell's stated commitment to CSR were about encouraging and ensuring that its actions were ethical, it is difficult to see how the company lived up to its own CSR rhetoric. None of the issues noted above indicate a strong commitment to CSR. Contrast this with the company's actions in other areas: In 2000, Shell established a charitable foundation aimed at supporting development and sustainability. The company also publishes a sustainable development report annually. It has invested in research and development into carbon capture and storage, as well as innovations in bio-fuel technology. In addition, Shell is a signatory to a range of international conventions and agreements, including the Mining industry Transparency

Initiative and the Universal Declaration of Human Rights. The company also provides considerable funds for social investment initiatives around the world.

There are numerous other examples of political risks in the mining industry. Much like Shell, Newcrest Mining Limited has been the subject of criticism on environmental grounds for years (Moody 2005). The company's CSR commitments did not prevent local community members from protesting against its mine in Indonesia's North Maluku province in 2004. The demonstrations turned violent when a paramilitary group allegedly attacked protestors. In The Philippines, Sagittarius Mines, a company well known for its stated commitment to sustainable development, has been unable to convince the South Cotabato provincial government of the merits of its open pit gold and copper mine, Tampakan. In South Australia, the State Government enacted legislation to prohibit Marathon Resources' attempt to establish a uranium mine at Mt Gee in the state's Arkaroola region, despite the project having the support of the local Adynamathanha Native Title holders.

The aforementioned cases are just several of many examples that demonstrate that CSR has not proven to be much help in growing reputational capital, or mitigating political risk. Of course, the mining industry's long history of devastating environmental disasters, and a perception that they failed in their social obligations for decades, will mean that it could take many more years to gain increased community acceptance for mining projects. However, this still does not account for the failure of CSR, in practice, positively to enhance the reputations of individual companies, or the industry as a whole. Thus, it is timely that the mining industry in particular seems intent on shifting its focus from the CSR paradigm to that of development partnerships. We are of the view that development partnerships potentially represent an advance on CSR, but with some important caveats.

The ICMM and the Development Partnerships Concept: Beyond CSR?

The ICMM is the peak global mining industry body. Founded in 2001, the organisation's goal is 'to act as a catalyst for performance improvement in the mining and metals industry.' It does this by promoting best practice among its members, advocating sustainable development, and seeking to improve the image and reputation of the industry as a whole. The organisation also undertakes research, publishing an array of relevant material, including reports and 'toolkits' on community engagement, mining and Indigenous Peoples, and mine closure and rehabilitation. One of the significant accomplishments of the ICMM is its understanding of the value of development partnerships to the industry and host communities. Principle nine of the organisation's ten principles of sustainability commits its members to 'contribute to the economic and institutional development of the communities in which [they] operate.' The ICMM's *Community Development Toolkit* and *Development Partnerships*

Toolkit identify opportunities for mining companies to contribute to development in their host communities. The following actions are proposed by the ICMM:

- Early and responsive engagement with those likely to be affected by mining activities,
- Ongoing dialogue with peoples affected by mining activities, including those from minority or marginalised groups, and in a culturally appropriate manner,
- Contributions to sustainable community development programs through collaboration with the host community,
- Promotion of cross-sectoral partnerships that lead effectively to well-designed and delivered community development programs, and
- Active identification of opportunities that may address poverty, thereby augmenting social and economic development.

The ICMM's 'partnerships for development' strategy is motivated by its members desire to become 'partners of choice for host countries and communities in the developing of mineral resources' (ICMM 2011). The strategy acknowledges that the ambit of a community development agenda is broad and that the host government must be the central partner, while the role of the mining company is to help ensure that the companies' 'investments in resources enhance social and economic development locally and nationally' (ICMM 2011).

Over a decade ago, Labonne (1999) and Hilson and Murck (2000) proposed partnerships as an approach to achieve the sustainable development objective of enhancing local community development for the mining industry. However, the concept only has gained currency more recently. Nelson (2012) explains that the private sector increasingly has become interested in the development sphere for a number of reasons. First, development opens up new investment opportunities in the form of labour and consumer markets near the 'bottom of the pyramid' (BOP), and, consequently, opens access to natural resources.[4] Moreover, for mining companies in particular, the development partnerships approach helps them to gain and maintain their social licence to operate. Second, development activities provide a philanthropic avenue for companies and their employees. Third, partnerships help companies to 'give back' to their home country, as development bonds represent a form of soft diplomacy. There also is potential for other partnerships beside those formed between mining companies and local communities to provide benefits; for example, collaboration between non-government organisations (NGOs), donor agencies, and researchers (Spagnoletti 2011; Nelson 2012). Fourth, partnerships can help to prevent or at least cushion the impact of failures in the market, regulatory governance, and the 'good intentions' failures of NGOs (Kolk et al. 2011). Fifth, development partnerships can assist stakeholders in their work toward mutually beneficial outcomes. Finally, partnerships may prevent companies from taking a top-down approach to their relationships with stakeholders.

Despite the ostensible benefits of the partnerships approach, it is not hard to imagine that mining companies and their stakeholders may approach the concept

[4] For an analysis of the BOP hypothesis, see Spagnoletti and O'Callaghan (2013).

with some trepidation. The complexity that surrounds the coordination of a sizeable group of stakeholders is daunting. Moreover, achieving consensus on what the partnership would achieve, how its activities would be financed, participants' expectations, and stakeholder identification would take considerable effort. For example, should international organisations be partners? What role should there be for regional economic organisations, such as the Asian Development Bank? To transition a mining company to be a *de facto* development agent would have ramifications for its core business activities. What, then, are the parameters of a company's role as an agent of development? These are mainly logistical issues, and large mining and other extractive companies have the human and financial resources to undertake such activities successfully. Consequently, our concerns are not with the logistics of achieving successful development outcomes; rather, they are philosophical. Fundamental to our concerns are issues pertaining to sustainable development, cross-cultural understanding, and the potential for inequalities of power to arise.

When the goals of a development partnership are proposed, inevitably, the topic of sustainable development will arise; especially the question of how sustainable development principles should inform business practice. The point addressed by Moody, earlier, is important. There is a widely held view that linking mining and, *ipso facto*, other extractive activities and the concept of sustainable development is problematic. The claim often made by industry advocates is that mining can be a sustainable activity. However, given that mineral and energy resources are finite and non-renewable—except perhaps for energy resources on a geological timescale—this is something of a perversion of the generally accepted definition of sustainable development, as enunciated in the Brundtland Commission's Report (Corder in this volume). Underpinning that definition is the view that resources should not be depleted so that they remain available for the benefit of future generations. Being finite, mineral and energy resources will be depleted if consumption continues at the present level.

The mining industry has used the cloak of sustainable development to advance its interests, but, in a development partnership, it is likely that some stakeholders would seek clarification from a company about its commitments to realising sustainable development goals. Accordingly, it is difficult to see how mining and other extractive companies could claim to be advancing the discourse of sustainable development without undergoing a significant behavioural shift.

Our second point relates to cross-cultural understanding. Historically, executives in the mining industry have had a reputation for exhibiting cultural insensitivities. In more recent years, this has become less problematic, as companies have hired staff to manage the cultural and community dimensions of their business activities. However, where land is sacred or there is a chance that a company might damage a culturally significant site, it may be difficult for a mining company to build trust with local communities, as recently occurred in Northern Australia. There, blasting by OM (Manganese) damaged a sacred site, angering the community. The company subsequently was fined AUD$150,000. A lesson, then, is that a development partnership may not proceed or realise its objectives if the host community perceives or experiences negative impacts.

Our principal concern with the partnerships approach is the potential for the partnerships to perpetuate inequalities of power between various stakeholders. Power is a difficult concept to define, though Feher makes an excellent attempt in the first section of this volume. Here, we mean simply that, in commercial and other transactions, differences in power can have a deleterious impact on the participants, especially the most vulnerable among them. Often, it is the mining company that has the capital and, so, the 'whip hand'. Where this is the basis of a partnership, the outcomes are likely to be poor. For development partnerships to succeed, stakeholders cannot take advantage of their relative power position. We suggest that Arnstein's theorisation of levels of citizen participation contains lessons for mining companies in this regard.

The development partnerships approach is, like the old adage, 'easier in theory than in practice'. While it is a practical strategy, it ignores the reality that it is not possible to take a wholly practical approach to anything. Theory lurks behind the curtain of all practical ideas. By examining Arnstein's ideas, we hope the mining industry might take a peek behind the theoretical curtain. This is what is missing in the ICMM's 'Development Partnership Toolkit'.

Arnstein's Ladder of Citizen Participation and Extractive Development Partnerships

Arnstein's ladder of citizen participation is one of the earliest examples of a conceptualisation of the relationship between citizens and those who hold power in contemporary society. Arnstein's paper first was published in 1969, during a period of great interest among political scientists regarding political rule. At the time, pluralists, elite theorists, and Marxists vied to enunciate the most accurate understanding of how politics works. Pluralists arguably won that debate, given the rise and increasing importance of interest groups to politics in America and elsewhere.

Arnstein's paper was a significant intellectual achievement. It began with the French conjugation of the verb 'to participate' (*participer*). Arnstein wrote her paper a year after the French student riots in May 1968, and the conjugation of *participer* came from a protest poster. Arnstein's final conjugation of *participer* was *íls profitent*. Clearly, this is not a conjugation of *participer*; instead, it means *they profit*. Arnstein's implication was that, while individuals may believe they are *participating* in decision-making in and about a community, strategically located individuals—bureaucrats, politicians, business leaders, and others who may benefit from unequal power relations—have a greater capacity to exercise power and, therefore, *profit* from the exercise of that power. Arnstein argued that the concept and meaning of participation is not self-evident. Indeed, powerful individuals often invoke the language of participation to mask the lack of consultation and transparency in their decision-making.

Table 17.1 An adaptation of Arnstein's ladder and its applicability to the mining industry

Urban citizen participation		Participation in the mining industry	
8. Citizen Control	Citizen power and genuine participation	High levels of partnership with an mining community	Community equity stake in mining projects with management responsibilities within the company: power differential is low
7. Delegated power		Medium levels of partnership with a mining community	No citizen control, but medium levels of engagement between the company and the community: power differential is medium
6. Partnership		Low levels of partnership with a mining community	No citizen control, but low levels of engagement between the company and the community: power differential is high
5. Placation	Tokenism	Good and bad faith negotiation	Instrumental approach to participation
4. Consultation		Good and bad faith consultation	Instrumental approach to participation
3. Informing	Non-participatory participation	Obligations to local communities	Company follows state regulation
2. Therapy		Public relations	Greenwash/'spin'
1. Manipulation		Coercion and manipulation of the community	Use of corporate will, force, manipulation and the exercise of power to achieve corporate goals

Source: Arnstein (1969)

According to Arnstein, the concept of participation can be divided into eight levels. The bottom three levels—manipulation, therapy, and informing (non-participatory participation)—treat participation as a cynical attempt to gain assent, and do not involve community members in a transparent way. Arnstein provides the example of placing community members on advisory boards with the sole purpose of indoctrinating them with the power-holders' epistemologies. These levels are about manufacturing consent. The two next levels are consultation and placation (tokenism). It only is after these stages that participation starts to become meaningful. In the sixth level, the community is engaged in a superficial way. In level seven, communities openly are engaged about local issues. Level eight places control of community projects and activities in the hands of the citizenry. It only is at levels six through eight that there is evidence of citizen empowerment. The concept of participation masterfully is deconstructed by Arnstein. However, as she notes, the levels are not distinct (Arnstein 1969). A 'bleeding' between the levels can, and does, take place. More importantly, though, through Arnstein's ladder, we can see why many community projects stall, are co-opted, or fail completely.

How is Arnstein's ladder relevant to the mining industry? We believe that it offers important insights into the difficulties of business conduct where multiple stakeholders are present. It provides a simple matrix through which an extractive company can assess the degree of its engagement with its host community, and what may be required for it to have a successful development partnership with that community. While the ladder may not obviate the need for some CSR activities, it is a more concrete way for a mining company to think about how it engages with its stakeholders. It also may provide a window into the complex issues around power and equity between mining companies and their stakeholders.

The first and second columns of Table 17.1 are a reproduction of Arnstein's ladder. The third and fourth columns are our adaptation of the ladder to the mining industry. The bottom of the ladder shows the capacity of extractive companies to achieve commercial outcomes through force, coercion, and manipulation. Such tactics unlikely will result in long-term, sustainable commercial success, however. Companies interested in building successful development partnerships with host communities should aim to be on the top two rungs of the ladder.

Corporate social responsibility can be a meaningful tool for mining companies if carried through with conviction. However, too often it is a form of greenwash. We believe that development partnerships offer a promising new approach to the difficulties that mining companies face when operating in challenging social, political, and cultural environments.

Conclusion

Strong corporate reputations are difficult to achieve. It takes a long and sustained effort by companies to live up to the expectations of stakeholders. This means being good corporate citizens by being socially and environmentally in tune with stakeholders and customers' values, and actively improving the communities in which they operate. Very few companies can claim to be acting in accordance with this higher order mandate (O'Callaghan 2016).

Mining companies in particular struggle in this regard. Despite the industries' rhetoric, mining companies are a key source of environmental pollution, have been accused of intimidating local communities, and, on occasion, negotiate in bad faith. The rhetoric of sustainable development and CSR is losing traction with various stakeholders.

The ICMM has played an important role in fostering self-regulation in the mining industry. However, as we have demonstrated in this chapter, the mining industry as a whole still have room for improvement. We believe that the broad criticisms of CSR are having a negative impact on the industry, and, therefore, a move beyond CSR is timely. In response to the criticisms, the ICMM has advocated development partnerships as a way forward, and we see merits—with some reservations—in such an approach. The commitment to sustainable development is hollow unless the industries define what they understand sustainable development to be. Issues, for example, with cross-cultural understanding, if not addressed appropriately and at the start of a potential venture, may hamper partnership projects going forward.

Power imbalances likely will be the most significant challenge for the mining industry in building successful development partnerships. Arnstein offers a way to understand the nuances in a participation discourse, as well as a way of understanding inequalities of power. However, more work remains to be done to demonstrate the applicability of Arnstein's ladder to the mining industry and, particularly, to partnerships between mining companies and their host communities.

References

Arnstein, S. (1969). A ladder of citizen participation. *Journal of the American Institute of Planners*, 35(4), 216–224.

Bakan, J. 2005. *The corporation: the pathological pursuit of profit and power*. New York: The Free Press.

Blowfield, M., Frynas, J.G. (2005). Setting new agendas: critical perspectives of corporate social responsibility in the developing world. *International Affairs*, 81(3), 499–513.

Brammer, S., Pavelin, S. (2004). Building a good reputation. *European Management Journal*, 22(6), 704–713.

Brammer, S., Pavelin, S. (2006). Corporate reputation and social performance: the importance of fit. *Journal of Management Studies*, 43(3), 435–455.

Burton, B. (2002). When corporations want to cuddle. In G. Evans, J. Goodman, J. N. Lansbury. (Eds.), *Moving mountains: communities confront mining and globalisation* (pp. 109–124). London: Zed Books.

Doorley, J., Garcia, H. (2007). *Reputation management: the key to successful public relations and corporate communication*. London: Taylor & Francis.

Ekatah, I., Samy, M., Bampton, R., Halabi, A. (2011). The relationship between corporate social responsibility and profitability: the case of Royal Dutch Shell Plc. *Corporate Reputation Review*. 14(4), 249–261.

Ellsworth, B. (2007). Conoco left out of Venezuela's Orinoco deals. Reuters. http://www.reuters.com/article/us-venezuela-nationalization-oil-idUSN2527844220070425. Accessed 18 June 2016.

Fombrun, C. (1996). *Reputation: realizing value from the corporate image*. Cambridge, Massachusetts: Harvard Business School Press.

Graetz, G. (2015). Energy for whom? Uranium mining, indigenous people, and navigating risk and rights in Australia. *Energy Research & Social Science*, 8, 113–126.

Grayson, D., and Hodges, A. (2004). *Corporate social opportunity!: 7 steps to make corporate social responsibility work for your business*. Sheffield: Greenleaf Press.

Griffin, A. (2008). *New strategies for reputation management: gaining control of issues, crises & corporate social responsibility*. London: Kogan Page Publishers.

Henderson, D. (2009) Misguided corporate virtue: the case against CSR, and the true role of business today. *Economic Affairs*, 29, 11–15.

Henderson, D. (2001). The case against "corporate social responsibility". *Policy*, 17, 28–32.

Hilson, G., Murck, B. (2000). Sustainable development in the mining industry: clarifying the corporate perspective. *Resources Policy*, 26(4), 227–238.

Hutchins, M. Walck, C., Sterk, D., Campbell, G. (2005). Corporate social responsibility. *Greener Management International*, 52(1). 17–30.

ICMM. (2012). *Community Development Toolkit*. http://www.icmm.com/document/4080. Accessed 4 February 2016.

ICMM. (2011). *Mining partnerships for development toolkit*. http://www.icmm.com/document/1945. Accessed 4 February 2016.

Jackson, K. (2004). *building reputational capital: strategies for integrity and fair play that improve the bottom line*. Oxford: Oxford University Press.

Jacobsen, J. (2010). Old problems remain, new ones crop up: political risk in the 21st century. *Business Horizons*, 53(5), 481–490.
Knox, S., Maklan, S. (2004). Corporate social responsibility: moving beyond investment towards measuring outcomes. *European Management Journal*, 22(5), 508–516
Kolk, A., van Dolen, W., Vock, M. (2011). Trickle effects of cross-sector social partnerships, *Journal of Business Ethics*, 94, 123–137.
Labonne, B. (1999). The mining industry and the community: joining forces for sustainable social development. *Natural Resources Forum*, 23(4), 315–322.
Larkin, J. (2003). *Strategic reputation risk management*. Houndmills, Basingstoke: Palgrave Macmillan.
Leith, D. (2002). *The politics of power: Freeport in* Suharto's Indonesia. Honolulu: University of Hawaii Press.
Mahon, J., Wartick, S. (2003). Dealing with stakeholders: how reputation, credibility and framing influence the game. *Corporate Reputation Review*, 6(1), 19–35.
Moody, R. (2007). *Rocks and hard places: the globalisation of mining*. London: Zed Books.
Moody, R. (2005). *The risks we run: mining, communities and political risk insurance*. The Netherlands: International Books.
Moran, T. (1998). *Managing international political risk*. London: Blackwell.
Nelson J. (2012). The power of development partnerships: an interview with Jane Nelson. www.usaid.gov/press/frontlines/fl_mar12_PPP_NELSON.htm\#top. Accessed 5 March 2016.
O'Callaghan, T. (2016). *Reputation risk and globalisation: exploring the idea of a self-regulating corporation*. London: Edward Elgar.
O'Callaghan, T. (2010). Patience is a virtue: problems of regulatory governance in the Indonesian mining sector. *Resources Policy*, 35(3), 218–225.
Parsons, R., Lacey, J., Moffat, K. (2014). Maintaining legitimacy of a contested practice: how the minerals industry understands its 'social licence to operate'. *Resources Policy*, 41(1), 83–90.
Peters, G. (1999). *Waltzing with the raptors: a practical roadmap to protecting your company's reputation*. New York: John Wiley & Sons.
Porter, M., Kramer, M. (2006). Strategy and society: the link between competitive advantage and corporate social responsibility. *Harvard Business Review*, 78–92.
Siltaoja, M. (2006). Value priorities as combining core factors between CSR and reputation – a qualitative study. *Journal of Business Ethics*, 68(1), 91–111.
Sluyterman, K. (2010). Royal Dutch Shell: company strategies for dealing with environmental issues. *Business History Review*, 84(2), 203–226.
Spagnoletti, B., O'Callaghan, T. (2013). Let there be light: a multi-actor approach to alleviating energy poverty in Asia. *Resources Policy*, 63(3), 738–746.
Spagnoletti, B. (2011). *Alleviating energy poverty in Asia: opportunities for development and commerce*. Unpublished Thesis. The University of South Australia.
Spagnoletti, B., O'Callaghan, T. (2011). Going undercover: The paradox of political risk insurance. *Asia-Pacific Journal of Risk and Insurance*, 5(2), 1–23.
Taylor, M., Supriatna, Y. (2013). Freeport can't say when Indonesia mine to reopen, confirms 28 dead. Reuters. http://www.reuters.com/article/us-indonesia-freeport-idUSBRE94L03420130522. Accessed 5 November 2015.
White, A. (2005). Fade, integrate or transform? The future of CSR. *Business for Social Responsibility*, August. New York. http://www.jussemper.org/Newsletters/Resources/BSR_Allen-White.pdf. Accessed 5 November 2015.
Whitmore, A. (2006). The Emperor's new clothes: sustainable mining? *Journal of Cleaner Production*, 14(3/4), 309–314.
Windsor, D. (2001). Corporate citizenship: evolution and interpretation. In M. McIntosh, J. Andriof (Eds.), *Perspectives on corporate citizenship* (pp. 39–52). Sheffield: Greenleaf Press.
Zadek, S. (2007). *The civil corporation*. London: Routledge.

Chapter 18
Mining and Climate Change

Vigya Sharma

Abstract This chapter argues that the relationship between a changing natural environment and the mining industry remains strong, but is greatly undervalued. In the absence of robust localised climate modelling and experiential knowledge that can help devise strategic climate management plans, ongoing performance as well as future viability of the mining sector remain under a significant climate threat worldwide. More research is needed to better understand, not only the differentiated nature of impacts from varied climatic changes (from droughts to floods to bush fires) across various stages of the mining lifecycle, but also how these impacts may lead to different consequences across various geographical regions. To this end, I identify five key areas to strengthen the mining industry's understanding of climate change. First, recognise the complexity and inter-linkages that underlie climate change. Second, recognise that climate is a relative, not an absolute risk. Three, leadership is important in order to promote a behavioural change in both employers and employees. Fourth, develop open and flexible institutions. Finally, begin to develop a repository of knowledge on climate change, its impact on the industry and successful mitigation strategies.

Abbreviations

BNE	Business and the Natural Environment
BoM	Bureau of Meteorology
GDP	Gross Domestic Product
GHG	Greenhouse Gases
GSP	Gross State Product
IPCC	Intergovernmental Panel on Climate Change
ICMM	International Council on Mining and Metals
OT	Oyu Tolgoi
QFCI	Queensland Flood Commission of Inquiry

V. Sharma (✉)
Centre for Social Responsibility in Mining, University of South Australia, Adelaide, SA 5001, Australia
e-mail: v.sharma@uq.edu.au

T. O'Callaghan, G. Graetz (eds.), *Mining in the Asia-Pacific*, The Political Economy of the Asia Pacific, DOI 10.1007/978-3-319-61395-6_18

SO	Southern Oscillation
SOI	Southern Oscillation Index
UNEP	United Nations Environment Programme

Introduction

A number of studies have undertaken in-depth assessments of the many social, economic and environmental impacts that a region experiences due to mining activity (Franks et al. 2010; Browne et al. 2011). However, relatively few studies have focused on the influence of the natural environment on the mining industry (Sharma et al. 2013). Considering that the location of a mining operation is directly tied to the location of the resource being extracted, the mining industry's relation to the natural environment is critical to its future viability. Indeed, the role that climate plays in providing a supportive natural environment can hardly be underestimated, especially given that many mining operations are either currently located, or are planned to be developed in some of the world's most ecologically fragile regions. Table 18.1 below provides a list of some of the key mining hot spots globally that are located in regions considered vulnerable to current and future changes in climatic conditions. Additionally, a changing climate may have the potential to influence irrevocably the mining industry's future. It can undermine investor confidence in mining activities that are planned in regions likely to face extreme climatic events (Freed 2012; International Council on Mining and Metals (ICMM) 2013). To retain investor confidence, then, it is important that mining companies develop climate-adaptive projects.

Over the last 5 years, research in Canada and Australia has examined the influence that climatic change may have on the future of the mining industry in resource-dependent economies (Ford et al. 2011; Pearce et al. 2011; Loechel et al. 2013). Yet the field remains largely under researched, particularly in low income countries, where mining projects promise significant pathways to development, but where technical and financial limitations constrain the extent to which these countries are able to adapt to climate change. Mining may further aggravate changes in natural environmental. This may, in turn, cause disruptions to resource-dependent traditional livelihoods, including, herding, agriculture, and fisheries (Holden 2013).

The last decade has witnessed some of the harshest climate-related natural disasters, including widespread flash flooding in Australia and Thailand, extended heat waves in Europe, bushfires and persistent droughts in Australia and the US, and increasingly powerful tropical cyclones around the world. It is not surprising in these circumstances that almost 40% of all Global 100 Index companies studied in a recent Standard and Poors inquiry believed they had already witnessed climate-influenced extreme weather events (Crawford and Seidel 2013).

In its Fifth Assessment Report, the Inter-governmental Panel on Climate Change (IPCC) (2014), notes that 'primary economic activities (e.g. agriculture, forestry,

Table 18.1 Global mining hotspots vulnerable to climate change

Mining hotspot	Place	Primary mineral(s) extracted	Regional vulnerability to climatic changes	Reference
Australia	Central Queensland; Pilbara, Western Australia	Coal, iron ore	High: (prone to droughts, floods and cyclones)	Sharma et al. (2013) and Loechel et al. (2013)
Chile	Atacama desert	Copper	High: (prone to droughts)	Romero and Kampf (2003)
West Africa	Niger, Ghana; Mali	Gold, uranium	High: (droughts in the Sahel Region; Ghana facing erratic rainfall)	Ayee et al. (2011), Kunateh (2011), and Watts (2012)
Mongolia	Gobi Desert	Coal, gold, copper	High: (prone to droughts and snowstorms *dzuds*)	Chuluunkhuyag (2008) and Sternberg (2014)
Caribbean	Guyana; Jamaica; Trinidad and Tobago	Gold, diamond, Tar sands, Bauxite-alumina	High: (sea level rise and cyclones)	United Nations Environment Programme (UNEP) (2008)
Canada and the Arctic		Coal, gold, diamond, oil	High: (fluctuations in water supply, warmer winter temperatures)	Ford et al. (2010) and Pearce et al. (2011)

fishing, mining) are particularly sensitive to the consequences of climate change, because of their immediate dependence on the natural environment. In some regions, these activities dominate the economy (Arent and Tol 2014). While it is established that the impacts of a changing climate may bear heavily on certain economic sectors such as energy, infrastructure, insurance and tourism, there remains a significant gap in knowledge in relation to 'potential climate impacts on other economic sectors, such as mining, manufacturing, and services … in particular assessments of whether these sectors are indeed sensitive to climate and climate change' (Arent and Tol 2014).

Implications of a Changing Climate for Business and the Economy

Businesses are deeply connected to the natural environment. Not only do they impact the environment in positive and negative ways through their operations, but they remain a critical component to solving problems such as climate change (Hawkins 2005). Over the last two decades, the scale of interaction between humans and the environment has resulted in business and the natural environment (BNE) emerging as a focuses of debate about climate change, pollution, habitat loss, desertification and changes in biodiversity (Hoffman and Georg 2013).

Global efforts to tackle these issues are very important. Possibly the most divisive of these is climate change (Hulme 2009). Australia's 2014 repeal of the Carbon Tax is an example of one of the world's most fossil fuel intensive economies refusing to embrace action on climate change, despite strong international and local pressure to do so (Whitmore et al. 2014). There are two possible reasons for a reluctance to take action on climate change. First, uncertainty about the nature, timing and severity of climate impacts fundamentally inhibits financial and human investments in building resilience across vulnerable socio-economic actors. Second, following on from this uncertainty, businesses, in particular, have long struggled to understand both the tangible and intangible risks from a changing climate, and the resulting implications for long-term viability (Hoffman 2010).

Notwithstanding this friction, recent research coming out of Europe has established that 'a region's physical and social sensitivity to extreme weather events is usually highly correlated with its economic sensitivity to climate change' (Luckenkotter et al. 2013). Although not exhaustive, Table 18.2 provides examples of how changes in climatic conditions may impact particular industries, in both the short and long-term. Potential ramifications for businesses have made several prominent business houses increasingly cognisant of the challenges that climate change may bring over the next several decades. Some companies have acknowledged the real threat that climate change poses to their operations, while also demonstrating their willingness to move beyond the business-as-usual models to embrace adaptive decision-making at both strategic and operational levels (Hoffman 2006).

It is vital to differentiate impacts of natural climactic variability from those that occur as a consequence of anthropogenic climate change. The distinction is important as there are different causal factors underlying both phenomena. These have different impacts and, consequently, require different response strategies.

Climate change encompasses changes that are a result of an accumulation of greenhouse gas emissions, due largely to human activities. These changes have generated considerable scientific uncertainty, primarily because the science is not well-understood (Sharma and Franks 2013).

Lessons from extreme events that result from natural climate variability may help build capacity and inform adaptation pathways to address longer-term anthropogenic climate change (Amundsen et al. 2010; Kates et al. 2012; Productivity Commission 2012). As a matter of fact, there is evidence that although adaptation planning may have occurred in the past, it is the extreme events and resulting loss of infrastructure, and economic and socio-ecological disruption that often provides the necessary 'trigger' to execute long-established adaptation plans (Leszczynska 2012). Extreme weather events thus provide a timely 'window of opportunity' to bring about changes in the way operations embrace climate changes more satisfactorily (Kates et al. 2012; Productivity Commission 2013).

Table 18.2 Impacts of climate change on various economic sectors

Economic sector/industry	Likely key impact(s)	Nature of impact
Arable farming/agriculture/seafood	Water scarcity, loss in productivity, reduced supply, rise in food prices, inequitable access to food, consequences for human health, positive impact on agriculture in colder regions (new crop species, higher plant growth)	Direct
Extractive sector/mining/energy production	Lost production, drop in revenues, reduced water availability, impacts on employee health and safety, impacts on hydropower production potential, increase in energy demand due to change in temperature	
Infrastructure	Repair costs for rebuilding transport and energy infrastructure, reduced accessibility, long-term need for reengineering road and rail; redesign of energy transmission; a rise in energy prices	Both direct and indirect
Manufacturing	Effect on supply and demand chains, reduced competitive edge	
Pastoral farming/forestry	Desertification, deterioration in livestock health, impact on supply, rise in dairy and meat prices, increase in forest fires	
Technology and communication	Disruption to telecommunication channels, inundation of facilities thereby disrupting operations and services	
Tourism	Loss of water and snow, affecting tourist attractions; impact on biodiversity and wildlife; indirect effects on transportation and cultural lifestyles across both source and destination regions	

Sources: Hoffman (2006), Sussman and Freed (2008), UNEP (2008), Marshall (2010), Murtinho and Hayes (2012), Crawford and Seidel (2013), ICMM (2013), Loechel (2013), and Luckenkotter et al. (2013)

Mining and Climate Sensitivity: Lessons Past and Current

The mining industry has a deep-rooted connection to the natural environment. Mining activities impact the air, land, and water of host communities, potentially leading to community discord (Rolfe et al. 2007). The climate of host countries may also influence the extent to which mining activities may successfully be undertaken (Sharma et al. 2013). In this sense, the link is unique. Manufacturing, financial and human services, and retail activities can easily be relocated should the environment become unsupportive of these activities. This is not the case with mining. High sunk costs and fixed infrastructure mean that mines are exclusively tied to the location of the resource to be mined (Linnenluecke et al. 2011). There is a growing recognition that, as the mining industry changes the natural environment, not only will this impact on extraction, production and shipping, but also impact the long-term viability of the sector (Pearce et al. 2011; ICMM 2013; Sharma et al. 2013).

Mining companies have, for many decades, dealt with a variable climate as part of an ongoing risk management strategy. According to several mining industry insiders, the sector has constantly manoeuvred around risky climatic conditions for decades, including switching from dry to wet conditions to sustain its businesses and remain competitive (ICMM 2013). Yet the idea of anthropogenic climate change as a 'game changer' is still to find substantive support industry-wide (Sharma et al. 2013). The industry is still coming to grips with the fact that, while extreme weather events have always occurred, their frequency and intensity is set to change as a result of anthropogenic climate change. This is a shift that calls for a proactive strategy in order for the mining industry to address the problem. Unfortunately, this is something that is not happening across the sector at present. This is an operational, organisational, and strategic issue. It is also a reputational issue. Individual mining companies that fail to address climate change in a demonstrable way will likely be seen as industry laggards, as the wider sector moves to confront the problem. Consequently, there is an early mover advantage for companies that factor climate change into their forward planning.

The remainder of this section presents two cases of mining industry development in Central Queensland and in Mongolia. Both regions are extremely vulnerable to climatic change and have had climate-influenced natural disasters in recent years. Also their socio-economic and institutional contexts are markedly different, as are their mineral development trajectories and expertise. So they make excellent examples of how climate change might impact, not only on mining companies, but also on the local communities in which they operate. The cases have been chosen to provide a broad understanding of the many ways in which climate may interact with the mining sector. At the end of the chapter I look at the implications for the mining sector in the Asia-Pacific Region more broadly.

Central Queensland, Australia

The dependence of the mining industry on the natural environment is of particular significance to Australia for two reasons. First, Australia's economy relies heavily on the mining sector for exports which, in turn, suggests a deep connection between Australian communities and the mining sector for long-term socio-economic development. Queensland is the world's largest exporter of coal, contributing to approximately 52% of the total coal exports worldwide (ABARES 2011). In addition, mining contributes to more than 20% of Queensland's economy and 13% of both direct and indirect employment (ABARES 2011; Queensland Flood Commision of Inquiry (QFCI) 2011). The Bowen Basin in Central Queensland is the largest

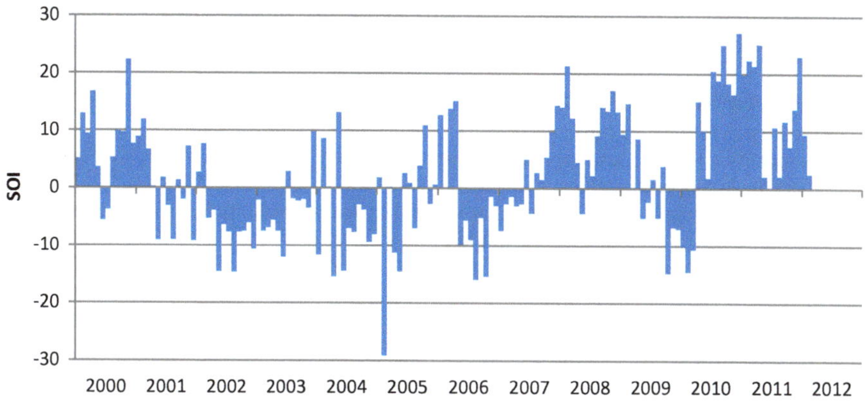

Fig. 18.1 Australia's Southern Oscillation Index from 2000–2012. Source: Sharma et al. (2013)

coal-reserve in Australia, and produces Queensland's high-grade coking coal, along with most of the exported thermal coal

Second, several characteristics inherent to Australia's climatic and oceanographic conditions have traditionally contributed to high natural climatic variability across the continent, thus causing an unstable climate that oscillates between extreme dry and wet conditions (Bureau of Meteorology (BoM) 2008). Climatic perturbations observed since 2000 demonstrate this variability quite well (see Fig. 18.1). From June 2002 to late 2010 a strong negative Southern Oscillation Index (SOI) resulted in an extended El Niño state causing severe drought conditions across several parts of Queensland, including the Bowen Basin. During the period of 2000–2011, four cyclones—Beni (February 2003), Ului (March 2010), Tasha (December 2010) and Yasi (February 2011)—were recorded that resulted in significant impacts on coal mining operations across the Bowen Basin. At the same time, major flood events occurred between the summer periods of 2007–2008 and 2011–2012 due to a positive SOI, resulting in a strong La Niña event.

In the most recent flood event of 2010–2011, several towns, both mining intensive and larger service hubs, in and around Central Queensland, were flooded. It took many days before normal services and mining operations were able to resume. In fact, the La Niña effect was one of Australia's strongest since 1917 (Nicholls 2011). This led to 'flooding of historic proportions' (ABARES 2011). Almost 80% of the entire state of Queensland—with more than 2.5 million people and several thousand kilometres of road and rail infrastructure—was declared flood-affected. The state's agricultural and mining sectors in particular, suffered huge losses both directly and indirectly due to prolonged floods in the region (Queensland Reconstruction Authority 2011; ABARES 2011). According to estimates, the 2010–2011 floods led to a total loss in excess of A$5 billion to Queensland's gross state product (GSP) that included more than A$2 billion in lost coal export earnings (QFCI 2011; ABARES 2011). The key impacts from drought and flood extremes that challenged the mining sector's capacity to conduct business as usual during and immediately after the events are listed below:

Droughts (reduced availability of water)
• Production falls below capacity; • Competition for water use negatively influenced industry's relations with downstream water users, including communities; • Scarce water supply leading to industry's psychological adaptability to a sudden change in climate leading to extreme wet conditions; • Rise in industry's operating costs due to increase in energy prices resulting from region-wide water scarcity; and • Rise in dust-related environmental hazards (for mine workers and local communities) from failure to undertake extensive dust suppression due to shortage of water.
Floods
• Lost revenue generation due to flooded pits limiting production for several months after the events; • Road and rail damage resulting in limited coal supply reaching ports for export to international markets; • Additional costs borne by the industry from hiring of pumping infrastructure for dewatering pits, and repairing and upgrading damaged mine site infrastructure; • Concerns over mine water releases leading to long-term damage to industry's relationships with communities located in the immediate vicinity of mining operations but also those further downstream; and • Lost time in production resulting in a significant drop in royalty revenues to the state government.

Source: Sharma et al. (2013)

The many lessons that mining companies have learnt after dealing with Queensland floods and droughts over the past decade warrant further discussion. First, the mining industry continues to experience high staff turnover rates, thereby resulting in an organisation-wide short-lived memory of extreme events, key impacts and responses undertaken at the time. What is needed is an established system that creates a repository of experiential knowledge, sound planning lessons and 'emergency' tips that can be revisited in the event of a future climate disaster. Unfortunately, the scope for 'reflective learning' in the industry remains rather limited. Ensham, an Australian mining company with coal operations in Central Queensland, provides a compelling example of how management's proactive thinking and the ability to apply knowledge from the past (experience from the floods of 2007–2008) did not only help prevent a repeat flood disaster in 2010–2011 but, in doing so, gained a significant competitive edge that allowed its operations to continue while other companies' mine sites remained closed or operated under-capacity following severe damage from the floods.

Second, the impacts of anthropogenic climate change are likely to interact with existing climate problems, and create a new wave of challenges underlined by increasing uncertainty and complexity (Palutikof 2010). Addressing these new challenges will require innovation and collaboration at a scale hitherto not experienced (Sharma et al. 2013). In the context of Central Queensland where agriculture and mining represent the two primary economic activities, a changing climate will further exacerbate challenges for the region's long-term sustainability. These might include, but are not limited to, compounding water stress due to the water-intensive nature of both mining and agriculture in areas likely to experience higher evaporation rates, and resulting conflict around resource access for various industries and stakeholders.

At the same time, extreme events can help drive innovation to ensure sustainability of resource regions (Kates et al. 2012; ICMM 2013). They 'force' stakeholders across government, industry and civil society to rethink current processes of risk management and apply new technologies to build collective knowledge that can transform crises into long-term advantages. While some of this innovation happens organically through the course of a climate-influenced natural disaster, the coming together of various actors provides a systematic opportunity to assess the nature of new knowledge needed, and how that can be processed, applied, and methodically updated. Such events are therefore 'catalysts' for the region by encouraging public and private stakeholders to be innovative in thinking about risks and resilience. Examples of innovation already in place due to climatic events experienced through the past decade in Central Queensland include industry's efforts to undertake reverse osmosis treatments and dry beneficiation during floods and droughts respectively to manage water surplus and scarcity and the establishment of joint water management actor groups, such as the Fitzroy Partnership for River Health (Sharma et al. 2013).

Finally, if the drought, flood and cyclone events witnessed in Queensland over the last decade are anything to go by, the future of the mining industry in Australia is under significant threat. Australia's recognition as one of the world's most vulnerable regions to anthropogenic climatic change adds a further layer of complexity to the mining-environment interface (Palutikof 2010). The enormity of impacts from extreme weather events in the last decade has clearly highlighted how climate change could stretch the socio-economic and institutional capacities of mining industry stakeholders operating in sensitive climatic conditions. These events have provided an opportunity to reflect on past experiences and better inform our understanding of what measures need to be undertaken to both assist recuperation of impacted operations and develop suitable adaptation options to prevent (or limit) impacts from such events in the future across Australia.

Mongolia

Mongolia provides a unique landscape of internal and external socio-cultural, economic, institutional and environmental forces that continue to challenge the country's long-term sustainability 'at multiple temporal and spatial scales' (Nixson and Walters 2006; Lioubimtseva and Henebry 2009). These forces include increasing climate variability; a host of context-specific factors such as remoteness, extremely low rates of population density, high dependence on the natural environment for livelihoods, and increasing urbanisation (the national urban population increased from 57% in 2000 to 68% in 2010). Moreover, a series of changes since 1991 including, democratisation after five decades of Soviet rule and subsequent opening of its economy to global market forces impacted the country significantly (Nixson and Walters 2006; Ochirsukh 2011; Reeves 2011; Sietz et al. 2011).

Traditionally, Mongolia's nomadic society, with strong dependence on a pastoral lifestyle, have dealt well with climate variability and aridity. However, the transition to a market economy in 1991 resulted in permanent changes to the nature and scope of traditional livelihoods (Sharma 2014). Mostly, these changes related to poor service delivery, limited government intervention to address climate-related water stress, and the effect of globalisation on changes in livestock composition, and subsequent negative impacts on the environment. Although around 40% of all Mongolians are involved in the agricultural sector (UN Statistics 2013), it has declined over the last two decades. Growing institutional allegiance to resource extraction as a fast track development tool, and increasing climatic variability manifested in higher rates of desertification, water stress, and extreme weather events (dzuds), may significantly threaten the country's herding-based traditional economy.

Mongolia's rapid resources boom based on copper, coal, gold, fluorspar, and uranium has led to a GDP rise of approximately 17.3% in 2011 (UN Statistics 2013). Recent census reveals that the two major mining centres in the South Gobi, Khanbogd and Tsogttsetsii, have witnessed an exponential growth in their population between 2000 and 2010 (approximately 201% and 229% respectively) (Ochirsukh 2011). In 2012, mining accounted for approximately 9% of all new employment opportunities created in Mongolia (Dalaibuyan 2013). Further, subject to China's ongoing demand for minerals, current trends project that Mongolia's GDP will continue to double every 2 years for the next 10 years (Fincenter 2011). The minerals sector already employs in excess of 14,000 people and contributes to more than 70% of Mongolia's total export earnings (Austrade 2011). Oyu Tolgoi (OT), besides being one of the world's largest copper mine, is expected to contribute approximately 34% to Mongolia's total GDP once fully operational in 2020 (Dalaibuyan 2013).

Mongolia is considered a 'climate hotspot', with more than 80% of the total territory vulnerable to climate extremes (Batima et al. 2005). Over the last 20 years or so, the country has experienced a number of devastating weather-related disasters (Tachiiri et al. 2008). As a case in point, desertification in Mongolia is currently estimated to affect '44–90% of the country's territory' (Reeves 2011). With strong changes in the frequency and intensity of both heat and cold waves, the country's population and natural resource-based economic enterprises may become more vulnerable to anthropogenic climate change. At the same time, loss of institutional support since Mongolia's transformation to a market economy has outstretched existing adaptive capacities. This has led to rising concern among policy-makers (Wang et al. 2013; Wachman Fall 2010).

Mongolia's institutions remain hopeful that the minerals boom will continue to contribute to national economic development (Ochirsukh 2011). However, its ability to do so is dependent on whether or not it is able to coexist with the Mongolia's strong custodial relationship with the land (Upton 2009). The nature of the relationship between the mining industry and rural, remote communities is an important one. Economic growth that disregards its impacts on the country's socio-ecological way of life is likely to lead to political risks for foreign mining companies.

Given the nature of Mongolia's traditional economy, both water availability (quantity) and quality remain central to the survival of several pastoralist communities (Sternberg 2008). The mining industry is becoming more water-intensive, at the same time as the country is experiencing reduced rainfall. There is a real possibility that further minerals development will exacerbate Mongolia's already tenuous water supply (Tiwary 2001; Combellick-Bidney 2012). Resulting impacts of any magnitude on pasture availability may become a potential source of long-term conflict between various actors operating in the region (Austrade 2011; Upton 2012; Wang et al. 2013). Most importantly, experience from other regions suggests that a conflict over water availability and quality may, in turn, cause irreparable damage to the industry's social license to operate (Lawson and Bentil 2013; Hart et al. 2008).

With both industries dependent upon access to minerals that are, in turn, likely to be under direct threat from climatic changes to natural environmental conditions, it is necessary to explore how mining and herding might coexist in a changing climate, as well as identify pathways to avoid conflict. The mining industry has access to technological innovation, financial resources and advanced knowledge of climate change. At the same time, pastoralists have a long history of coexisting with extreme weather events. Indeed, there may be scope to develop strategies based around both these unique perspectives.

Table 18.3 below deconstructs some of the complexity underlying the mining-herding-climate nexus in Mongolia. From an ecological perspective, water availability, dust, and availability of good pasture, may have a direct bearing upon Mongolia's future sustainability. Mining has led to increase in traffic through areas that had limited vehicular movement previously. While increased traffic has, in turn, generated excessive amount of road dust, mining operations also contribute large amounts of coal dust around their operations. In addition, mining brings with it an increase in traffic and infrastructure development. These activities have also aggravated the rate of pasture degradation across several parts of Mongolia. An already arid environment undergoing extreme climatic variability manifested in shorter, buvt more intense rainfall events, suggests that further pasture degradation, erosion and over-grazing are likely.

A number of these impacts are affecting the health of herder families and their livestock, but also their livelihoods in numerous ways (e.g. increased prevalence of asthma and lung disorders among herders and their herds, and dust settling on livestock leading to reduced returns from sale of animal fur and cashmere). To that extent, mining and herding in the presence of growing climate variability is creating a landscape of multiple cause and effect relationships that form an emerging area of concern for Mongolia's policymakers.

Finally, three important observations may be drawn in the context of Mongolia's future sustainability. First, considering that the steady minerals boom has put Mongolia at an important development crossroads, it is crucial that the boom is managed in a way that supports and does not undermine the traditional herding-

Table 18.3 The mining-herding-climate nexus in Mongolia

Impact area/driver of change	Water	Dust	Pasture
Mining (socio-economic)	Water-intensive industry; potential impact on water quality through unregulated mine-affected water discharge	Coal dust from mining; road dust from increased traffic	Pasture fragmentation and loss due to mining activities (including traffic movement, extraction and infrastructure development)
Herding (socio-economic)	Relies on water availability of good quality	Excessive grazing-induced erosion and desertification	Over-grazing and changes in livestock composition (increase in goat ownership due to market demands for cashmere)
Climate (bio-physical/natural)	More intense rainfall over a shorter duration; increased run-off; drying up of rivers and streams	Lack of moisture in an arid environment	Desert landscape, takes longer to regrow; loss of good quality pasture due to extended droughts followed by severe cold (as in a dzud)

Sources: Chuluunkhuyag (2008), Sternberg (2008, 2014), Tachiiri et al. (2008), Lioubimtseva and Henebry (2009), Upton (2012), Wachman (2010), and Batima et al. (2005)

based lifestyles of its people. Second, at the same time, it is important for Mongolia's policymakers to acknowledge that the mining industry remains vulnerable to a changing climate as well, particularly in light of Mongolia's increasing water stress and aridity. Finally, the interdependence between herders and mining companies and the implications of climate change, collating accurate climate data is necessary. This data is necessary for impact attribution and management. In the likelihood of multiple mining operations intensifying over certain regions of Mongolia (such as in the South Gobi), cumulative impact attribution and assessment is vital to developing a robust response strategy, both operationally as well as at the policy level.

Where to Next? Future Prospects for Resource Development in a Changing Climate

Most businesses have adapted, through the course of their existence, to a range of external factors: market volatility, technological transitions, financial crises, changing consumer awareness and demand parameters, and unstable regulatory environments. Yet, businesses are generally unwilling to accept that climate unpredictability is a potential driver of change that may impact negatively of their long-term growth (Acclimatise 2010). This reluctance can be attributed to a number of factors. These include a lack of confidence in climate science, the distant temporal nature of projected impacts, and a short-term profit agenda. Such an approach does not allow for long-term planning and performance monitoring.

Although the last few decades have seen a better integration of how business influences and is, in turn, influenced by the natural environment, there is still a long way to go for certain sectors, including the mining industry (ICMM 2013). Very little research currently exists in this space, most of which is drawn from developed countries, such as Canada and Australia. There is an urgent need for greater financial and time investment from all mining stakeholders to develop broad, yet robust, practical guidelines about how mining companies best identify and address climate thresholds that may threaten their long-term viability in both emerging and developed economies (Hodgkinson et al. 2010; Pearce et al. 2011; Sharma and Franks 2013; Arent and Tol 2014).

To identify new thresholds in light of looming uncertainty over the exact nature of future climate impacts coupled with an industry-wide lack of experience in addressing anthropogenic climate change, past experiences dealing with natural extreme weather events may provide invaluable insights. In particular for the mining industry, where a strong reliance on stakeholder engagement exists, the role of 'reflective learning' and 'sense making' is important (Nass et al. 2005; Madsen 2009). On the one hand, it may influence industry behaviour and generate a wider appreciation for proactive planning and adaptive risk management strategies. On the other hand, it can create incentives for various actors to function along a collaborative platform to develop a coherent adaptation policy for future climatic events that is cognisant of local context, opportunities and priorities. Further, weather-related natural disaster events provide a useful trigger to bring about innovation and a change to the way policy-makers and practitioners address and manage climate change (Sharma et al. 2013).

Research from other business organisations demonstrates that climate-sensitising business modelling, as well as organisational psychology, are important elements that allow mining companies to foster opportunities for improved competitiveness, innovation and economic longevity (Hoffman 2006; Hoffman 2010). The mining industry too, is demonstrating positive developments that are gradually challenging the sector's deep-seated scepticism around climate change. For example, Rio Tinto and other mining giants are already undertaking actions aimed to help their organisations better understand, and adapt to, the likely impact of anthropogenic climate change on their future business models (Sussman and Freed 2008; Freed 2012; ICMM 2013). Considering the highly competitive nature of the business, it is possible that early leadership from these actors may encourage climate 'laggards' to follow suit (Herremans et al. 2009).

An integrative approach that allows managing a rapidly expanding risk landscape and building climate resilience across the mining lifecycle will not only boost business reputation, stakeholder relationships, thus providing direct socio-economic benefits, but also foster investor confidence for future business growth (ICMM 2013). Such an approach will be founded on the following five cornerstones of effective adaptation.

Recognise the Complexity and Inter-Linkages that Underline Climate Change

Climate science is characterised by plurality and on-going uncertainty (Dessai et al. 2004) that often translates into inaction or scepticism in relation to adapting to climate change. The underlying complexity is further aggravated when climatic changes interact with existing stressors to create additional challenges and compound impacts that question conventional risk management strategies (Acclimatise 2010).

In the mining sector, as has already been noted in numerous studies, diverse pre-existing challenges span the breadth of development (Rolfe et al. 2007; Browne et al. 2011). Understanding how these interact with a changing climate is invaluable for developing guidance on impact attribution and management.

Recognise Climate as a Relative, Instead of an Absolute Risk

Unlike ozone depletion, which is an absolute change to the environment, climate change is relative in that it may have both positive and negative implications across different regions (Hulme et al. 1999; Hulme 2008). Its impacts also rely, to a large extent, on social and individual perceptions and values. In light of the relative nature of its impacts, climate change management, including resilience-building and adaptation requires a nuanced approach, not only to assessing the *costs of inaction* that the industry may bear over the long-term, but also the best strategy to capitalise on the opportunities that changing natural conditions provide to drive innovation and mineral industry longevity.

Effective Leadership

Leadership across all levels is a central tenet of effective adaptation (Hulme 2009; Hoffman 2010). The ‘reinforcing loop’ is recognised as playing a particularly important role where a committed work culture strengthens strong leadership that, in turn, promotes behavioural change in both employee and employer (Hoffman 2006). Further, policy incentives for leaders and penalties for laggards must apply, as they do in some other industries (Hoffman 2010).

Flexible, Open Institutions

Given the competitive nature of the mining industry, restricted information sharing within the industry affects the mining sector’s flexibility in adapting to climate change (Hoffman 2006). In addition, there is an established need for government and its regulators to be open to robust policy making in relevant areas of environmental regulation, such as greenhouse gas emissions (GHG) and climate mitigation (Sharma et al. 2013).

Knowledge Repository

According to Hulme (2008), '[c]limate is an idea which encapsulates the immersion of the physical with the cultural, in which local and global dynamics interweave and where the memory of the past meets the possibilities of the future'. Experiences of extreme weather events associated with anthropogenic climate change, can provide a great deal of information to guide 'future thresholds and adaptive behaviour' (Dessai et al. 2004). Understanding what strategies worked well in managing and coping with past events, how local conditions may have changed since the event, and identifying what additional efforts may be required as a result need to be better understood. Also the experiences from events occurring across diverse socio-economic and institutional settings may further enrich this knowledge repository. Such an iterative database may capture a wide range of information including, site-level experiences and responses to extreme events, organisational decision making processes and bottlenecks during a natural disaster and, particular events that may have acted as catalysts to bring about a modest change in behaviour towards climate preparedness.

Finally, this chapter has argued that the relationship between a changing natural environment and the mining industry remains strong, but is greatly undervalued. In the absence of robust localised climate modelling and experiential knowledge that can help devise strategic climate management plans, ongoing performance as well as future viability of the mining sector remain under a significant climate threat worldwide. More research is needed to better understand, not only the differentiated nature of impacts from varied climatic changes (from droughts to floods to bush fires) across various stages of the mining lifecycle, but also how these impacts may lead to different consequences across various geographical regions.

Conclusion

Although the nature and scope of impacts from a changing climate may vary, all kinds of economic activities remain vulnerable to climate change. Climate change has become a buzzword for business over the last decade, in much the same way that 'environment' was in the 1950s and 1960s. The seminal work of Carson was an important catalyst here (Carson 1962). From international financial companies like Citigroup and Goldman Sachs, to industrial players like Royal Dutch Shell, Rio Tinto and DuPont, businesses are actively engaging with the climate debate and using the opportunities to explore the possibility of innovative climate-sensitive approaches to business.

Significant research has been undertaken to establish the impacts of a changing climate on the future of some of our key primary industries such as agriculture, fishing and forestry. Little sound knowledge, however, currently exists to help understand how these critical multi-dimensional dependencies might play out for the mining industry worldwide. Some research from Canada and Australia provide a limited understanding of the relationship between climate change and the future of competitiveness of mining companies, or what implications this might have on the social and economic landscape of regions that host mining activities.

There are several characteristics specific to the mining sector which further accentuate its vulnerability to a changing climate. These include, but are not limited to: the location of some of the world's major mining operations in remote, highly climate-sensitive environments, direct dependence on suitable natural environmental conditions for the industry to extract resources, no option to relocate elsewhere should the climate become unsuitable for further operations, a conventional market-orientated profit-based approach to decision-making at the cost of ignoring climate change and, last, but not least, high staff turnover across the sector that further limits in-house knowledge accumulation on early resilience building and long-term climate-adaptive planning.

This chapter has taken a step further in the discussion on mining and climate change by providing an assessment of how climate impacts may interact with mining operations in diverse host regions. Australia and Mongolia are at critically different stages in their minerals-based economic development pathways. Yet, the mining industry in both regions remains under-resourced in relation to building their adaptive capacities. It is important that further research is undertaken with mining stakeholders across different world regions to allow a comprehensive synthesis of what processes and strategies are currently underway to build both site and organisation level resilience to future climate change, as well as identify collaborative and partnership opportunities across diverse stakeholder groups in the mining sector to boost innovation in long-term adaptation planning.

References

Australian Bureau of Agricultural and Resource Economics and Sciences. (2011). The impact of recent flood events on commodities. ABARES special report. Canberra: Australia. http://resources.news.com.au/files/2011/01/21/1225992/459658-flood-report.pdf. Accessed 4 June 2012.

Acclimatise. (2010). Building business resilience to inevitable climate change. http://www.slideshare.net/BlakeLapthorn/building-business-resilience-to-inevitable-climate-change-seminar. Accessed 12 August 2012.

Amundsen, H., Berglund, F., Westskog, H. (2010). Overcoming barriers to climate change adaptation question of multilevel governance? *Environment and Planning C: Government and Policy,* 28, 276-289.

Arent, D., Tol, R. (2014). Key economic sectors and services. *Fifth assessment report: Impacts, Adaptation, and Vulnerability*. Cambridge: Cambridge University Press, 659-708. https://www.ipcc.ch/pdf/assessment-report/ar5/wg2/WGIIAR5-Chap10_FINAL.pdf. Accessed 4 December 2014.

Australian Trade Commission. (2011). Mongolian mining projects. http://www.austrade.gov.au/Australian/Export/Export-markets/Countries/Mongolia/Market-Profile/mining-projects-report. Accessed 5 February 2012.

Ayee, J., Søreide, T., Shukla, G., et al. (2011). Political economy of the mining sector in Ghana. Policy research working paper, 5730. Washington D.C.: The World Bank. http://elibrary.worldbank.org/doi/abs/10.1596/1813-9450-5730. Accessed 6 April 2012.

Batima, P., Natsagdorj, L., Gombluudev, P. et al. (2005). Observed climate change in Mongolia. AIACC working paper No. 12. http://www.start.org/Projects/AIACC_Project/working_papers/Workingper cent20Papers/AIACC_WP_No013.pdf. Accessed 6 April 2012.

Bureau of Meteorology. (2008). Report on Queensland floods. http://www.bom.gov.au/hydro/flood/qld/fld_reports/reports.shtml. Accessed 21 September 2011.

Browne, W., Franks, D., Kendall, G. (2011). The foundations for responsible mining in Cambodia – suggested approaches. Phnom Penh, United Nations Development Programme, Cambodia. https://www.csrm.uq.edu.au/Portals/0/docs/02_Policy-Brief-En-V5.pdf. Accessed 5 July 2011.

Carson, R. (2002). *Silent spring*. New York: Houghton Mifflin Company.

Chuluunkhuyag, S. (2008). The impact of climate change and human activity on Mongolian water resources. Third WEPA International Forum on Water. http://www.wepa-db.net/pdf/0810forum/paper32.pdf. Accessed 24 October 2012.

Combellick-Bidney, S. (2012). Mongolia's mining controversies and the politics of place. In J. Diekerks (Ed.). *Change in democratic Mongolia: social relations, health, mobile pastoralism, and mining* (pp.271-296). Leiden: Brill.

Crawford, M., Seidel, S. (2013). Weathering the storm: building business resilience to climate change. Centre for Climate and Energy Solutions. Arlington: Virginia. http://www.c2es.org/initiatives/business-resilience. Accessed 4 May 2014.

Dalaibuyan, B. (2013). Scoping the potential for enterprise facilitation in Mongolia. Brisbane, International Mining for Development Centre. https://www.csrm.uq.edu.au/publications/scoping-the-potential-for-enterprise-faciliation-in-mongolia. Accessed 1 August 2013.

Dessai, S., Adger, W., Hulme, M., et al. (2004). Defining and experiencing dangerous climate change. *Climatic Change* 64(1/2), 11-25.

Fincenter. (2011). Brief information: current economic situation of Mongolia. http://www.slideshare.net/FinproRy/brief-information-about-mongolia. Accessed 20 October 2013.

Ford, J., Pearce, T., Prno, F., et al. (2011). Canary in a coal mine: perceptions of climate change risks and response options among Canadian mine operations. *Climatic Change*, 109 (3-4), 399-415.

Ford, J., Pearce, T., Prno, J., et al. (2010). Perceptions of climate change risks in primary resource use industries: a survey of the Canadian mining sector. *Regional Environmental Change*, 10(1), 65-81.

Franks, D., Brereton, D., Moran, C. (2010). Managing the cumulative impacts of coal mining on regional communities and environments in Australia. *Impact Assessment and Project* Appraisal, 28(4), 299-312.

Freed, J. (2012). It will hit trade and investor confidence, say miners. *Australian Financial Review*. http://www.afr.com/p/national/it_will_hit_trade_and_investor_confidence_TvLN34xlr4PGXK2PC1UbWM. Accessed 20 November 2014.

Hawkins, P. (2005). *The ecology of commerce: A declaration on sustainability*. New York: Collins Business.

Hart, B., Greenfield, P. Pascoe, M. (2008). Report to Queensland Premier: review of the Fitzroy River water quality issues. Department of Environment and Resource Management. Queensland.

Herremans, I., Sandy Herschovis, M., Bertels, S. (2009), Leaders and laggards: the influence of competing logics on corporate environmental action. *Journal of Business Ethics*, 89(3), 449-472.

Hodgkinson, J., Littleboy, A., Howden, M., et al. (2010). Climate adaptation in the Australian mining and exploration industries. Climate adaptation flagship working paper No. 5. CSIRO http://csiro.au/resources/CAF-working papers.html. Accessed 5 January 2011.

Hoffman, A. (2006). *Getting ahead of the curve: corporate strategies that address climate change*. Ann Arbor, Michigan, University of Michigan.

Hoffman, A., Georg, S. (2013). A history of research on business and the natural environment: conversations from the field. In S. Georg and A. Hoffman (Eds.). *Business and the environment: critical perspectives in business and management* (pp.1-58). Oxford: Routledge.

Hoffman, A. (2010). Climate change as a cultural and behavioural issue: addressing barriers and implementing solutions. *Organizational Dynamics* 39(4), 295-305.

Holden, W. (2013). Neoliberal mining amid El Niño induced drought in the Philippines. *Journal of geography and geology*, 5(1), 58-77.

Hulme, M. (2009). *Why we disagree about climate change: understanding controversy, inaction and opportunity*. New York: Cambridge University Press.

Hulme, M. (2008). Geographical work at the boundaries of climate change. *Transactions of the Institute of British Geographers*, 33(1), 5-11.

Hulme, M., Barrow, E., Arnell, N., et al. (1999). Relative impacts of human induced climate change and natural climate variability. *Nature*, 397(6721), 688-691.

International Council on Mining and Metals (2013). Adapting to a changing climate: implications for the mining and metals industry. http://www.icmm.com/page/92086/adapting-to-a-changing-climate-implications-for-the-mining-and-metals-industry. Accessed January 3 2014.

Intergovernmental Panel on Climate Change. (2014). *Climate Change 2014: Impacts, Adaptation, and Vulnerability* (pp.659-708). New York: Cambridge University Press.

Kates, R., Travis, W., Wilbanks T. (2012). Transformational adaptation when incremental adaptations to climate change are insufficient. *Proceedings of the National Academy of Sciences*, 109(19), 7156-7161.

Kunateh, M. (2011). Climate change threatens Ghana's food security. http://thechronicle.com.gh/climate-change-threatens-ghanaper centE2per cent80per cent99s-food-security/. Accessed 20 January 2015.

Lawson, E., Bentil, G., (2014). Shifting sands: changes in community perceptions of mining in Ghana. *Environment, Development and Sustainability*, 16(1), 217-238.

Leszczynska, A. (2012). Adaptation to climate change as a process of organizational learning. Proceedings of the Management, Knowledge and Learning International Conference, International School for Social and Business Studies, Celje, Slovenia. issbs.si/press/ISBN/978-961-6813-10-5/papers/ML12_106.pdf. Accessed 3 March 2016.

Linnenluecke, M., Stathakis, A., Griffiths, A. (2011). Firm relocation as adaptive response to climate change and weather extremes. *Global Environmental Change*, 21(1), 123-133.

Lioubimtseva, E., Henebry, G. (2009). Climate and environmental change in arid Central Asia: impacts, vulnerability, and adaptations. *Journal of Arid Environments*, 73(11), 963-977.

Loechel, B. (2013a). Mining companies are underprepared for climate change. The Conversation. http://theconversation.com/mining-companies-are-underprepared-for-climate-change-13091. Accessed 29 April 2013.

Loechel, B., J. H. Hodgkinson and K. Moffat (2013b). Climate change adaptation in Australian mining communities: comparing mining company and local government views and activities. *Climatic Change*, 119(2), 465-477.

Luckenkotter, J., Langeland, O., Langset, B., et al. C. (2013). Economic impacts of climate change on Europe's regions. In P. Schmidt-Thome and S. Greiving (Eds). *European climate vulnerabilities and adaptation: a spatial planning perspective* (pp. 95-112). Oxford: Wiley-Blackwell.

Madsen, P. (2009). These lives will not be lost in vain: organizational learning from disaster in U.S. coal mining. *Organization Science*, 20(5), 861-875.

Marshall, N. (2010). Understanding social resilience to climate variability in primary enterprises and industries. *Global Environmental Change*, 20(1), 36-43.

Murtinho, F., Hayes, T. (2012). Adaptation in resource-dependent communities: a call for greater methodological clarity in adaptation field research. *Society and Natural Resources*, 25(5), 513-522.

Nass, L., Bang, G., Eriksen, S., Vevatne, J (2005). Institutional adaptation to climate change: flood responses at the municipal level in Norway. *Global Environmental Change*, 15(2), 125-138.
Nicholls, N. (2011). The strongest La Niña on record. The Drum Opinion. ABC News. http://www.abc.net.au/unleashed/42858.html. Accessed 6 October 2011
Nixson, F., Walters, B. (2006). Privatization, income distribution, and poverty: the Mongolian experience. *World Development*, 34(9), 1557-1579.
Ochirsukh, Y. (2011). Mongolia 2010 population census: main findings. http://www.slideshare.net/Ochiro/mongolia-2010-population-census-main-findings. Accessed 20 October 2013.
Palutikof, J. (2010). The view from the front line: adapting Australia to climate change. *Global Environmental Change*, 20(2), 218-219.
Pearce, T., Ford, J., Prno, J., et al. (2011). Climate change and mining in Canada. *Mitigation and Adaptation Strategies for Global Change*, 16(3), 347-368.
Productivity Commission (2013). Barriers to effective climate change adaptation: final report. http://www.pc.gov.au/inquiries/completed/climate-change-adaptation/report. Accessed 14 March 2013.
Queensland Flood Commission of Inquiry. (2011). Interim Report. http://www.floodcommission.qld.gov.au/__data/assets/pdf_file/0006/8781/QFCI-Interim-Report-August-2011.pdf. Accessed 5 October 2011.
Queensland Reconstruction Authority. (2011). Rebuilding a stronger, more resilient Queensland. Brisbane, Queensland Government. http://qldreconstruction.org.au/publications-guides/resilience-rebuilding-guidelines/rebuilding-astronger-more-resilient-queensland. Accessed 3 March 2012.
Reeves, J. (2011). Mongolia's environmental security. *Asian Survey*, 51(3), 453-471.
Rolfe, J., Miles, B., Lockie., et al. (2007). Lessons from the social and economic impacts of the mining boom in the Bowen Basin 2004-2006. *Australasian Journal of Regional Studies*, 31(2), 134-153.
Romero, H., Kampf, S. (2003). Impacts of climate fluctuations and climate changes on the sustainable development of the arid Notre Grande in Chile. In H. F. Diaz and B. J. Morehouse (Eds.). *Climate and water* (pp. 83-115). The Netherlands: Kluwer Academic.
Sharma, V. (2014). Can traditional livelihoods and mining co-exist in a changing climate: strengthening public-private partnerships in Mongolia to reduce risk and address loss and damage. http://www.apn-gcr.org/resources/items/show/1938#.VNBn-ukcSUl.
Sharma, V., Franks, D. (2013a). In situ adaptation to climatic change: mineral industry responses to extreme flooding events in Queensland, Australia. *Society and Natural Resources*, 26(11), 1252-1267.
Sharma, V., van de Graaff, S., Loechel, B., et al. (2013b). Extractive resource development in a changing climate: Learning the lessons from extreme weather events in Queensland, Australia. Gold Coast, National Climate Change Adaptation Research Facility. https://www.nccarf.edu.au/publications/extractive-resource-development-changing-climate.
Sietz, D., Ludeke, M., Walther, C. (2011). Categorisation of typical vulnerability patterns in global drylands. *Global Environmental Change*, 21(2), 431-440.
Sternberg, T. (2014). Transboundary hazard risk: the Gobi Desert paradigm. *Natural Hazards*, 72(2), 533-548.
Sternberg, T. (2008). Environmental challenges in Mongolia's dryland pastoral landscape. *Journal of Arid Environments*, 72(7), 1294-1304.
Sussman, F. Freed, J. (2008). Adapting to climate change: a business approach. Arlington, Virginia. The Centre for Climate and Energy Solutions. http://www.c2es.org/publications/business-approach-adaptation. Accessed 15 October 2015.
Tachiiri, K., Shinoda, M., Klinkenberg, B., Morinaga, Y. (2008). Assessing Mongolian snow disaster risk using livestock and satellite data. *Journal of Arid Environments*, 72(12), 2251-2263.
Tiwary, R. K. (2001). Environmental impact of coal mining on water regime and its management. *Water, Air, and Soil Pollution*, 132(1-2), 185-199.
United Nations Statistics. (2013). 2013 World Statistics Pocketbook Country Profile: Mongolia. http://unstats.un.org/unsd/pocketbook/PDF/2013/Mongolia.pdf. Accessed 5 July 2013.

United Nations Environment Programme. (2008). Climate Change in the Caribbean and the challenge of adaptation. Panama City, United Nation Environment Programme (UNEP) Regional Office for Latin America and the Caribbean, Panama. http://www.pnuma.org/deat1/pdf/Climate_Change_in_the_Caribbean_Final_LOW20oct.pdf. Accessed 5 July 2013.

Upton, C. (2012). Mining, resistance and pastoral livelihoods in contemporary Mongolia. In J Dierkes (Ed.). *Change in Democratic Mongolia: social relations, health, mobile pastoralism, and mining* (pp. 223-248). Leiden, Brill.

Upton, C. (2009). 'Custom' and contestation: land reform in post-socialist Mongolia. *World Development*, 37(8), 1400-1410.

Wachman, A. (2010). Suffering what it must? Mongolia and the power of the 'weak'. *Orbis*, 54(4), 583-602.

Wang, J., Brown, D., Agrawal, A. (2013). Climate adaptation, local institutions, and rural livelihoods: a comparative study of herder communities in Mongolia and Inner Mongolia, China. *Global Environmental Change*, 23(6), 1673-1683.

Watts, R. (2012). Managing climate change and conflict in Mali (Case study 13). https://www.ids.ac.uk/files/dmfile/LHcasestudy13-Mali.pdf. Accessed 19 October 2013.

Whitmore, J., Hopkin, M., Hall, S. (2014). Carbon tax repealed: experts respond. http://theconversation.com/carbon-tax-repealed-experts-respond-29154. Accessed 20 January 2015.

Chapter 19
Mining and Infrastructure

Tom Johnson

Abstract This chapter argues that the provision of reliable, accessible and affordable infrastructure is one of the main challenges facing mining projects in the Asia-Pacific Region. Infrastructure is one of the most important features of a mining project, as it is the means for transporting minerals and metals to market, and for maintaining access to a mine by staff. However, the high cost of infrastructure and its associated economic, social and political risks means it can strongly influence the feasibility of a mining project. Infrastructure provision is an even bigger problem for junior mining companies, who may not have the budget or experience to implement the infrastructure they need to progress a mining project. The chapter begins by explaining infrastructure and the growth of infrastructure investment in the Asia-Pacific Region in recent years. The chapter looks at the risks associated with infrastructure investment, along with the ways in which governments seek to attract this type of investment. The chapter concludes with an analysis of the present and future challenges of infrastructure investment in the mining industry of the Asia-Pacific Region.

Abbreviations

AFC	Asian Financial Crisis
APEC	Asia-Pacific Economic Cooperation
BOO	Build-Operate-Own
BOT	Build-Operate-Transfer
BTO	Build-Transfer-Operate
CEO	Chief Executive Officer
FMG	Fortescue Metals Group
Mt	Million Tonnes
PPP	Public-Private Partnership
SOE	State Owned Enterprise

T. Johnson (✉)
Centre for International Risk, School of Communication, International Studies and Languages, University of South Australia, Adelaide, Australia
e-mail: Tom.Johnson@sca.com.au

T. O'Callaghan, G. Graetz (eds.), *Mining in the Asia-Pacific*, The Political Economy of the Asia Pacific, DOI 10.1007/978-3-319-61395-6_19

Introduction

Infrastructure is an integral part of each stage of a mining project, from on-site utilities for staff to the transportation of commodities to market. However, the development of the mining industry across the Asia-Pacific Region is impeded by inadequate, unreliable, or unaffordable infrastructure. The Ernst and Young (2014) report, *Business Risks Facing Mining and Metals 2014–2015*, found infrastructure access to be the seventh greatest risk facing the mining industry. In fact, the report revealed the cost of developing infrastructure to be almost 75% of the total project cost in some cases. Across the region, there are various regulatory, economic, and social barriers preventing mining companies from addressing the infrastructure deficit.

The infrastructure 'problem' negatively affects the mining industry in three ways. The first is its effect on business confidence, best shown by concerns stated by the heads of multiple mining companies. In July 2014, Rio Tinto Chief Executive Officer (CEO), Sam Walsh, called on the public and private sectors to focus on infrastructure development as a way of encouraging the development of large mining projects (Dagge 2014). Likewise, BHP Billiton CEO, Andrew Mackenzie, described infrastructure as 'pivotal' for boosting economic activity in the mining sector (Hepworth 2014). The second effect of the infrastructure deficit is how it hampers the development of the mining industry across the Asia-Pacific Region. An example is Indonesia's South Sumatra coal mines losing a potential 35 million tonnes (Mt) of coal production annually after the cancelation of the construction of a $1.7 billion railway to service the mine. Another example is Coal India Limited losing 300 million tonnes of annual coal production due to a lack of transport and evacuation infrastructure in its mines (Das 2015). The third effect of the infrastructure deficit is its impact on competition in the industry. An example is the decision by Fortescue Metals Group (FMG) to charge rival Brockman Mining AU$576 million per year to access part of its rail line in the Pilbara Region (Validakis 2013). FMG's decision was reversed only after Brockman Mining invoked Western Australia's State Agreements Act governing third party rail access (Australian Broadcasting Corporation 2014).

The infrastructure deficit in the Asia-Pacific Region is sizeable. The Asian Development Bank (2009) report, *Infrastructure for a Seamless Asia*, revealed Asia needs $8 trillion in investment into infrastructure between 2010 and 2020. However, the PricewaterhouseCoopers (Rathbone and Redrup 2014) report, *Developing Infrastructure in Asia-Pacific: Outlook, Challenges and Solutions*, finds that, with the exception of China, very little progress has been made toward this goal so far. The report revised the region's required infrastructure investment target to be $1.3 trillion annually to 2020.

This chapter examines the problem of infrastructure provision in the mining industry across the Asia-Pacific Region, identifying why the development of infrastructure is so challenging for mining companies. Through the chapter, I argue that infrastructure provision is one of the greatest challenges facing mining companies in the region. I begin the chapter by explaining the role of infrastructure in the region's mining industry. Following this, I look at the ways in which governments and private companies invest in mining-related infrastructure and the risks associated with these investments. I conclude by assessing the future challenges of infrastructure investment in the mining industry across the Asia-Pacific Region.

Infrastructure used in the mining industry includes: electrical generators, roads, railways, airports, seaports, telecommunication networks, information and communication technology, water and sanitation networks, education hubs, housing, and medical facilities. It also includes the infrastructure associated with mine construction as well. As these components contribute toward the operation of a mining project, one can appreciate the difficulties faced by mining companies needing to develop their infrastructure requirements. The importance of infrastructure is best described by, Flyvbjerg, Bruzelius, and Rothengatter (2003), who argue that infrastructure is at 'the very core' of production and consumption. This is because advances in infrastructure provision equate to a shrinking of space through improved productivity. In the mining industry, this means the quality and availability of infrastructure directly affects the feasibility and profitability of a mining project.

Private investment is an important contributor to infrastructure development in the Asia-Pacific Region. The zenith of private infrastructure investment was during the early 1990s, before infrastructure asset prices plunged as a consequence of 1997 Asian Financial Crisis (AFC). However, investment in mining infrastructure is booming once again. Using the most recent available data from the World Bank (2014), the private sector invested US$19.4 billion into infrastructure projects in the East Asia and Pacific Region in 2013. That year saw the biggest annual financial investment into infrastructure projects in the region since 2007, with most investment being in energy infrastructure projects. Since 1990, private investors have spent US$383 billion on infrastructure projects in the region. In Australia, private investors currently are committed to nine mining infrastructure projects valued at AU$11.7 billion (Minerals Council of Australia 2013). There are more projects on the way, including an AU$6 billion railway and port to transport coal from the Galilee Basin in Queensland for export. This is expected to be the most expensive infrastructure project in that state's history (Manning and Remeikis 2013).

There are a variety of recently completed and ongoing infrastructure developments in the mining industry across the Asia-Pacific Region. One example is the haul road constructed for the Goro Nickel Project in New Caledonia (GHD n.d.). Another is FMG's construction of a 620 km railway in Western Australia's Pilbara Region, completed in 2007. This is the fastest and heaviest haul line in the world (FMG n.d.). Also in the Pilbara, a 300 km railway connecting the Roy Hill mine to Port Hedland has recently been completed (Lannin 2014). The Anvil Hill coal mine in the Hunter Valley, New South Wales, is complemented by the construction of access roads, offices, and a compound with staff amenities (Bailey 2008). There also are a number of mining infrastructure projects awaiting government approval. One such project is a proposed power plant for the Phulbari coal mine in Bangladesh, which has been waiting for government approval for more than 8 years (Khan 2014).

Private investment into mining infrastructure projects can occur, and are funded, in a number of ways. Common investment structures include build-operate-transfer

Table 19.1 Common investment structure for infrastructure projects

BOT	'A project company finances the building of an infrastructure facility and operates it for a fixed period, after which ownership over the asset is transferred to the host government'. An example of a BOT is the Russey Chrum Krom hydropower plant in Cambodia, which was built under a 35 year BOT contract by China Huadian Corporation and began operations in 2015 (Larson 2015)
BOO	'Similar to a BOT, except that the asset is not transferred to the host government. However, operating concessions typically are on a fixed-term basis, after which the operating rights can be transferred to other private service providers'. An example is Hong Kong-based China Everbright International, signing a BOO contract with the government of China to build a US$51 million biomass power generation facility in Sichuan province (Messenger 2015)
BTO	'A new facility is built on a turnkey basis with private capital, and the ownership title is transferred to the host government after construction completion. The private contractor operates the facility for a fixed term under a separate agreement'. An upcoming BTO project is the construction of the first subway transport system in Manila, Philippines, due to begin in 2015 (Peña 2015)
PPP	'The host government provides capital, typically in exchange for an equity position, to support the commercial viability of a project'. An example of a PPP is the Alice Springs to Darwin extension of the Adelaide to Darwin railway in Australia. The extension was an AU$1.2 billion PPP between company, FreightLink, and the South Australian, the Northern Territory, and the Commonwealth Governments. Three mines currently use the railway to transport commodities (Om 2010)

Source: Sader (2000)

(BOT), build-operate-own (BOO), build-transfer-operate (BTO), and public-private partnerships (PPP) (see Table 19.1).

Private investment in infrastructure projects has become more common over the past twenty years. Prior to the 1990s, investment in infrastructure typically was seen as the responsibility of the public sector. This was because private investors saw the large, upfront investments required for infrastructure projects as too risky, along with a broad perception that affordable prices for public and private consumers is best achieved through public ownership. Additionally, the standard of technology at the time meant a monopoly service provider often was the most effective means of service provision.

During the 1990s, a shift to private investment in infrastructure occurred, in conjunction with the development of newer lower cost, and decentralised technology, especially in the telecommunications sector. A growing dissatisfaction with state-owned enterprises (SOE). The inefficient operation of SOEs and lack of cash-flow and reinvestment caused service quality to deteriorate and strained government budgets. This opened the opportunity for private investors successfully to enter the market. Although the 1997 Asian Financial Crisis and 2008 Global Financial Crisis damaged investor confidence, there has been a revival of infrastructure investment as governments across the region realise the economic links created by affordable and accessible infrastructure (Public-Private Infrastructure Advisory Facility 2009). Indeed, government leaders attending the 2014 Asia-Pacific Economic Cooperation (APEC 2014) summit showed an ongoing commitment to private investment in infrastructure projects by endorsing the *Multi-Year Plan on Infrastructure Development and Investment (Annex D - Action agenda on promoting infrastructure investment through public-private partnership (PPP) 2014)*. This plan requires governments across the region to develop more transparent and business-friendly investment guidelines, to improve project finance mechanisms, and to encourage more PPP infrastructure projects.

Mining Infrastructure Projects and Risk in the Asia-Pacific Region

Although governments in the Asia-Pacific Region have signalled a commitment to private investment in infrastructure projects, there are many risks associated with these investments. In the mining industry, Ramamurti and Doh (2004) argue that mining-related infrastructure is exposed to the same risks as the mining project itself. This is because, like a mine, infrastructure cannot easily be redeployed for other uses, and are fixed at their location. This makes infrastructure a sunk cost and vulnerable to an 'obsolescing bargain' should a host government attempt to nationalise its privately-owned assets (Vernon 1971). Additionally, infrastructure is a politically sensitive component of economic activity. This puts infrastructure under greater political scrutiny and risk of nationalisation than assets in other sectors (Sader 2000). This is a form of political risk, defined by McKellar (2010) as 'potential harm to a business operation arising from political behavior'. Other political risks that have the potential to affect mining infrastructure projects include domestic unrest, terrorism, corruption, political instability, and civil war. Helm (2009) notes that even if a government does not own infrastructure assets, it still can wield significant control over these assets. Indeed, Helm (2009) remarks, it is often easier (for governments) to impose control in the absence of ownership'. For these reasons, mining companies must approach infrastructure investments with the same care and consideration they would for a mining project.

Besides the risks imposed by government policy, investors must consider the risks associated with the large sunk costs of mining infrastructure projects. Addison and Anand (2012) argue that mining companies that invest in infrastructure face significant upfront costs and long periods of maintenance, which require considerable financial and technical resources. This may divert a mining company's resources away from its main asset, the mine itself. Flyvbjerg et al. (2003) suggest other risks associated with infrastructure investments include the potential for construction cost overruns, increased financing costs from fluctuating interest and exchange rates, and less-than-expected revenues, which can be caused by falling commodity prices. Another problem for private investors is the cost effectiveness of the infrastructure in relation to the expected revenue of a mine. A mining project is unfeasible if the projected value of the commodities, and lifespan of the mine, is unable to justify the cost of the associated infrastructure.

When considering the risks of mining infrastructure projects, it is understandable that mining companies seek to internalise the benefits derived from such projects. This can have negative effects on competition in the mining industry, especially for junior miners, who cannot afford to develop their own infrastructure. A lack of infrastructure access affect mining companies when excessive usage fees are charged for access to privately owned infrastructure. Even with legislation intending to allow reasonable third-party access to mining infrastructure, there still may be the need for expensive and time-consuming legal disputes over the matter. The lengthy legal battle between FMG and Brockman Mining over access to the Fortescue railway line in the Pilbara is a case in point.

Although there are many risks associated with mining infrastructure investments, there are certain circumstances under which projects more likely will succeed. One circumstance identified by Stern and Holder (1999) is the existence of a regulatory framework conducive to infrastructure investment. In the mining industry, this would involve checks to ensure changes in government policy do not affect the ownership or functioning of the infrastructure assets, along with support for contracts enabling a mining company to derive a reasonable rate of return on the infrastructure project. This may involve longer contract timeframes or a streamlined approval process. Additionally, Levy and Spiller (1994) note the importance of an independent judiciary, credible and trustworthy legislation, a degree of flexibility in lawmaking, and competent government regulators as factors contributing to the success of infrastructure investments.

Australia is a major mining producer in the Asia-Pacific Region where mining infrastructure investment likely will succeed. Australia was ranked fifth on the 2015 Index of Economic Freedom, with South Korea (27th) and Malaysia (29th) ranked the next highest in the region (Heritage Foundation 2015). Australia's attributes, identified by Levy and Spiller's regulatory framework is highly conducive to infrastructure investment (O'Callaghan and Vivoda in this volume). According to the 2014 Corruption Perceptions Index, Australia is the 11th least corrupt country in the world, with Malaysia (ranked 50) the next least corrupt major mining country from the region (Transparency International 2014). Using data from the World Bank (2013), Australia is ranked seventh in the world for regulatory quality and tenth in the world for rule of law—the highest of the major mining countries in the Asia-Pacific Region.

In addition to Australia, other countries in the Asia-Pacific Region are working toward creating more attractive investment climates. Sader (2000) argues that a key step in attracting infrastructure investment is having cooperation between government agencies and a clear process for obtaining all necessary government approvals. Committees for streamlining the approval process and potentially authorising public financing of projects have been introduced in Thailand and the Philippines. Another way countries are committing to infrastructure investments is by introducing legal frameworks dedicated to facilitating concession or BOT-type investments. For example, the Philippine BOT Law provides clear guidelines for infrastructure investment and signals an explicit commitment to the private sector's involvement in infrastructure projects (The Government of the Philippines 2013). Partly thanks to this Act, the Philippines has become one of the most successful destinations for infrastructure investments in the region, with mining infrastructure projects planned for the coming years including a power station, port, and pipeline for the Tampakan Copper-Gold Project, and a port facility for the Canatuan mine (AECOM n.d.; Mindoro Resources 2013).

Future Infrastructure Challenges for Mining Companies in the Asia-Pacific Region

Although there is evidence of improvement in the regulatory frameworks governing mining infrastructure investment in the Asia-Pacific Region, some challenges remain. One challenge is posed by local government regulation, which affects the construction of infrastructure. One example is the preferential treatment of local suppliers (KPMG 2015). A survey of 25 Australian mining companies in the 2011–2012 financial year revealed $32.1 billion was spent on local contractors and suppliers, which exceeded the industry's tax and royalty payments of $21 billion (Heber 2013). Rio Tinto's Argyle diamond mine, in the Kimberly Region of Western Australia, is required to demonstrate how Indigenous suppliers and workers will be included in contracts valued over $250,000 (Esteves et al. 2014). Similarly, Indonesia has introduced legislation requiring mining companies to use local and national suppliers and mining service companies.

Social opposition to infrastructure projects is also a problem for mining companies across the region. Environmental concerns are a common reason for galvanising resistance. In 2015, protestors blocked vehicles seeking to clear forests for the construction of rail infrastructure at the Maules Creek coal mine, in northwest New South Wales (Sturmer 2015; Woods 2015). In 2013, local residents stormed the Martabe gold mine in North Sumatra, in response to the construction of a pipeline (Bland 2013). In the Philippines, the displacement of local communities has also been met with anger (Sader 2000). The spread of technology, in particular social media, demonstrate that a project needs a social licence to operate. This may further fuel opposition to mining infrastructure projects in the future.

The depletion of raw material reserves in the Asia-Pacific Region presents a further challenge for mining companies, particularly junior mining companies. According to Andrew Keith, transportation costs are one of the largest items of expenditure in a mining project. Keith predicts these costs will increase as more mines are established further away from inhabited areas, and in more difficult terrain (Duffy 2012). The need to mine less accessible deposits is costlier and requires more expertise than conventional deposits. It may also be the case that ore grades in less inhabited areas may be lower, and have significantly more overburden to remove. This is because these areas may lack facilitating infrastructure, thereby requiring the construction of new associated facilities. For example, Rio Tinto has had to construct a power plant because its Oyu Tolgoi mine is in an area which has no power available. In addition, junior mining companies, either must raise additional capital to invest in infrastructure investments, of be forced into a merger with a larger company.

Conclusion

The Asia-Pacific Region's mining industry faces numerous infrastructure challenges. Due to the integral role of infrastructure in a mining project, it is imperative for mining companies to understand these challenges and identify solutions. These challenges are best dealt with, either on a case-by-case basis, or by collaborating with other companies in order to share costs. As noted earlier, there is also potential to become involved in public-private partnerships with local, regional and national governments. The reality is, however, that each mining project faces unique financial constraints, differing regulatory frameworks, and various social and political risks. Some overarching concerns for miners in the Asia-Pacific Region include: the need for more streamlined regulations for infrastructure investment and improve the cost-efficiency of infrastructure projects, particularly as mines are located in increasingly remote areas.

Further research is required into how governments can implement more effective regulatory frameworks for infrastructure investment. This especially is important for countries that lack a strong legal system or that score poorly in ease of doing business surveys. There also are opportunities to develop new ways to improve third party access to mining-related infrastructure. As the need for more cost-effective infrastructure solutions becomes apparent, and in order ideally to avoid lengthy and bitter legal disputes over access to privately owned infrastructure, there is an opportunity to research new ways of improving third party access to infrastructure. However, the infrastructure deficit is widening, despite private investors spending billions of dollars each year on projects in the region. The chapter has outlined the principal reasons why infrastructure investment is so difficult for mining companies, and the ways in which governments can increase the likelihood that infrastructure projects will succeed. Until governments improve the investment climate for projects, infrastructure will remain one of the most pressing concerns for mining companies operating in the Asia-Pacific Region.

References

Addison, T., Anand, P. (2012). Aid and infrastructure financing: emerging challenges with a focus on Africa. United Nations University, Working Paper No. 2012/56, Helsinki. http://papers.ssrn.com/sol3/papers.cfm?abstract_id=2524546. Accessed 9 June 2013.

AECOM (n.d.) Tampakan Copper-Gold Project, Philippines - Asia - AECOM. http://www.aecom.com/Where+We+Are/Asia/Mining/_carousel/Tampakan+Copper-Gold+Project,+Philippinespercent3E. Accesed 7 May 2015.

Asia-Pacific Economic Cooperation. (2014). Annex D - action agenda on promoting infrastructure investment through public-private partnership (PPP). APEC Ministerial Meeting, Beijing. http://www.apec.org/Groups/Committee-on-Trade-and-Investment/~/link.aspx?_id=AFC4C880508F440180F2155C6E29CECD&_z=z. Accessed 6 October 2015.
Asian Development Bank. (2009). Infrastructure for a seamless Asia. Tokyo. http://www.adb.org/publications/infrastructure-seamless-asia. Accessed 7 March 2015.
Australian Broadcasting Corporation. (2014). Brockman Mining wins legal bid to access Fortescue Metals Group's rail infrastructure. http://www.abc.net.au/news/2014-09-27/brockman-wins-legal-bid-to-access-fmg-rail-infrastructure/5773538. Accessed 7 May 2015.
Bailey, H. (2008). Anvil Hill mine modification to project approval: environmental assessment, Xstrata Mangoola. http://www.mangoolamine.com.au/EN/approvals/Earlypercent20workspercent20modificationpercent20documentation/06_0014_mod1_anvil_hill_ea_mainreport.pdf. Accessed 7 October 2015.
Bland, B. (2013, January 2). Environment protests hit Sumatra gold mine. *Financial Times*. http://www.ft.com/intl/cms/s/0/7bce1a92-37a7-11e2-a97e-00144feabdc0.html#axzz44r7o4VFn. Accessed 12 March 2015.
Dagge, J. (2014, July 17). Rio boss Sam Walsh says business leaders must find how to fund more projects. *Herald Sun*. http://www.heraldsun.com.au/business/rio-boss-sam-walsh-says-business-leaders-must-find-how-to-fund-more-projects/story-fni0dcne-1226992772849. Accessed 27 February 2015.
Das, A. (2015, February 18). CIL willing to increase capex for infrastructure investment. *Mining Weekly*. http://www.miningweekly.com/article/cil-willing-to-increase-capex-for-infrastructure-investment-2015-02-18. Accessed 27 February 2015.
Duffy, A. (2012). Mining infrastructure: from pit to port. *Mining Australia*. http://www.miningaustralia.com.au/features/mining-infrastructure-from-pit-to-port. Accessed 27 February 2015.
Ernst & Young. (2014). Business risks facing mining and metals 2014-2015. London. http://www.ey.com/Publication/vwLUAssets/EY-Business-risks-facing-mining-and-metals-2014percentE2percent80percent932015/$File/EY-Business-risks-facing-mining-and-metals-2014percentE2percent80percent932015.pdf. Accessed 30 June 2015.
Esteves, A., Ogorodnikova, V., Putz, C. et al. (2014). Increasing domestic procurement by the mining industry in Central Asia. World Bank. http://www-wds.worldbank.org/external/default/WDSContentServer/WDSP/IB/2014/06/18/000470435_20140618091347/Rendered/PDF/ACS88340WP0P1331680Box385243B00PUBLIC0.pdf. Accessed 21 March 2015.
Flyvbjerg, B., Rothengatter, W., Bruzelius, N. (2003). *Megaprojects and risk: an anatomy of ambition*. New York: Cambridge University Press.
Fortescue Metals Group. (n.d.). Port and rail infrastructure http://www.fmgl.com.au/our_business/Port_and_Rail_Infrastructure. Accessed 5 March 2015.
GHD. (n.d.). Infrastructure for the mining industry. http://www.ghd.com/PDF/InfrastructurefortheMiningIndustry.pdfpercent. Accessed 5 March 2015.
Heber, A. (2013). Mining sector community spending exceeds $34 billion. *Australian Mining*. http://www.miningaustralia.com.au/news/mining-sector-community-spending-exceeds-$34-billi. Accessed 12 March 2015.
Helm, D. (2009). Infrastructure investment, the cost of capital, and regulation: an assessment. *Oxford Review of Economic Policy*, 25(3), 307-326.
Hepworth, A. (2014, November 12). BHP boss urges investment in trade-related infrastructure. *The Australian*. http://www.theaustralian.com.au/business/bhp-boss-urges-investment-in-traderelated-infrastructure/story-e6frg8zx-1227119989213. Accessed 27 February 2015.
Khan, S. (2014, March 5). Tricky to decide, easy to keep hanging. *Daily Star*. http://www.thedailystar.net/tricky-to-decide-easy-to-keep-hanging-37401. Accessed 25 May 2014.
KPMG. (2015). Future of procurement in Asia-Pacific: keeping pace with challenges in the energy & natural resources sector. https://www.kpmg.com/AU/en/IssuesAndInsights/ArticlesPublications/Documents/future-of-procurement-in-asia-pacific.pdf. Accessed 3 May 2016.

Lannin, S. (2014, May 16). First tracks laid in Roy Hill project in WA's Pilbara. *Australian Broadcasting Corporation*. http://www.abc.net.au/news/2014-05-16/first-tracks-laid-in-roy-hill-project/5457764. Accessed 13 March 2016.

Larson, A. (2015). Cambodia's largest hydropower plant begins operation. *Power Magazine*. http://www.powermag.com/cambodias-largest-hydropower-plant-begins-operation/. Accessed 6 March 2015.

Levy, B., Spiller, P. (1994). The institutional foundations of regulatory commitment: a comparative analysis of telecommunications regulation. *Journal of Law Economics and Organisation*, 10(2), 201-246.

Manning, P., Remeikis, A., (2013, March 13). Single rail deal to Galilee Basin. *Sydney Morning Herald*. http://www.smh.com.au/business/single-rail-deal-to-galilee-basin-20130311-2fwcn.html. Accessed 5 March 2015.

McKellar, R. (2010). *A short guide to political risk*. Surrey, England: Gower.

Messenger, B. (2015). Major new waste to energy & biowaste projects for Everbright in China. *Waste Management World*. http://www.waste-management-world.com/articles/2015/01/major-new-waste-to-energy-biowaste-projects-for-everbright-in-china.html. Accessed 6 March 2015.

Mindoro Resources. (2013). TVI Pacific Inc. and Mindoro Resource Ltd. announce robust economics for the Agata High-Iron DSO project. http://www.mindoro.com/s/PressReleases.asp?ReportID=602317. Accessed 15 September 2015.

Minerals Council of Australia. (2013). Submission to the Productivity Commission's inquiry into public infrastructure. http://www.pc.gov.au/__data/assets/pdf_file/0020/131735/sub070-infrastructure.pdf. Accessed 5 May 2015.

Om, J. (2010, June 10). Taxpayer funds sought from new rail owner. *Australian Broadcasting Corporation*. http://www.abc.net.au/worldtoday/content/2010/s2923550.htm. Accessed 6 March 2015.

Peña, Z. (2015, March 5). P370-B PPP projects up for P-Noy OK. *Philippine Star*. http://www.philstar.com/business/2015/03/05/1430021/p370-b-ppp-projects-p-noy-ok. Accessed 6 March 2015.

Public-Private Infrastructure Advisory Facility. (2009). 2009 Annual Report. Washington. http://documents.worldbank.org/curated/en/2009/01/11362156/public-private-infrastructure-advisory-facility-ppiaf-annual-report-2009. Accessed 6 March 2015.

Ramamurti, R., Doh, J. (2004). Rethinking foreign infrastructure investment in developing countries. *Journal of World Business*, 39(2), 151-167.

Rathbone, M., Redrup, O. (2014). Developing infrastructure in Asia-Pacific: outlook, challenges and solutions, PricewaterhouseCoopers, Singapore. http://www.pwc.com/sg/en/capital-projects-infrastructure/assets/cpi-develop-infrastructure-in-ap-201405.pdf. Accessed 4 January 2015.

Sader, F. (2000). Attracting foreign direct investment into infrastructure: why is it so difficult? Occasional Paper No. 12, Foreign Investment Advisory Service. http://elibrary.worldbank.org/doi/abs/10.1596/0-8213-4602-4. Accessed 5 February 2014.

Stern, J., Holder, S. (1999). Regulatory governance: criteria for assessing the performance of regulatory systems. *Utilities Policy*, 8(1), 33-50.

Sturmer, J. (2015, February 15). Protestors chain themselves to land clearing equipment at Maules Creek coal mine site. *Australian Broadcasting Corporation*. http://www.abc.net.au/news/2015-02-15/activists-disrupt-maules-creek-coal-mine/6109164. Accessed 12 March 2015.

The Government of the Philippines (2013). The Philippine Amended BOT Law R.A. 7718 and its revised implementing rules and regulations (IRR). http://ppp.gov.ph/wp-content/uploads/2015/01/BOT-IRR-2012_FINAL.pdf. Accessed 5 April 2015.

The Heritage Foundation. (2015). Country rankings: world and global economy rankings on economic freedom. http://www.heritage.org/index/ranking. Accessed 11 March 2015.

Transparency International. (2014). 2014 Corruption Perceptions Index. http://www.transparency.org/cpi2014/results. Accessed 11 March 2015.

Validakis, V. (2013, May 31). FMG puts $576m price tag on rail access. *Australian Mining*. http://www.miningaustralia.com.au/news/fmg-puts-$-576m-price-tag-on-rail-access. Accessed 27 February 2015.

Vernon, R. (1971). *Sovereignty at bay: the multinational spread of U.S. enterprises*. New York: Basic Books.

Woods, G. (2015). Maules Creek protestors blockade to stop mine construction. Greenpeace. http://www.greenpeace.org/australia/en/mediacentre/media-releases/climate/Maules-Creek-protestors-blockade-to-stop-mine-construction1/. Accessed 7 October 2015.

World Bank. (2014). East Asia and Pacific Region - private infrastructure projects. http://ppi.worldbank.org/explore/ppi_exploreRegion.aspx?RegionID=2. Accessed 4 March 2015.

World Bank. (2013). Worldwide governance indicators. The World Bank DataBank. http://databank.worldbank.org/data/views/variableselection/selectvariables.aspx?source=Worldwide-Governance-Indicators. Accessed 11 March 2015.

[illegible]

Chapter 20
Legacy Issues and Abandoned Mines

Corinne Unger

Abstract The chapter summarises the principal problems associated with current approaches to closure planning. Then it discusses the critical relationship between effective mitigation and management of mining legacies and abandoned mines. In the third section, I review global and Australian initiatives on mining legacies. Then, drawing upon global leading practices, I develop a maturity model to guide jurisdictions as they move toward more mature management of abandoned mines. In the fifth section, I present a case study of the abandoned Mount Morgan gold mine in Queensland, Australia. Throughout the chapter, I draw attention both to the importance of fully accounting for liabilities to internalise costs, and to the benefits that can flow to industry, governments and communities when beneficial post-mining land uses are created.

Abbreviations

AMD	Acid Mine Drainage
AMWG	Abandoned Mines Working Group
ANZMEC	Australian and New Zealand Minerals and Energy Council
ARD	Acid Rock Drainage
ASM	Artisanal and Small-scale Mining
AusIMM	Australasian Institute of Mining and Metallurgy
BCCCSP	British Columbia Crown Contaminated Sites Program
CLM	Contaminated Land Management
DTIRIS	Department of Trade and Investment, Regional Infrastructure and Services
EIS	Environmental Impact Statement
EPA	Environmental Protection Agency
FA	Financial Assurance
GRI	Global Reporting Initiative
ICMM	International Council on Mining and Metals

C. Unger (✉)
School of Business, University of Queensland, St Lucia, QLD 4067, Australia
e-mail: c.unger@uq.edu.au

T. O'Callaghan, G. Graetz (eds.), *Mining in the Asia-Pacific*, The Political Economy of the Asia Pacific, DOI 10.1007/978-3-319-61395-6_20

IIED	International Institute for Environment and Development
IUCN	World Conservation Union
LMU	Legacy Mine Unit, NT
MCA	Minerals Council of Australia
MCMPR	Ministerial Council on Mineral and Petroleum Resources
MMSD	Mining, Minerals and Sustainable Development
NOAMI	National Orphan/Abandoned Mine Initiative
OECD	Organisation for Economic Co-operation and Development
UNEP	United Nations Environment Programme

Introduction

Abandoned mines are mines for which mining leases or titles no longer exist, and for which responsibility for their rehabilitation cannot be allocated to any individual, company or organisation that has undertaken mining activities. Mining legacies can be positive or negative in their impact; however, the focus of this chapter is on the negative legacies—what factors have caused them, who is responsible for them, and what is needed to mitigate and manage them now and in the future. While there is value in promoting the positive legacies of extractive developments, given that human society is dependent upon the products of mining, comparatively few resources have been dedicated to the study of negatives legacies and, subsequently, to disseminating information so that knowledge about best practices can be shared. In this chapter, I review progress in this regard, by examining leading practices that may address mining legacy challenges. Also I identify the risks of inaction, as well as the opportunities that exist to mitigate more effectively and manage legacy sites to generate beneficial post-mining land uses.

The 2002 Mining, Minerals and Sustainable Development (MMSD) report of the International Institute for Environment and Development (IIED) noted that 'some [legacy] impacts can be long-term [and] society is still paying the price for natural capital stocks that have been drawn down by past generations'. If they are left unmitigated, mining legacies can pose risks to human health and safety, the environment, and additionally may precipitate negative socio-economic impacts. The reputation of industry participants and governments also may be affected by the negative legacies of abandoned mines.

Largely unquantified, poorly mapped and often remote, stakeholders have limited understanding of the challenges that abandoned mines pose—economically, socially, politically and environmentally. Abandoned mines and mining legacies often inhabit a space in which legislation is absent or ambiguous. Similarly, the boundaries of responsibilities and liabilities of stakeholders are not clear. Research has highlighted the important role of the office of the Auditor-General in Australia and Canada in drawing attention to liabilities and inadequate governance of mining operations at all stages of the mine life-cycle. Audits of contaminated land programs and environmental regulation have drawn attention to the need effectively to account for and manage existing mining liabilities from abandoned mines (Victorian Government Auditor-General's Office 2011). In the pages that follow, I discuss the urgent need for effective regulation to prevent potential future mining environmental liabilities from becoming the responsibility of the state and community. I draw on examples from several states in the Asia-Pacific Region in order to identify the risks, challenges and opportunities for improved mitigation and management of mining legacies.

The chapter proceeds as follows: First, I summarise the principal problems associated with current approaches to closure planning. Second, I discuss the critical relationship between effective mitigation and management of mining legacies and abandoned mines. In the third section, I review global and Australian initiatives on mining legacies. Fourth, drawing upon global leading practices, I develop a maturity model to guide jurisdictions as they move toward more mature management of abandoned mines. In the fifth section, I present a case study of the abandoned Mount Morgan gold mine in Queensland, Australia. The case study speaks to the magnitude of legacy mine liabilities and shows that technical solutions exist. The case additionally demonstrates how ambiguity regarding regulatory and corporate responsibility is holding back progress on remediation. Throughout the chapter, I draw attention both to the importance of fully accounting for liabilities to internalise costs, and to the benefits that can flow to industry, governments and communities when beneficial post-mining land uses are created.

Problems with Closure Planning

The causes of abandoned mines are complex, and may be unique to one site. However, in this chapter, I identify the common themes that lead to negative legacies. The absence of appropriate regulatory frameworks is only one aspect. Another involves a focus on short-term needs by industry to reduce the regulatory burden for approval of mining projects, which may result in insufficient attention being paid to the mitigation of long-term legacy risks during planning and design phases. Mines can be operated over many decades and by multiple owners. Consequently, it can take many years before liabilities and potential legacy risks become apparent. Due diligence assessments during mergers and acquisitions may not reveal the magnitude of legacy risks. This occurs when poorly prepared closure plans overlook complex risks, and when regulatory requirements for financial assurance do not reflect the full costs of closure. Closure assumptions may not be fully questioned and quantified unless M&A teams include closure specialists. Internal closure provisions incorrectly may assume that expenditure for closure only is required at the time of closure, instead of understanding the information needs and progressive remediation required throughout the mine life-cycle. Mining engineers are taught how to develop a mine, but they are not taught how effectively to close a mine that transitions to an alternative post-mining land use. It has been said: '[n]o mine manager wants to have a mine closure on his or her resume'. However, at some point, a resource will be exhausted. When mines are planned, both industry and government assume they will close when the resource is depleted, yet as few as 25% of 800 mines surveyed in one study (Laurence 2006) closed due to this reason.

Designing for closure requires multidisciplinary teams that are able to integrate closure considerations into key aspects of design and operations. Early in the life of a project, there are opportunities to ensure that negative legacies are mitigated; however, as mining progresses, options close off, and the likelihood of residual legacies increases.

Laurence's (2006) study of why mines close identified the most significant environmental impact from closed mines as Acid Rock Drainage (ARD), followed closely by tailings, waste dumps and open pit-related impacts. ARD refers to drainage affected by the products of sulphide oxidation, including acid, neutral, and saline drainage. The generation, release, mobility, and attenuation of ARD is a complex process, which is governed by a combination of physical, chemical, and biological factors. Characterisation of sources and pathways is a key to the prediction, prevention, and management of drainage affected by the products of sulphide oxidation at mine sites.

The inappropriate placement of waste landforms in the landscape is a product of systemic problems at the feasibility study, approvals, and design stages (McCarthy 2014). During feasibility studies for a mining project, tensions already are in place between the owner, who/which wants to reduce capital and operating costs, and engineers, whose first priority is to design a mine that maximises technical success. Significant decisions that affect whole-of-mine life, including environmental management and legacy risks, usually are not well addressed at this stage. They often are considered operational matters, which 'will be addressed later' (McCarthy 2014). This, combined with discounted cash flow to quantify the value of a project—the most common financial modelling tool of mining companies—diminishes the liabilities associated with closure, by assuming closure costs can be deferred until the end of the mine's life. The result is flawed assumptions, which lead to budget deficiencies throughout the project life when investment in closure activities is needed to mitigate significant end-of-life liabilities (McCarthy 2014). Greater emphasis on investment in life-of-mine (including closure) design and associated investigations therefore must commence prior to, and continue during, operations. Some examples include:

- Understanding topographic and hydrological (surface and groundwater) contexts to determine where in the landscape waste landforms should be located to address engineering requirements, as well as to mitigate life-of-mine environmental impacts;
- Characterisation and verification of ore and waste geochemistry to ensure effective mitigation and management of acid mine drainage (AMD);
- Design of stable landforms, impoundments and covers that encapsulate wastes for hundreds, if not thousands, of years (for AMD and uranium mill tailings);

- Prediction and management of ground and surface water interactions (clean and contaminated waters), as part of an operational water management system that will facilitate the transition to a non-polluting system after mining ceases in order to protect environmental values in the long-term; and
- Engagement with communities on agreed post-mining landforms and land uses, and development and proof of sustainable revegetation methods via progressive remediation.

Life-of-mine planning for new mines offers many opportunities to optimise the handling of waste. Such planning can reduce the volumes of waste and/or minimise the development of mine drainage containing high concentrations of metals, and likely will reduce overall project costs. While prevention is best, once oxidation and other natural weathering processes are well advanced. This can occur during the active mining (operational) period, where prevention no longer is realistic.

Mining companies that operate sites with AMD problems may find they cannot gain regulatory relinquishment through a 'walk away' solution to mine closure and, subsequently, that they are required to maintain a presence on site to manage and/or treat water in the long-term (Peck et al. 2005). When negative legacies persist long after mining operations have ceased, there is a risk that any socio-economic benefits that accrue from a mining project will be overshadowed by the environmental impacts and associated costs to manage and mitigate those impacts post-closure. As an example, Byrne (2013) found that 25% of sites undertaking closure had accepted they would need to manage water in perpetuity.

In some instances, where mining companies come to understand the scale and complexity of the closure task, and their under-estimated closure provision, they sell or transfer ownership to another company. If the new company has the capacity to further extract value from the mine or the financial resources to close the site to an accepted standard, this is a good outcome. However, when smaller, less well-resourced companies with limited capacity to deal with the scale of closure risks take over the site, the mining operation rapidly can transition to insolvency. Closure/legacy risks may occur as a consequence of the following:

- Closure objectives and assumptions for mine planning are not supported by evidence;
- Failure of closure risk assessment processes to detect significant risks (e.g., due to a lack of data, or the lack of relevant cross-disciplinary expertise that have been engaged in risk assessment);
- Investigations and discoveries close to the end of the mine life yield expensive solutions (e.g., double handling of waste materials for major landform design changes);
- Delay of progressive remediation to cut costs in the short-term, and/or failure to realise the need to trial and demonstrate remediation methods over the life of a project in order to develop effective solutions;
- Underestimation of the predicted costs and timeframes for closure;
- Inadequate stakeholder engagement leading to a mismatch between community expectations and remediation standards and/or post-mining land use(s);

- Poor understanding of the connection between the withdrawal of social licence to mine in regions where un-remediated abandoned mines exist, even when they are not the direct responsibility of the current mining companies;
- Assumption that compliance with regulatory standards will facilitate effective mine closure;
- Assumption that all mines and commodity types have a 'walk-away' solution;
- Failure of regulatory authorities to mitigate negative legacies through the development of evidence-based policy;
- Inadequate engagement with Indigenous communities to understand cultural heritage and the relevance of connection to land in mine closure design (industry) and abandoned mine management (governments).
- Assignment of non-specialised environmental officers to mining regulatory roles, with insufficient knowledge to understand the long-term legacy implications of the proposals they are reviewing. This can occur when graduates are equipped with templates and checklists as a surrogate for more highly paid and experienced technical experts within government;
- Assumption that, when a mine closes, another company will mine what is left of the deposit in the future, and failure to establish systems to evaluate the likelihood of such an occurrence;
- Where the Mines Department (or responsible agency) administers mining tenure and becomes, by default, responsible of mining legacies and abandoned mines, failure to recognise the need effectively to engage the relevant functions of other departments (e.g., environment, heritage, tourism, agriculture, regional development), to ensure good governance, programs and post-mining land uses;
- Assumption that the mining company will mine and meet all of its initial closure commitments;
- Assumption that governments have developed guidelines (across the multiple departments and different legislation) for: (a) approval of rehabilitation; (b) mine closure approval; (c) lease relinquishment; and (d) approval of new/alternative post-mining land uses;
- Inability to quantify the link between social licence and mining legacies, and abandoned mines;
- Government confusion between *responsibility* with *liability* for legacy sites; that is, responsibility may reside with the current land owner, however, liability may resides with the government as regulator;
- Agreement by governments to unrealistic timeframes to review mining applications, thereby applying pressure on regulatory personnel to approve quickly. (The regulatory approvals process is a critical stage that allows for the detection of fundamental flaws in closure design and long-term legacy impacts. Significant cost savings—environmental, financial and social—can be achieved if flaws are detected early in the design process, and the design is subsequently modified
- Limited ability to identify and mitigate cumulative environmental impacts;
- Communities' assumption that regulators are learning about the causes of mining legacies from existing abandoned mines and actively are translating that learning into robust policy for current mining activities;

Table 20.1 Estimated numbers of orphaned, abandoned or derelict mine sites (outside Australia)

Country	Number of sites	Information quality
Canada	10,100	Good
Japan	5500	Good
South Africa	8000	Average
Sweden	1000	Average
United Kingdom	11,700	Average
United States	600,000+	Poor
Rest of the world	Likely millions	Very poor

Source: Worrall et al. (2009)

Assumption that financial assurance (FA) cost calculator tools provided by governments for projects take account of all aspects of closure.

While a lack of funding often is cited as the reason abandoned mines inadequately are addressed, it is clear that this argument is flawed. For example, twice as much money is spent on abandoned mines in the Australian State of Queensland each year compared with British Columbia in Canada. However, unlike the British Columbia Crown Contaminated Sites Program (BCCCSP 2012, 2014), there is no evidence from Queensland that environmental impacts are being regulated, nor that environmental liabilities are being reduced. In the next section, I examine the global initiatives to cope with legacy issues.

Global Initiatives on Mining Legacies

The report of the World Commission on Environment and Development (United Nations 1987) defined Sustainable Development as ‘development that meets the needs of the present without compromising the ability of future generations to meet their own needs’. This definition embeds the concept of inter-generational equity, which is relevant to mining legacies. Contained within the definition are two key concepts: needs and limitations. Mineral and energy resources are extracted without being replaced; hence, this aspect of mining is not sustainable. However, there are ways of mitigating impacts and generating wealth for governments and communities from mining; for example, by investing in community development and the regeneration of mined landscapes to create positive legacies.

According to a study by Worrall et al. (2009), there is average to poor data quality on the number of orphaned, abandoned or derelict mines outside Australia, with the exception of Canada, Japan and Sweden (Table 20.1). Quantifying the number of sites is only the first stage in gaining an understanding of the risks and impacts of legacy sites. The commendable and necessary efforts to apply sustainable development principles in the mining sector unfortunately are undermined by the existence of so many negative mining legacies globally. Further, the inconsistent application of intervention measures to prevent the creation of future negative environmental legacies, and the often reactive approaches to the management of abandoned mine programs, highlights the need for global leadership in this regard.

From an economic perspective, countries that set aside wealth from the resources sector through sovereign funds prolong the benefits derived from mining for future generations. Perhaps the best example is Norway's sovereign wealth fund, the world's largest, which uses petroleum revenues to carry forward the benefits of resource extraction to future generations.

Sovereign wealth funds (and Future Funds) are pertinent when we consider the growing community expectations for environmental restoration of mine sites and the current ages and phases of mining. Even with modern environmental regulations, there remain a significant number of mines that commenced operations under weaker environmental regulatory regimes and that are unlikely to meet community expectations when they close, unless there is considerable new investment and leadership.

In the United States, the Office of Surface Mining introduced The Surface Mining Control and Reclamation Act of 1977 in order to address the cumulative impacts of coal mines by requiring mining companies to return land to agricultural productivity with requirements to backfill open cut mine voids. The legislation also addressed funding for abandoned mines and associated research facilities.

In 2001, the Chilean Copper Commission and the United Nations Environment Programme (UNEP) hosted an abandoned mine forum to address policy challenges for decision-makers. As a consequence, UNEP began to collect information using national or regional inventories in addition to a global survey. The objective was to start the discussion to define the current state of knowledge of abandoned mines in each jurisdiction, criteria for prioritisation, and how to apportion costs and responsibilities. However, this process appears to have come to a halt as there is no active UNEP webpage on abandoned mines. UNEP observed that some governments have reacted more effectively in preventing current mines from becoming abandoned sites. For example, in Japan, where a national survey found 5500 abandoned mines, the government has taken a two-pronged approach to address the problem. Where the original owner exists, they remain responsible for the safety of the mine and the prevention of all pollution. Where the original owner cannot be found, or in the case of bankruptcy, the local and national governments jointly assume responsibility.

Elsewhere in the Asia-Pacific Region, pressure for land in China has propelled land reclamation and environmental concerns from mining into prominence. Mining waste stockpiles currently cover over two million hectares of desperately needed land. Moreover, research on reclamation and post-mining closure in Indonesia used the United States as a benchmark against which to analyse the current mining closure and reclamation framework in the country (Paramita and Asri 2011). The recommendations of the study highlighted the importance of a social licence through community engagement and improved regulation. Recommendations include:

- The need for greater transparency in permitting processes and bond/financial assurance processes, including the reporting and release of bonds;

Table 20.2 Towards a sustainable future for the mining and minerals sector

Worst of the past	Vision for the future
A legacy of ghost towns, poverty and pollution	Integrated planning for sustained post-closure environmental, social and economic benefits
Minerals development as a threat to protected areas and biological diversity	Mineral development in appropriate places and as a source of revenue to ensure the protection of areas critical to biodiversity

Source: IIED (2002)

- The need to encourage the public to become involved in reclamation projects;
- The need to ensure compliance to raise corporate social and environmental performance;
- The need to improve coordination, planning and supervision of agencies involved in reclamation and post-mining land use(s);
- The need to improve current standards/regulation whilst also developing specific laws for mining activities on smaller islands and coastal areas;
- The need to build the capacity and resources of regulatory agencies; and,
- The need to strengthen law enforcement for violations regarding provisions related to reclamation and mining closure.

The MMSD project, *Breaking New Ground* (IIED 2002), focussed global debate on how the mining industry could maximise its contribution to sustainable development. It was through this process that research first was undertaken and a global dialogue on mining legacies commenced. The project report noted different forms of legacy—riverine tailings, legacies of conflict and heritage values—as well as the positive legacy of education and skills training, which continues when mining ceases.

Importantly for this chapter, the project report observed that most countries with a long history of mining had little data on the environmental legacies of their mines; however, there was enough information to know that problems were widespread. The report described the worst of the past and presented a vision for the future of the mining industry across 13 characteristics, with the two particularly relevant characteristics included here (see Table 20.2).

Common closure scenarios are listed below, of which only the first three are considered acceptable (Peck et al. 2005):

- Ordered advance and completion;
- Strategic revision of final operational period;
- Ordered retreat;
- Retreat in disarray; and,
- Dereliction of duty.

The International Council on Mining and Metals (ICMM) is the peak global body tasked with representing the interests of the world's largest mining companies. The Council additionally facilitates the self-regulation of the mining industry, and develops guidance and protocols to ensure that its member companies—and aspiring members—adhere to sustainable development principles. For example, the ICMM has developed good practice guidance regarding mining and biodiversity and integrated mine closure (ICMM 2006, 2008). At the World Summit on Sustainable Development in Johannesburg in 2002, the ICMM, in conjunction with the World Conservation Union, launched a joint dialogue on mining and biodiversity (IUCN-ICMM 2006; IUCN-ICMM and Post-Mining Alliance 2008). Legacy issues subsequently were identified as important to the industry.

In 2007, mining legacy stakeholders were surveyed (Whitbread-Abrutat 2008) in preparation for a mine legacy roundtable that was held in Toronto in March 2008 (IUCN-ICMM 2008). The results of the survey identified funding/finance, partnerships and stakeholders, and legislation as priority topics for the forum, followed by case studies, knowledge transfer, and leadership issues. Approximately 65% of respondents were based in the broader Asia-Pacific Region (encompassing Australasia, Asia, North America and Latin America). When respondents ranked the most important environmental issues accruing from mining legacies, acid rock drainage (ARD) was first, followed by contaminated land, public safety and biodiversity loss. The local economy was the largest socio-economic impact identified.

The most important stakeholders with regard to planning for the regeneration of legacy sites were identified as local communities, local government agencies and mining companies, followed by Indigenous Peoples, state/provincial governments, national governments, industry bodies, and inter-governmental agencies. The survey also identified barriers to effective regeneration. Poor enforcement of legislation was the most important factor, followed by:

- A lack of good closure planning and bonding, including insufficient surety calculation and provisioning;
- A lack of peer pressure within the mining industry to 'do the right thing';
- Legal liability for those willing to address legacy concerns, and the need for a remediation fund: (a) for when new mines are established; and (b) to encourage regional cooperation by companies to address legacy issues;
- Poor formulation of the original agreement between relevant government agencies and the mining company;
- A lack of 'environment protection' attitude;
- Poor company planning and commitment;
- Corporate greed; and,
- An unwillingness to invest in remedial action.

The forum was effective in bringing together legacy site and program managers from government, industry and civil society. Some of the relationships built during the forum have sustained knowledge sharing in the years since (Unger 2009; The Centre for Mined Land Rehabilitation 2012; Unger et al. 2015). A number of recommendations were made during the forum, but perhaps the most important was the agreement that there was a need for an organisation formally to address and lead discussions on legacy site issues at national and international levels. The forum found that:

> The organisation needs to ensure that legacies site are well placed in the international environmental agenda so that industry and governments will recognise the significance of legacy sites throughout both the developed and developing world. Furthermore, the organisation should act as a lead agency for raising awareness, identifying priority sites, collecting and disseminating information on toolkits, best practices and funding mechanisms, and portal to exchange information among like-minded organisations and individuals (IUCN-ICMM 2008).

The need for tools to be developed and made available to the international community by such an organisation also was identified. However, in the absence of an implementation mechanism or lead entity from the forum, no such organisation was established. This is typical of the discontinuous nature of initiatives on mining legacies.

The report, Minerals, Mining and Sustainable Development +10, (IIED 2012) reflected on a decade of mining and sustainable development since the initial study in 2002. A key finding was that the ICMM has implemented many of MMSD's recommendations for industry and has been the impetus for collective action from the sector. However, complementary measures for government, the small-scale mining sector, and communities were lagging behind. Also noted by the report was that there had been little advancement on the environmental issues surrounding legacy sites where legal responsibility remained unclear.

One aspect of relevance to mining legacies, which is a focus of the World Bank, is financial assurance (FA). This is the funding mechanism required by governments to cover the costs of mine remediation. A requirement for FA may be accompanied by a closure plan prior to commencement, and updated during the life of a mine to reflect closure requirements and costs. A guideline developed in 2009 (Sassoon 2009) addressed FA for mining closure in the oil, gas and mining policy division of the World Bank Extractives Industries for Development by evaluating and comparing the various forms of surety. This provides a useful reference point for jurisdictions when developing their financial assurance policies. Effective financial assurance policies provide incentives for industry to implement good closure design from the outset, and encourage progressive remediation during the life of the mine. This is an important an indicator of environmental liabilities; however, the amount held should not be relied upon as being a reflection of full liability. Independent third party evaluations of closure costs are likely to be more robust. Even with the best intentions in the world, accurately forecasting closure costs is extremely difficult; accordingly, the best that might be expected is an approximation. Some jurisdictions provide a financial assurance cost calculation tool for use by industry. One such tool is the Nevada Standardised Reclamation Cost Estimator (Poulin and Jacques 2009). Use of the tool provides a consistent estimation method. However, the tools can omit key costs or have discounts applied by governments. Costs for long-term mine water management and extended time periods for closure are not routinely included in Australian cost estimation tools, for example. This is despite an indication that time frames for closure regularly are underestimated (Byrne 2013). Mining companies may have their own cost calculation tool, which may be more or less robust depending upon corporate policy.

The International Atomic Energy Agency (2014) has contributed a significant body of knowledge to mine remediation through its development of guidelines, technical documents and country-specific reports, and provision of access to environmental databases on abandoned mines (Kelly and Slater 1994). Technical and non-technical lessons learned from environmental remediation programs also have been documented by the organisation.

Uranium mine remediation has much in common with the remediation of other commodities. Sites with ARD risks require similar containment strategies (including secure impoundments and effective covers) to encapsulate wastes in order to limit oxygen and water ingress. The Rum Jungle former copper and uranium mine in Australia's Northern Territory is one example where remediation of the environmental impacts of ARD impacts also will address uranium-specific impacts. The very long time frames for waste impoundment design (200–1000 years in the case of uranium mill tailings storage) are now applied to other types of tailings and residues in large dam design guidance (Australian National Committee on Large Dams 2012). These higher standards reflect what has been learnt from incidents where failed impoundments have had catastrophic impacts on humans and the environment.

In the last decade, there also has been far greater awareness and understanding of artisanal and small-scale (ASM)) mining in developing countries (ICMM 2009; Scholz in this volume).[1] In Mongolia, ASM threatens the environment, as well as the safety and health of small-scale miners. In 2009–2010, an inventory of abandoned areas (due to mining and other activities) was performed in 15 provinces of Mongolia (Buya 2013). The inventory found that 3984 hectares of land had been abandoned across 500 unit areas of 56 sub-provinces in 15 provinces. Approximately 60% of these areas were abandoned by formal licence holders, while the remainder were abandoned due to informal ASM activities.

Whilst ASM can exacerbate the environmental legacies of mining, they also have the potential to be part of the solution to remediation if they can be engaged effectively. The production of mining wastes, which are accessible to illegal miners, is only one part of this challenge. If illegal miners are not effectively managed, then remediation projects are likely to fail, particularly where remediation projects are carried out in the same catchment as illegal mining activities. There needs to be greater emphasis on and resources directed at how remediation of legacy sites can build capacity and provide employment in remote communities. Informal miners also may provide access to a body of knowledge that could be of value to the exploration activities of formal miners.

[1] They are also sometimes referred to as illegal or informal mining.

Mining Legacies in Australia

In Australia, the Mt Lyell mine near the west coast of Tasmania deposited tailings in the King River up until the early 1990s. Investigations of impacts and recommendations on remediation involved the Australian government's Supervising Scientist, as well as the Tasmanian government. New mining operations continue at the site with a tailings impoundment; however, since the 1990s, there is no evidence of further remediation works in the river delta of the Macquarie Harbour. ARD continues to flow into the King River from other sources of the historic operation, with the Tasmanian Government Environmental Protection Agency finding that:

> [a] biologically dead lower King River and biologically depauperate ecosystem in Macquarie Harbour, Australia's second largest bay, due to high metal levels (copper and aluminium) in the river and high copper levels in the harbour.

Because water treatment was not cost-neutral, it was not undertaken. The government actively is seeking proposals for the extraction of metals in the water. Queensland's Mount Morgan gold mine (discussed in further detail below) closed in the early 1990s, just prior to changes to environmental regulation of mines under the Queensland Mineral Resources Act (1989), which required projects to have financial assurance for rehabilitation. It is common for governments in Australia either to be surprised or to enter a state of denial when the magnitude of liabilities is realised. The complex challenges and costs often are postponed until the next government is elected. The Australian Commonwealth government evades responsibility for negative mining legacies and abandoned mines, claiming that their management is a matter for states and territory governments, as part of their regulation of mining activities. This section explores the roles of both levels of government.

The Organisation for Economic Co-operation and Development (OECD) undertook Environmental Performance reviews on integration of environmental concerns in the global mining industry. Regarding Australia, the OECD noted:

> Several specific mining objectives set in the 1992 National Strategy for Ecologically Sustainable Development have yet to be met, notably with respect to community involvement and mine rehabilitation. The problems posed by Australia's legacy of abandoned mines are being addressed in a piecemeal fashion, without any clear knowledge of their overall extent and nature, or of the priorities and resources needed for rehabilitation (OECD 1998).

The *Commonwealth's Environment Protection and Biodiversity Conservation Act 1999*, which came into effect in July 2000, established a nationally consistent framework for environmental assessment of new projects, based on consultative agreements between the Commonwealth and state and territory governments. Conditions relating to mine closure are included in the act's approvals where relevant. The mining industry responded to changing regulatory/legislative standards by developing a Code for Environmental Management. An Australian strategic framework for mine closure also was developed by the Australian and New Zealand Minerals and Energy Council 2000).

While the *EPBC Act* may be applicable for conservation of biodiversity where endangered species exist on or near abandoned mines (Thomson et al. 2005), there is no evidence that the legislation is applied to enforce remediation of abandoned mines where biodiversity of 'national significance' currently is affected. Ecological values of national significance are affected by abandoned mines where AMD has impacts on aquatic ecosystems, birds, and other fauna reliant on these water sources. At a jurisdiction level, regulatory black holes and ambiguities regarding the responsibilities of respective stakeholders have been reported, with recommendations made (Keliher 2007; Emerson 2010). Yet, despite having knowledge of these black holes and ambiguities, the Commonwealth government has not required that the environmental standards of state/territory governments be enforced in order to protect the environment from harm.

At a state/territory level, there is evidence in some jurisdictions of avoidance of regulatory responsibility by one department (the Environment Department) with regard to its oversight of the department responsible for the management of abandoned mines (the Mines Department). Avoidance of the issue likely is contributing to the lack of learning from the past, allowing governance gaps to be perpetuated. Meanwhile, degraded environments remain un-remediated, affecting biodiversity and Indigenous cultural heritage values. The impacts of abandoned mine pollution on food production areas are poorly mapped and quantified in Australia as well.

In 2001, UNEP noted that approximately 150 tonnes of mercury were lost from gold mine diggings in Australia to the environment, due to inefficient historic mining methods. More recent estimates indicate that there are approximately 600,000 tonnes of mercury in the environment in Australia attributable to past gold mining activities (Alpers et al. 2005). Australia, like many other states in the Asia-Pacific, signed the Minimata convention in 2013, which commits governments to the removal of mercury from the environment. Remediation of sites by recovering mercury and other valuable metals from wastes and waterways not only could help to address global environmental and human health impacts, but also could offset remediation costs. Moreover, during downturns in the mining industry, remediation activities could help to provide employment and community capacity building projects in regional Australia.

The application of sustainable development principles is important; however, there exists a perception at the national level in Australia that we have transcended ecological sustainable development. Indeed, there is a sense of complacency about the application of sustainable development principles in the country. As a consequence, knowledge has been lost by the industry and regulatory agencies, with socio-political interactions a contributing factor.

Australia points to good examples of land rehabilitation and closure in bauxite and sand mining. However, we have few good case studies for other mine types. It is more challenging to rehabilitate and close a metalliferous mine. Research has highlighted the important role of novel ecosystems for difficult sites (Doley and Audet 2014), where the aim is not to duplicate pre-existing natural ecosystems on mine landforms (e.g., waste rock dumps and tailings dams), but to develop sustainable and compatible ecosystems for those landforms. Further research and documentation of case studies on Australian mine rehabilitation and closure is needed, especially for metalliferous mines, in order to mitigate mining legacies.

The value of establishing wiki-based knowledge management systems for sharing information about the management of mine legacies also has been recognised. A mine closure site that is being developed by the Geological Survey of Finland provides a valuable benchmark for the resourcing and subsequent publication of research by governments at a national level in order to address the challenges of identifying sustainable solutions for mine closure. This is a model that Australia could benefit from.

In some jurisdictions in Australia, mines have opened and operated under site-specific legislation, which specifically excluded the requirement to undertake rehabilitation works and prepare closure plans. While this may have facilitated a short-term need for governments and industry, this approach does not serve the environment or communities well from a mining legacy perspective. Examples include the former Queensland Special Agreement Act mines and current State Agreement Act mines in Western Australia. Changing expectations have required the relevant Queensland mines to transition to current regulatory processes. In Western Australia, the relevant mines have been included for the first time in draft Mine Closure Guidelines. The guidelines outline specific regulatory processes for evaluation, which involves the EPA, not only the Department of Minerals and Petroleum.

Learning from Legacies

Of the 73 mine closure plans in Australia that Byrne (2013) studied, he found that 91% had a closure duration of more than 21 years, with the responsible companies still working toward relinquishment. He noted that long-term water management typically is ignored in mine closure planning. Byrne (2013) also identified that there is a gap between what is being allowed in closure plans and the reality of closure. His research identified a number of factors leading to the inherent optimism of the closure plans, including unrealistic assumptions, pressures to reduce closure cost provisions, and the technical complexity of many ARD issues. Whilst there is a large body of science on ARD management, the understanding of technical solutions still is evolving. However, if integrated with mine planning early in the mine life-cycle, the impacts for mining companies, governments, communities and the environment can be reduced. As Byrne notes:

> Of the closure plans that were reviewed only one site had allowance for long term (>50 years) water treatment. Of the sites in actual closure 25 per cent have reported that they expect such long term water treatment, with many acknowledging that in perpetuity treatment is probable (Byrne 2013).

Byrne (2013) additionally observed that most closure plans are littered with assumptions—sometimes documented, but sometimes inferred or hidden in an appendix. Typical assumptions include:

- That there will be no requirement for water treatment after production ceases;
- That waste rock dump and tailings storage facility caps will eliminate poor water quality discharges (even if there has been little modelling or design for those caps);
- That regulators and communities will accept an acidic pit lake; and,
- That water quality issues will be resolved during the closure execution phase.

A summary of costs for remediation associated with pumping and treating groundwater at legacy uranium mines provides an indication of the potential costs for companies/responsible for remediation, if closure is not well planned and executed. These examples highlight the uncertainties that exist around closure duration and costs when closure is poorly planned for during mine design.

Tailings containment strategies increasingly are requiring longer-term time frames for design, reflecting risk management evaluation of case studies overseas of tailings storage facilities and failures. The most recent edition of the tailings dam design guidance in Australia noted that closure design tends to be defined on a geological time-scale. ANCOLD (2012) notes that, for Australia, '[t]he period of 1000 years is considered reasonable, given that in Europe there are currently examples of tailings storages in excess of 800 years old that are being actively monitored'.

Research on 'Designer Tailings' (Edraki et al. 2014) aims to develop techniques that will result in geochemically benign wastes, with fewer legacy risks. Recent research on denser and more stable paste tailings in the aluminium sector also will help with mitigation of long-term geotechnical stability risks and contamination from draining tailings water (Boger 2012). These methods already are being applied at some sites to ensure long-term stability of wastes and, in so doing, are mitigating the creation of negative mining legacies.

Laurence's (2006) review of 800 mine closures in Australia over 25 years defined key causes of mine closure and the associated risks. The majority of mine closures were in Western Australia (60%), followed by New South Wales (20%) and Queensland (10%). The most commonly closed mine has been gold mines, followed by coal mines. Not all of these mines necessarily are abandoned, and some may have recommenced operations. Leaving this caveat to one side, only 25% of the mines closed are due to resource exhaustion/depletion. Most closures are unplanned and a consequence of falls in commodity prices. The most significant environmental impact from mine closure is ARD, confirming earlier observations that successfully closing metalliferous mines is a significant challenge. The data confirm the importance of early closure design, regular review and updating of closure plans, full financial provisioning for closure at any time during the mine's life, progressive remediation to reduce closure risks, and the need for research to address knowledge gaps.

Kuipers et al. (2006) confirms the challenges the mining sector has in predicting and managing ARD, and in applying appropriate regulations to minimise negative legacies. They reviewed predicted and actual water quality in hard rock mines, drawing upon information provided in Environmental Impact Statements (EIS) and subsequent publicly available data. Their study identified two primary modes of failure: geochemical characterisation and mitigation. In 64% of cases examined (16 case study mines), mitigation measures failed, while in 44% of cases (11 case study mines), the ability accurately to predict mine water geochemistry failed.

If closure risks are to be well understood, industry and governments must gather and analyse data from multiple sites, which is not easily accessible. Regulatory approval processes that focus principally on short-term/transient impacts with insufficient focus on closure risks contribute to the problem. Removal of requirements that support a long-term view, for example, the former Queensland Liberal National Party government's 'greentape reduction legislation', could increase the risk of future negative legacies or permanent site management. There currently is no legislative requirement for a mine closure plan in Queensland, as a consequence of the removal of the former rehabilitation planning process—which had been included as part of the Environmental Management Plan—from legislation in 2013 (Unger et al. 2013). However, during 2014, the Queensland government provided additional guidance, through the introduction of a rehabilitation cost calculator tool similar to that of NSW and Victoria.

Where closure guidance is weak or absent, the responsibility falls more heavily on the mining industry to reduce uncertainty for closure, lease relinquishment and post-mining land use. Companies must design for closure applying leading practices, and engage with the host community and other stakeholders regarding desired future land use(s). Companies that embrace this reality likely will track toward a successful outcome. Those that focus only on short-term compliance requirements may find themselves unable to close and achieve regulatory release.

Moreover, mining companies that combine the health, safety and environmental management and community relations functions into one role may contribute to the creation of legacy risks. Professionals with a health and safety focus in such a role naturally will focus on these risks. The productivity and legal liabilities of a serious H&S incident more easily are quantified, with impacts additionally being experienced in the short-term. Only those operations with experienced and senior environment managers are likely to apply sufficient focus to mitigation and management of environmental legacy risks. Those that engage teams of remediation researchers over the long-term likely will be able effectively to identify, evaluate and mitigate legacy risks.

Responsibility for Abandoned Mines in Australia

Worrall et al. (2009) report that there are at least 32,600 abandoned mines in Australia; however, there is average to poor data quality with unknown numbers in two jurisdictions. The authors also report that the high-level ICMM SD principles, the Minerals Council of Australia's (MCA) Enduring Value process, and the Global Reporting Initiative (GRI) provide frameworks that are suited to large-scale operations, but these do not translate well to small mines. Such frameworks additionally fail to address mining legacies in a meaningful way. There are a number of issues that need to be considered.

National Policy and Progress

The need for national coordination of abandoned mine management in Australia was recognised during a 'Management and Remediation of Abandoned Mines' forum held in Brisbane over a decade ago (Bell 2003). An overview of the National Orphaned/Abandoned Mine Initiative, which commenced in Canada in 2001, was presented as context for a national strategy in Australia. An Abandoned Mines Working Group was established in 2005–2006 under the Ministerial Council on Mineral and Petroleum Resources. In late 2010, the Strategic Framework for Managing Abandoned Mines in the Minerals Industry (Strategic Framework) was published.[2] It was developed to promote convergence on:

- Site inventories and site data management;
- Improved understanding of liability and risk relating to abandoned mines;
- Improved performance reporting;
- Standardisation of processes and methodologies; and,
- Sharing of knowledge and skills across jurisdictions (MCMPR/MCA 2010).

There are five chapters in the Strategic Framework: (1) valuing abandoned mines; (2) data collection and management; (3) risk assessment and management; (4) resourcing and partnership opportunities; and (5) information sharing and 'leading practice' (MCMPR 2010).

Estimates of mining legacy liabilities are $1 billion each for the Northern Territory and Queensland. These estimates do not include the gap between estimated liabilities and financial surety held by governments, which are recorded for some jurisdictions where audits have been performed. In the case of WA, the Auditor-General reported that 'financial securities held against poor end-of-mine outcomes accounted for less than 25% of potential rehabilitation costs'. The WA government initiated new policies for rehabilitation and closure since 2011, as described earlier.

The Multiple Land Use Framework 'developed in recognition of the conflict arising over land access and land use' is another nationally developed framework. However, it is not clear how this is being implemented into jurisdictional planning frameworks, nor how it integrates the requirements of the Strategic Framework (MCPR/MCA 2010). In Australia, the dialogue on mine closure at a national level has ground to a halt.

---

[2] There are a number of Australian strategic frameworks; however, only the National Mine Safety framework appears to have progressed to an implementation plan. The abandoned mine Strategic Framework does not have an implementation plan.

It is evident that where national governments administer mining tenure and regulate environmental aspects of mining, such as in Chile, there is greater progress (compared with Australia) on implementation of good governance for mine closure, remediation, and legacy site management. Chile has had mine closure legislation—Law 20.551—since 2012 (Olivari 2014).

In Australia, by contrast, each state/territory has its own legislation for environmental regulation of mining. In order to compete for mining developments, state and territory governments have embarked on a wave of 'greentape' reduction, implying that environmental requirements have been removed. Perhaps as a consequence of the defunding of environmental and mine regulatory agencies, the data required to inform effective policy-making and regulatory oversight increasingly are unavailable. There is only one rehabilitation performance measure available in the public domain in Queensland; that is, the proportion of disturbance that has been rehabilitated compared with total mine footprint in the state between 1990 and 2006 (Unger et al. 2013). The data showed an increase in the proportion of coal mine rehabilitation (reported by companies) from the early 1990s, when regulatory reforms for rehabilitation were introduced. However, since 2002, the proportion of coal mine rehabilitation has declined. Metalliferous mine rehabilitation was fairly constant until about 1998. Rehabilitation declined after that, with the data set ending in 2006. Increasingly, the availability of good quality spatial data and innovative interpretive tools provide mechanisms for policy development based on more effective evaluation of both quantitative and qualitative data on mine rehabilitation.

In response to the development of Australia's Strategic Framework, the Australasian Institute of Mining and Metallurgy (AusIMM—representing mining industry professionals) actively embarked on an abandoned mine discussion paper and survey (van de Graaff et al. 2012). The resultant policy endorsed education of mining industry professionals on effective closure design, as well as advocacy and engagement with governments responsible for the management of abandoned mines in order to improve rehabilitation (AusIMM 2013).

In 2011, a map of Australian abandoned mines was developed to provide information on the location and nature of Australia's abandoned mine records (Figure 20.1) (Lechner et al. 2011; Unger et al. 2012). In 2011, only three states had formal abandoned mine programs (Queensland, New South Wales and Tasmania), while other jurisdictions had programs for individual sites. For example, the Brukunga mine in South Australia. Western Australia and the Northern Territory have developed new programs to address abandoned/legacy mines (Unger et al. 2014).

Data quality in each jurisdiction is highly variable, with each state defining their mining legacies differently. In the absence of cross-jurisdictional data consistency for abandoned mines, it is not possible to identify priorities at a national scale. This level of data availability would benefit planning for addressing mercury risks from historic gold mines in line with Minamata Protocol requirements. Also useful would be an understanding of AMD impacts on biodiversity of national significance (Ramsar-listed wetlands). Geosciences Australia provides spatial data on mine fea-

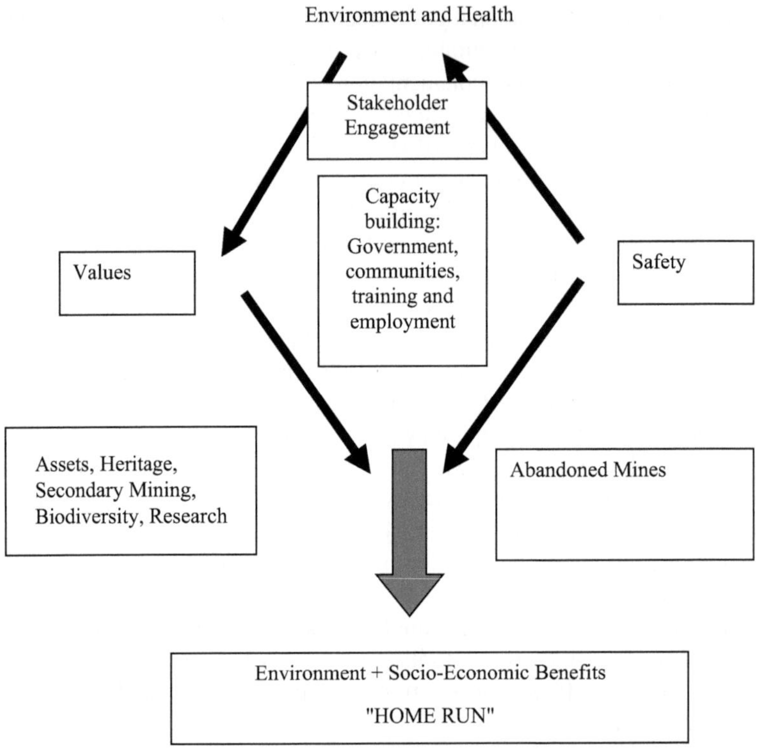

Fig. 20.1 Leading practice elements of abandoned mine regeneration

tures, with the data last updated in January 2012. It is not clear how state-based data sets are integrated with national data sets on abandoned mines. One of the few abandoned mine inventories in Australia is the Western Australian Inventory of Abandoned Mines, which specifically was developed to locate and describe abandoned mines across the state. This occurred after the 1997 death of a member of the public who fell into a vertical shaft. It took about 15 years for the WA government to commence policy development to prioritise sites for management based on the data collected from the inventory. It is intended that remediation of abandoned mines will be funded from interest raised from a recently established mine rehabilitation fund.

The management of abandoned mines and legacy sites clearly is not a priority for the Australian government, which views the implementation of the aforementioned Strategic Framework as a state/territory matter. The Australian government has not led an Abandoned Mine Working Group meeting since completion of the national policy (MCMPR/MCA 2010). In the absence of leadership on the issue, stakeholders are not effectively engaged and the energy of past initiatives may be lost.

Increased Awareness of Liabilities and Regulatory Black Holes

Auditors-General in several Australian jurisdictions recently have undertaken performance audits on environmental compliance of mining and contaminated land management, which have included mention of abandoned mines, mining liabilities and legacy issues. These audits have identified accountability and legislative gaps.

In Victoria, there is no single agency that serves as a contact point regarding abandoned mines. Instead, the department that manages the Crown Land upon which there is an abandoned mine is responsible for the site. An audit of contaminated land (including some abandoned mines) identified that both planning and environmental legislation are relevant. However, the responsible agencies and local councils are not effectively managing contaminated sites and, consequently, cannot demonstrate that they are reducing to acceptable levels potentially significant risks to human health and the environment. The complex regulatory framework governing abandoned mines has gaps, and key elements lack clarity. Inaction in dealing with contamination was driven in part by an undue emphasis on avoiding legacy and financial liability, rather than concerns about the need to protect human health and the environment.

In New South Wales, derelict mines arguably represent the State's largest category of contamination liability, with 500 abandoned mines not regulated under the Contaminated Land Management program (CLM) and its associated Act. A more recent audit of contaminated land recommended that the EPA be notified about seven high-risk derelict mine sites on Crown Land, which are managed by the Department of Trade and Investment, Regional Infrastructure and Services (DTIRIS). As DTIRIS has known about these sites for a number of years, it could be failing its duty to report under section 60 of the CLM Act.

DITIRIS does not view itself as the landowner or manager of these mines, and thus does not believe that it has an obligation to notify the EPA regarding how abandoned mines can be abandoned more than once. Indeed, in the first instance, a mine may be abandoned by the mining company following poor regulatory guidance. A mine may be abandoned for a second time—by the (ostensibly) responsible government department in order to avoid responsibility, and allow liabilities to flow through to the environment and community.

Government policies outlining roles and responsibilities for abandoned/legacy mines are difficult to find in Australian jurisdictions. However, Tasmania is performing better than its Australian peers in this regard. That state has clear publicly available information that outlines the functions and funding mechanisms for its abandoned mines program. In contrast, in Queensland, the Mines Safety Department manages abandoned mines. However, legislative 'black holes' mean that environmental impacts of abandoned mines are not regulated by the state's Department of Environment and Heritage Protection; accordingly, roles within government are unclear. Moreover, abandoned mines with potential environmental and human health impacts are not recorded on the Environmental Management or Contaminated Land Registers. Queensland spends between four and six million dollars per year on more than 15,000 abandoned mines, which have a total estimated liability of $1 billion (Interim Commission of Audit Report 2012). Avoidance of responsibility and liability is a clear priority of the state government:

> DNRM is directly responsible for the management of approximately 3500 of these sites … DNRM has not recognised in its financial statements a provision for costs associated with abandoned mines sites on or under state controlled land. The DG has advised that the state does not have a legal or constructive obligation to restore, either partially or in full, any of the 3500 abandoned mines for which it is directly responsible (QAO 2013).

The Western Australian Auditor-General (2011) noted that ownership and responsibility for the 11,000 abandoned mines in that state was not clear, especially with regard to rehabilitation. Performance audits on environmental regulation of mining identified that levels of financial security or bonds required from operators leaves states exposed to potentially significant financial risk in the event of inadequate mine closure. In Western Australia, the Audit Office also found that mines operating under a state agreement were not generally required to pay any bond. There is no standard mechanism to provide compensation to the State in the event of unacceptable rehabilitation in such cases.

Where abandoned mines fall into legislative ‘black holes’ or become a future liability, the environment and community may not be adequately protected from harm. Governments are slow to rectify these problems as is evidenced by the recommendations of a number of commissions/studies, which remain unimplemented (Emerson 2010; Keliher 2007).

The Fraser Institute (2013) mining industry survey places Western Australia at the highest of all jurisdictions in Australia on its Policy Perception Index, and in the top 10 globally. This measure is described as a report card for governments on the attractiveness of their mining policies. WA in particular is recognised for its investment attractiveness. The survey notes that the ‘new rehabilitation fund/bond retirement scheme’ is an exemplary policy. By reducing the financial surety it requires, the Department of Mines and Petroleum (DMP) accounted for both the needs of industry and its own need to gather funding for mining legacies through its pooled funding model. The department estimates ‘the total mine rehabilitation and closure costs for all mines operating under the Mining Act in Western Australia to be between $4 billion and $6 billion (actual data does not exist as there is no requirement for mine sites to report estimated closure costs to DMP)’ (Leybourne 2014). The new 1% bond for mines is deposited with the Mining Rehabilitation Fund (MRF), which enables a capital base to be accumulated and interest to be earned on this base, which subsequently can be directed to the management of historic abandoned mines. However, 25 mining projects managed under State Agreement Acts were exempted from the legislation (the Mining Act) and consequently are not required to contribute to the MRF. Elsewhere in Australia, in 2014, the Northern Territory’s Legacy Mine Unit (LMU) estimated liabilities to be $1 billion. The LMU was established to administer the 1% mining legacy levy, which was introduced in 2013 under the Mining Management Act (van Holthe 2013). Unlike Western Australia, the Northern Territory seeks a 100% rehabilitation security, plus a 15% contingency fee.

As is clear from this discussion, each jurisdiction in Australia has a unique bond/financial assurance system, as well as discrete environmental rehabilitation and closure policies. Evidence-based policy requires on-going monitoring of industry performance on key closure indicators. It would be wise for state and territory governments to undertake reviews of their bond policies and closure legislation to evaluate the effectiveness of mine closure and abandoned mine policies in the context of global leading practices to ensure that the costs of mining and historic legacies are not externalised.

Case Study: Mount Morgan

The Mount Morgan Mine is the largest abandoned mine in Queensland. The mine is located adjacent to the Dee River in the Fitzroy River catchment—the largest catchment flowing to the east coast of Australia. It was the largest producing gold mine and fourth-largest copper producer in Australia, and operated for over 100 years before closing in 1990.

During the course of the mine's lifecycle, approximately 7.5 million ounces (250 tonnes) of gold and 360,000 tonnes of copper were extracted, generating 134 million tonnes of waste rock and tailings (Wels et al. 2006). The Mount Morgan site is the primary source of ARD within the Fitzroy River catchment. The context for rehabilitation planning integrated with secondary mining at Mount Morgan was documented in 2003 (Unger et al. 2003). The objectives of rehabilitation planning were to:

- Improve water quality downstream;
- Avoid having to manage ARD interception and treatment indefinitely;
- Manage the site in accordance with its significant mining heritage; and,
- Develop and apply best-practice rehabilitation and management of the site.

Remediation estimates by the Queensland Government in 2003 indicated up to $100 million is required (Unger et al. 2003). The site actively is managed to collect seepage and to treat and evaporate water, which costs approximately $3.5 million per annum. Costs to remediate the mine today are likely to be greater than the 2003 estimate, due to the greater inventory of pit water and inflation. If there are found to be sufficient mineral resources in the site wastes and a company can be found that is able to undertake some of the remediation, this would go some way toward offsetting remediation costs.

Mount Morgan is referred to as an abandoned mine; however, the company left the site having met the agreed regulatory obligations at the time. Significant changes to environmental legislation were soon to be introduced under the *Minerals Resources Act 1989*. This included requirements for financial assurance for rehabilitation to be linked to a Plan of Operations. In addition, the requirement for an Environmental Management Overview Strategy was introduced. The company that last held the licence to mine at Mount Morgan, knew it would not be able to meet the new environmental conditions as it neared the conclusion of a tailings reprocessing phase. Accordingly, it undertook works (including establishing a seepage interception system), before leaving the site. Four companies in succession over more than 20 years have acquired the mining leases to evaluate mineral resources. However, to date, no project has been successful. While it is ideal to have the full extent of resource value achieved at abandoned mines, there is very little publicly accessible information on this subject in Australia. Savage River in Tasmania is one example where there was a successful agreement for further mining at an abandoned mine (Government of Tasmania (1996); Goldamere Pty Ltd (Agreement) Act 1996).

With support from the Australian government and the Fitzroy Basin Association, funding of $3.4 million was provided to build the Mount Morgan mine water treatment plan under the National Action Plan for Salinity and Water quality (Queensland Government 2006). Additional funds have been spent on the expansion of water treatment capacity and the installation of water cannons for enhanced evaporation.

Water treatment helped to reduce the volume of water contained in the open cut pit, however, the Queensland government chose not to fund major remediation works. Significant rainfall events in the summers between 2010 and 2013 caused the pit to overflow, with water quality impacts from ARD in the Dee River experiences many kilometres downstream (ABC Radio National 2013).

Water treatment reduces the volume of ARD water stored in the pit by treating and discharging water to the Dee River. Rainfall runoff in the form of overland flow from ARD-generating wastes cannot be intercepted, so this water flows into the Dee River. There are no data or reports available in the public domain on the water quality objectives for the Dee River in line with Australian and New Zealand Environment and Conservation Council's (2000) water quality guidelines. Expenditure of around $3.5 million a year for site management, including water treatment and management, may reduce the safety risks and volume of ARD entering the Dee River in the short-term, but it does not reduce the overall remediation liability of the site.

Mount Morgan Mine is not simply a mine safety project. The site is technically complex and requires multi-disciplinary guidance. The knowledge exists on how to address the technical environmental challenges. The Eden Project Post-Mining Alliance (2008) highlighted how management of mining legacies requires an economic and socio-political focus in addition to the obvious environmental focus. Policies that define cross-functional engagement by more than one government department and that have clear legislative requirements are just a few of the ingredients needed to address significant risks to, and enhance opportunities that arise from, abandoned mines, so that beneficial post-mining land uses can be achieved.

The Mount Morgan community has had its fair share of optimism and disappointment since the mine closed 25 years ago. The township had the highest unemployment rate of any community in Queensland at 23.4% in the 2001, according to the Australian Bureau of Statistics census (2001). Not only are environmental issues poorly addressed by inadequate leadership on abandoned mines, so too are social impacts. As a heritage-listed site of significance in Queensland, the opportunity remains to develop heritage tourism and alternative land uses at the Mount Morgan mine. The Queensland government hopes for a solution to its environmental legacies in the form of a new mining project. While waiting for this to occur, opportunities for alternative and concurrent beneficial post-mining land uses, employment, training and community capacity building need to be fully explored.

Leading Practices from Successful Case Studies

The cumulative impacts and liabilities of abandoned mines at a jurisdiction and national scale are even more challenging than individual sites. Such problems have complex social interactions, with no simple solution. As noted earlier, multiple agencies and stakeholders are involved, often with ambiguous responsibilities. Due to the complexity of such challenges, planning for abandoned mines falls into the category of 'wicked problems' (Rittel and Webber 1973). Mining legacies and abandoned mines have no clear single solution, and context is critical. Abandoned mines likely will require site-specific or jurisdiction-specific solutions. Because the legacy mine issue is prone to changes over time, participatory and process-oriented approaches are needed, as these provide mechanisms for policy learning, adaptation and constant improvement in result-oriented goals (Melish et al. 2012).

In order to develop an understanding of how successful projects have been completed at this larger, more complex scale, the present author undertook a review of leading practices for abandoned mine management and post-mining land use in 2009 (Unger 2009). Leading practices were drawn from:

- The International Atomic Energy Agency's global abandoned uranium mine program, which is coordinated from Vienna, Austria;
- A former East German uranium mine remediation project headquartered in Chemnitz;
- The Eden Project Post-Mining Alliance and other initiatives in Cornwall, England;
- The National Orphan/Abandoned Mine Initiative in Canada; and,
- The British Columbia Crown Contaminated Sites Program in Canada (BCCSP 2014).

The core elements of successful programs were identified and conceptualised using a softball pitch analogy (Figure 20.1). The first base, safety, often is the first aspect to be addressed by those managing legacy sites in order to ensure the public and site investigators are safe to assess site impacts. For those abandoned mines with environmental and health impacts, these are addressed next (second base). Concurrent with these processes, the values of the site (third base) are explored in order to maximise the opportunity for new economies to be integrated with closure planning. Successful projects identify and engage stakeholders actively from the outset of planning, yielding optimal socio-economic benefits. The 'home-run' signifies that when key risks and opportunities have been identified and addressed in the planning process—in concert with stakeholder engagement—outcomes most likely will be sustainable. Whitbread-Abrutat et al. (2013) argue that assisting communities and alleviating socio-economic impacts should be at the heart of closure planning. Figure 20.1 can be applied at both site and jurisdiction levels.

Leading practice programs overseas highlight the important role different stakeholders play in successful projects (Unger 2009). External stakeholders, including industry, researchers and NGOs, also need to be engaged to support successful programs. Effective engagement of stakeholders, as well as greater transparency, typified successful programs overseas.

Communities and other stakeholders cannot rely solely upon governments to address the wicked problem of mining legacies and abandoned mines. The history of unimplemented recommendations on abandoned mines suggests that the challenges are too complex and long-term in nature for governments to manage alone.

The Maturity Model Applied to the Strategic Framework

Expanding on the core elements of leading practice projects and programs, a risk-based model (The Maturity Model) was developed by Unger, in conjunction with collaborators. This model provides a measure of performance or maturity for abandoned mine management (Unger et al. 2012, 2015). The approach is proposed as a tool to support implementation of Australia's national policy on abandoned/legacy mines, as well as abandoned mine programs elsewhere in the Asia-Pacific Region.

The maturity model draws upon the risk management 'journey model' approach in the safety literature. Figure 20.2 illustrates the journey that management systems can take from vulnerable to resilient.

The leading practice review (discussed above) generated 14 elements of mature programs (Table 20.3), which were aligned with the five chapters of the Strategic Framework (MCMPR 2010; Unger et al. 2012, 2015).

A review of web-accessible data using this maturity model compared jurisdictions in Australia with each other and with jurisdictions overseas. Figure 20.3 shows that Western Australia is more mature in the area of data and information management than any other criteria (Unger et al. 2014). All jurisdictions were included in the web-based review and were compared with the British Columbia Crown Contaminated Sites Program (BCCCSP), where about 80% of the program is focused on abandoned/orphaned mines (Fig. 20.4). This was used as a benchmark for a mature program as it has been in place for about a decade. The program provides biennial performance reports on its website. These reports quantify liabilities and show how liabilities have been reduced by projects undertaken in a transparent and technically rigorous program (Unger 2009; BCCCSP 2012, 2014). Figure 20.4 shows that for almost all elements, the BCCCSP was more mature than the initiatives of Australian jurisdictions.

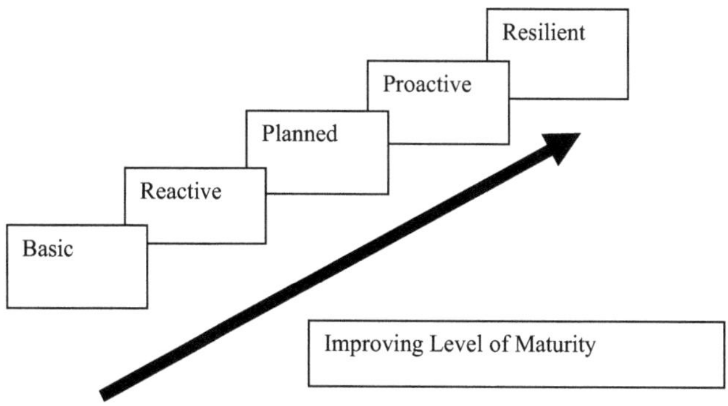

Fig. 20.2 Improving level of maturity model

Table 20.3 Elements of a mature abandoned mine program applied to Australia's strategic framework

Item	Mature program conceptual model	Strategic framework
1	Data/information management	Chapter 2: Data collection and management
2	Jurisdiction-wide knowledge of health, safety, environment, and socio-economic impacts	
3	Site-specific rehabilitation and management plans for high-risk sites	
4	Leadership, legislation, policy and guidance to address abandoned mines	Chapter 3: Risk assessment and management
5	Legislation, policy and guidance to prevent new abandoned mines	
6	Risk assessment and prioritisation of programs	
7	Abandoned mine program leadership and capacity/ skills	
8	Funding sources, mechanisms and resources	
9	Focus on beneficial post-mining land/water uses	Chapter 1: Valuing abandoned mines
10	Heritage conservation; Indigenous, cultural and Industrial	
11	Secondary mining opportunities	
12	Resourcing in partnership	Chapter 4: Resourcing and partnership opportunities
13	Stakeholder engagement at jurisdiction level	
14	Communication and networks	Chapter 5: Information sharing and 'leading practice'

Source: Unger et al. (2015)

Conclusion and Future Developments

The MMSD project recommended that 'one way to create a credit in the current natural capital account would be to deal with the worst environmental problems at abandoned mine sites' (IIED 2002). It further stated: '[i]mproving these sites could create benefits, which could offset or perhaps even exceed any deficits attributable

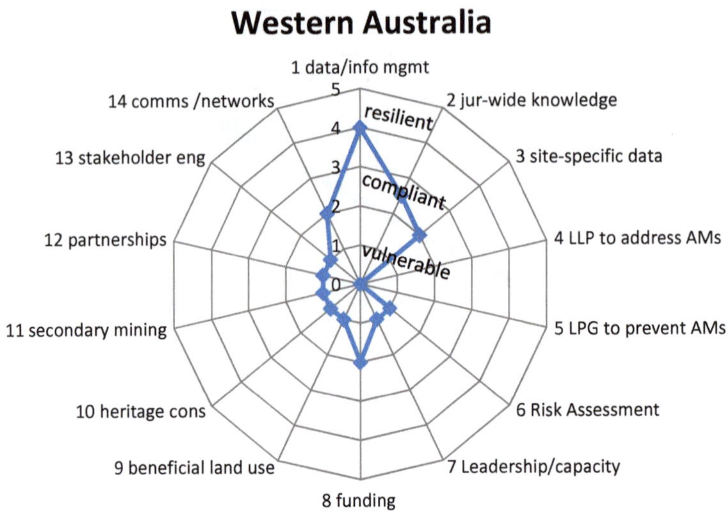

Fig. 20.3 Maturity of abandoned mine program in Western Australia. Source: Unger et al. (2014)

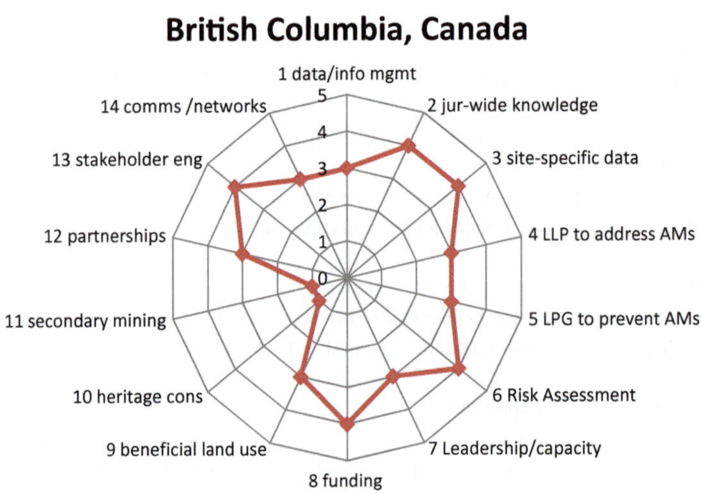

Fig. 20.4 Maturity of crown contaminated sites program in British Columbia, Canada. Source: Unger et al. (2014)

to current operations. And at some of these sites even a relatively small investment can have a big environmental payback' (IIED 2002). This presents both a challenge and an opportunity for the global mining industry.

It has been argued that society as a whole contributed to the creation of today's mining legacies; indeed, many of these mines met the regulatory requirements of the time (Pearman 2009). It follows that governments, industry, research institutions and civil society all must be involved in addressing them. Recommendations from audits, global forums and initiatives that have engaged stakeholders and that have clear direction should be revisited and implemented.

Companies must demonstrate that they can successfully close mines and generate positive legacies in order to obtain and maintain a social licence to operate. Where positive mining legacies successfully have been established, knowledge about these successes can and should be shared so that lessons can be learned and best practices established.

Leadership by the majors fundamentally is important to demonstrating that positive legacies from mining can be achieved. However, the roles that smaller companies, governments, NGOs, and other stakeholders can play requires further research (Pearman 2009; Gladwell 2013). Some companies actively are seeking ways to demonstrate to host communities and governments how they are different to other mining companies with regard to remediation, closure, stakeholder engagement and establishing post-mining land use capabilities. For example, some companies have developed their own internal closure planning mechanisms and have made them publicly available in multiple languages. This encourages the use of such tools and builds trust through transparency. However, the opportunity exists for more companies to publicise their remediation activities, closure planning and post-mining land use standards, and to engage more widely with host communities about the closure design process.

Opportunities also exist for greater creativity regarding post-mining land uses. The Eden Project has drawn attention to this through the publication of its book, *101 Things to do with a Hole in the Ground* (Pearman 2009). It is a leading example of inspiring post-mining land use and innovation in community engagement and education. Creative post-mining land uses in quarries already has provided opportunities for greater engagement with society post-mining (Hine et al. 2014). Broadly, the quarry sector has made more progress in achieving beneficial post-mining land uses, perhaps due to their more benign geology and proximity to urban development.

MMSD+10 (IIED 2012) showed that while progress has been made by industry on sustainable development initiatives, governments have lagged behind. However, MMSD+10 also noted was that industry, represented by the ICMM, has chosen not to lead on mining legacies. Thus, the roles of each stakeholder group need to be further explored as they each play a role in improved mitigation and management of mining legacies and abandoned mines globally. Accordingly,

National, state and territory governments need to:

- Ensure the application of evidence-based mine remediation, closure policies and legislation to support mitigation and management of long-term environmental impacts;
- Ensure that significant closure commitments are carried through the life of a project irrespective of the owner of the mine or changes in regulatory personnel;
- Allow for improvement and refinement of closure plans over the life of a project through greater understanding of the biophysical environment and community needs;

- Ensure a whole-of-government approach to mitigating and managing mining legacies;
- Ensure there are agreed standards for closure planning/closure with mechanisms to evaluate closure designs to which all relevant departments have contributed to and agree upon.

For legacy mines, national, state and territory governments need to:

- Ensure that there is cross-functional integration to solve complex challenges, analyse site risks and opportunities, and incorporate lessons into policy;
- Address legislative obstacles for mining of brownfields sites where remediation outcomes can be improved through further mining or reprocessing (e.g., 'Good Samaritan' legislation);
- Address responsibility and legislative 'black holes' for management of mining legacies and abandoned mines;
- Understand the role of effective remediation as part of a reconciliation process with Indigenous Communities;
- Engage more effectively with Indigenous and other communities in closure regulation and abandoned mine management;
- Apply consistent good practice standards to government departments managing legacy and abandoned sites;
- Clarify the role of contaminated land legislation ('polluter pays') for mitigating and managing mining legacies;
- Ensure that policies for regulation and remediation address the need to internalise environmental costs of mining in the long-term;
- Ensure effective, quarantined funding and financial mechanisms to mitigate/manage long-term legacies;
- Develop guidelines so that good practices are applied to legacy and abandoned mines, as well as the active mining sector;
- Fully account for legacy and abandoned mine environmental liabilities, ensuring regular updating of liabilities and liability reduction when works are undertaken to demonstrate transparency and accountability;
- Provide knowledge management toolkits to share good practice knowledge and case studies on mitigation/management and remediation of mining legacies; and,
- Engage communities and other stakeholders in closure policy development.

Industry needs to:

- Fully account for closure costs and integrate costs into design, consultation and progressive remediation from the outset;
- Prioritise the integration of SD principles for remediation and closure into mining even when compliance is in order to prevent new legacies from being created;
- Create 'wow' sites to demonstrate leading practice post-mining land use;
- Share knowledge with stakeholders so that capacity can be built in governments and civil society;

- Engage Indigenous Land Holders and communities in closure design to ensure their knowledge is used and needs are incorporated;
- Encourage small mining enterprises to be developed during the life of the mine that can sustain and support remediation during and after mining;
- Raise mining industry professionalism so that community and environmental aspects more fully are understood across the mining disciplines of mining, civil and geotechnical engineering, minerals processing and geology;
- Collaborate with other stakeholders and 'unusual partners' (Pearman 2008) on remediation of legacy sites;
- Explore the opportunity for ecological offsets, including former mines;
- Contribute to closure planning at a cumulative impact/regional scale (with neighbours and other stakeholders) for good environmental outcomes at a regional scale; and,
- Demonstrate that approved remediation and closure commitments will be met and not passed onto another company/operator who lacks the capacity to follow through.
- NGOs.

The broader community needs to:

- Build capacity locally through collaboration and training;
- Involve regional catchment groups and landholders in the gathering and sharing of data, through healthy waterways studies and partnerships for river health;
- Identify legacy sites where historic databases and inventories have gaps;
- Raise awareness of land and water values and remediation priorities;
- Applying for funding and providing oversight for community-based monitoring and research, potentially in collaboration with government;
- Expand Indigenous training and employment through remediation projects and the creation of new businesses; and
- Participate in groups working toward sustainable post-mining land uses, such as catchment groups, mining heritage conservation, interpretation, tourism and other businesses.

Finally, research can provide:

- Independent evaluation and a deeper understanding of issues, risks, challenges and opportunities;
- Mechanisms to analyse complex 'wicked problems';
- Novel approaches in the application of social, environmental and economic models to derive innovative diagnostic tools;
- New toolkits and resources for use by specific—or all—stakeholders;
- Quantification of impacts and risks to inform current and future business cases for mining projects;
- New knowledge of leading practices at cumulative and site specific levels;
- Access to cost effective research by tapping into undergraduate and research higher degree students;

- Linkages to wider specialist knowledge networks globally and within their research institutions;
- Multi-disciplinary approaches to problem solving;
- Expert panels to review data and plans for abandoned mines;
- Forums focussed on specific challenges and engagement processes; and,
- Peer reviewed publications and reports.

The recommendations of the Legacy Mine Roundtable forum (IUCN-ICMM 2008) remain valid and should be implemented; however, this requires leadership at a global level. Similarly, the national policy for abandoned mine management in Australia (MCMPR 2010), also remains to be implemented. The Australian government additionally should update its Strategic Framework for mine closure (ANZMEC 2000) and develop an effective mechanism for its implementation by state and territory governments. The current frameworks for the management of existing abandoned mines—and the prevention of new legacy sites—are ineffective in the absence of legislation and resources.

Ownership through multiple stakeholders will be needed to drive these processes. Refreshed engagement processes also will be required. An Australian network to connect governments and industry facing the challenges of mining legacies should be established, based on the NOAMI model from Canada, to connect stakeholders to each other. Future leadership from Australia in this regard could support leadership in a wider Asia-Pacific network.

Ultimately, the mining sector and its stakeholders will benefit from mining in a more comprehensive manner. The provision by communities of social licences for future mining activities will be based on companies' demonstration of a far greater level of success with regard to sustainable mine remediation and closure supporting beneficial post-mining land use. Host communities also will benefit from reduced levels of conflict from impacts on the environment, human health, and land and water use. The environment, under increasing demands from growing populations, additionally will benefit from the remediation of the physical, chemical and biological impacts precipitated by mining activities.

References

Alpers, C., Hunerlach, M., May, J., Hothem, R. (2005). Mercury Contamination from Historical Gold Mining in California. *US Geological Survey*, November 2005. http://pubs.usgs.gov/fs/2005/3014/. Accessed 2 April.

ANCOLD. (2012). Guidelines on Tailings Dams – Planning, Design, Construction, Operation and Closure. http://www.ancold.org.au/?page_id=334. Accessed 5 May 2014.

ANZECC. (2000). Australian and New Zealand guidelines for fresh and marine water quality - Volume 1. The guidelines. Australian and New Zealand Environment and Conservation Council. http://www.environment.gov.au/water/quality/publications/australian-and-new-zealand-guidelines-fresh-marine-water-quality-volume-1. Accessed 7 March 2014.

Australian and New Zealand Environment and Conservation Council and Agriculture and Resource Management Council of Australia and New Zealand. (2000). Australian Water Quality Guidelines for Fresh and Marine Waters, National Water Quality Management Strategy. Australian and New Zealand Environment and Conservation Council, Canberra. http://erin.gov.au/water/quality/nwqms/index.html External Link. Accessed 14 April 2014.

Audit Office of NSW. (2014). Performance Audit - Managing contaminated sites, Environment Protection Authority Department of Trade and Investment, Regional Infrastructure and Services. http://www.audit.nsw.gov.au/ArticleDocuments/336/01_Managing_Contaminated_Sites_Full_Report.pdf.aspx?Embed=Y. Accessed 14 August 2014.

Audit Office of NSW. (2011). Financial Audit: Volume Six 2011: Focus on Environment, Water and Regional Infrastructure, Audit Office of New South Wales. http://www.audit.nsw.gov.au/ArticleDocuments/226/01_Volume_Six_2011_Full_Report.pdf.aspx. Accessed 14 August 2014.

AusIMM. (2013). Abandoned Mines Policy Statement, AusIMM and Discussion paper http://www.ausimm.com.au/content/docs/policy/abandoned_mines_policy_June_2013.pdf. http://www.ausimm.com.au/content/docs/policy/abandoned_mines_discussion_paper_June_2013.pdf. Accessed 23 April 2014.

Australian and New Zealand Minerals and Energy Council/Minerals Council of Australia (ANZMEC/MCA). (2000) Strategic Framework for Mine Closure. http://www.sernageomin.cl/pdf/mineria/cierrefaena/DocumentosRelacionados/Strategic-Framework-Mine-Closure.pdf. Accessed 12 March 2014.

Australian Broadcasting Commission. (2015). Protesters chain themselves to land clearing equipment at Maules Creek coal mine site. http://www.abc.net.au/news/2015-02-15/activists-disrupt-maules-creek-coal-mine/6109164. Accessed 15 March 2015.

Australian Bureau of Statistics. (2001). 2017.3 - Census Of Population And Housing: Selected Education And Labour Force Characteristics For Statistical Local Areas, Queensland. http://www.abs.gov.au/ausstats/abs@.nsf/productsbytopic/58CE9B2ACC286C90CA256D01007F02EC?OpenDocument. Accessed 14 February 2014.

Australian Government, Department of Resources, Energy and Tourism. (2009). Evaluating performance: monitoring and auditing. LPSD program for the mining industry. http://www.industry.gov.au/resource/Documents/LPSDP/EvaluatingPerformanceMonitoringAuditing_web.pdf. Accessed 5 April 2014.

BCCCSP. (2014). Crown Contaminated Sites Program, 2014 Biennial report. Crown Land Opportunities and Restoration Branch, Ministry of Forests, Lands and Natural Resource Operations, Victoria, British Columbia. www.agf.gov.bc.ca/clad/ccs/cabinet/reports/2014_CCSP_Biennial_Report.pdf. Accessed 5 July 2014.

BCCCSP. (2012). Crown Contaminated Sites Program, Biennial Report. Crown Land Opportunities and Restoration Branch, Ministry of Forests, Lands and Natural Resource Operations, Victoria, British Columbia. www.agf.gov.bc.ca/clad/ccs/cabinet/reports/12_CLORB_Biennial_Report.pdf. Accessed 14 February 2014.

Bell, L, (2003). Management and remediation of abandoned mines. Australian Centre for Environmental Research (ACMER), Brisbane. http://www.jktech.com.au/sites/default/files/knowledge-transfer/other/AbandonedMinesTableofContents.pdf. Accessed 23 April 2014.

Boger, D. (2012). Rheology and the resource industries. Proceedings of the Life-of-Mine 2012: Maximising rehabilitation outcomes conference, 10-12 July, Brisbane, Queensland, The AusIMM/CMLR Published by the Australasian institute of Mining and Metallurgy, Publication Series No 6/2012.

Buya, T. (2013). Government policy on developing environmental friendly mining: success and challenges in MIRECO Korean Society Mineral and Energy Resources KSMER, 4th ISMR intercontinental Seoul COEX Hotel, Seoul Korea 29 May.

Byrne, G. (2013). Long-term water management – the forgotten legacy of mine closure. *Proceedings Water in Mining 2013*, The Australasian Institute of Mining and Metallurgy, Melbourne, 163-168.

Centre for Mined Land Rehabilitation. (2013). Submission to inquiry: Environment Protection Biodiversity Conservation Amendment Bill 2013 [Provisions].

DMP, 2014. http://www.dmp.wa.gov.au/documents/Cuidelines_Closure_Plan.pdf. Accessed 24 May 2014. Accessed 2 February 2014.

Chilean Copper Commission. (n.d.). http://www.cochilco.cl/english/. Accessed 4 February 2014.
Doley D., Audet P. (2014). Changing restoration priorities in the 21st century – opportunities for novel ecosystem design in mine closure. Proceedings, Life of Mine conference, Brisbane, 16-18 July.
Eden Project Post Mining Alliance. (2008). Background paper -ICMM Roundtable on the restoration of legacy sites. Eden Trust. Bodelva, Cornwall. www.postmining.org/documents/Roundtablebackgroundpaper.pdf. Accessed 4 April 2014.
Edraki M., Huynh, T., Baumgartl T., et al. (2014). Designer tailings – an integrated model for tailings management. Proceedings, Life of Mine 2014 conference, Brisbane, 16-18 July 2014.
Emerson, S. (2010). Speech on the mines and energy legislation amendment bill. *Queensland government hansard*. Wednesday 14 April 2010.
Fraser Institute. (2013). Survey of Mining Companies, 2012/2013 Fraser Institute Annual Survey. http://www.fraserinstitute.org/uploadedFiles/fraser-ca/Content/research-news/research/publications/mining-survey-2012-2013.pdf. Accessed 4 May 2014.
Gladwell, M. (2013). *David and Goliath: Underdogs, misfits and the art of battling giants*. New York: Back Bay Books.
Government of Tasmania. (1996). Goldamere pty. ltd. (agreement) act. http://www.thelaw.tas.gov.au/tocview/content.w3p;doc_id=30++1996+AT@EN+20111122000000;rec=0. Accessed 3 May 2014.
Hine, A., Kirsch, P., Amizlev, I. (2014). Red mud: art and the post-mining landscape. *Artlink Magazine*, 33(4).
IAEA. (2014). Lessons learned from environmental remediation programmes. http://www-pub.iaea.org/MTCD/Publications/PDF/Pub1630_web.pdf. Accessed 19 January.
ICMM. (2009). Workshop on Artisanal and Small-Scale Mining (ASM) and Mining Companies. http://www.icmm.com/document/919. Accessed 5 May 2014.
ICMM. (2008). Planning for integrated mine closure: toolkit http://www.icmm.com/document/310. Accessed 5 May 2014.
ICMM. (2006). Good practice guidance for mining and biodiversity. http://www.icmm.com/page/1182/good-practice-guidance-for-mining-and-biodiversity. Accessed 5 May 2014.
IIED. (2012). MMSD+10: Reflecting on a decade of mining and sustainable development, sustainable markets Discussion paper: June. http://pubs.iied.org/pdfs/16041IIED.pdf? Accessed 5 May 2014.
International Network for Acid Prevention (n.d.). Global acid rock drainage guide. http://www.gardguide.com/index.php?title=Chapter_4#4.1_Introduction.
IIED/WBCSD. (2002). Mining, minerals and the environment. In *Breaking new ground report of the mining, minerals and sustainable development (MMSD) Project*. http://pubs.iied.org/pdfs/G00902.pdf. Accessed 23 May 2014.
IUCN-ICMM and Post-mining alliance. (2008). Roundtable on restoration of legacy sites. The Old Mill Inn, Toronto Canada. http://www.icmm.com/document/511. Accessed 23 May 2014.
IUCN-ICMM (2006). Dialogue discussion report. http://cmsdata.iucn.org/downloads/2006_april_meeting_mining_eval.pdf. Accessed 23 May 2014.
Keliher, L. (2007). Review of the roles and responsibilities of the Department of Natural Resources, Mines and Water, Environmental Protection Agency and Department of Primary Industries and Fisheries. Service delivery and performance commission, Queensland Government. http://www.parliament.qld.gov.au/documents/tableOffice/TabledPapers/2007/5207T1077.pdf. Accessed 4 April 2014.
Kelly, G., Slater, R. (1994). The mining environment database on abandoned mines, acid mine drainage, and land reclamation. Proceedings of the international land reclamation and mine drainage conference and third international conference on the abatement of acidic drainage. Volume 4: Abandoned mine lands and topical issues – SP 06D-94. https://inis.iaea.org/search/search.aspx?orig_q=RN:26040794. Accessed 2 April 2014.
Kuipers, J., Maest, A., MacHardy, K., Lawson, G. (2006). Comparison of predicted and actual water quality at hardrock mines: the reliability of predictions in environmental impact statements. http://www.pebblescience.org/pdfs/ComparisonsReportFinal.pdf. Accessed 29 March 2014.

Laurence, D. (2006). Why do mines close? In Mine Closure 2006, proceedings of the first international seminar on Mine closure 13-15 September. Australian Centre for Geomechanics, Western Australia. http://www.acg.uwa.edu.au/__data/page/3735/Table_of_Contents.pdf. Accessed 17 July 2014.

Lechner, A., Unger, C., Edraki, M., Fletcher, A., Mulligan, D. (2011). Development of national abandoned mines inventory for risk modelling, Centre for Mined Land Rehabilitation. Poster, 7th Australian Workshop on Acid and Metalliferous Drainage, 21-24 June 2011, Darwin. https://www.researchgate.net/publication/236900953_DEVELOPMENT_OF_NATIONAL_ABANDONED_MINES_INVENTORY_FOR_RISK_MODELLING_National_Abandoned_Mines_Inventory_Future_work_Development_of_risk_model. Accessed 10 September 2014.

MCMPR/MCA. (2010). Strategic framework for managing abandoned mines in the minerals industry. Ministerial Council on Mineral and Petroleum Resources, Canberra. www.industry.gov.au/resource/Mining/Documents/StrategicFrameworkforManagingAbandonedMines.pdf. Accessed 23 April 2014.

McCarthy, B. (2014). The domains concept – 20 years on retrospective, obstacles and opportunities for effective life-of-mine implementation. Proceedings of AusIMM/CMLR Life-of-Mine conference, Brisbane, Queensland.

Ministerial Council on Mineral and Petroleum Resources. (2009). National mine safety framework, implementation report. Canberra. www.industry.gov.au/resource/Documents/Mining/NMSF-Implementation-Report-June2009.pdf. Accessed 14 May 2014.

Organisation for Economic Co-operation and Development. (1998). OECD environmental performance reviews, Australia: conclusions and recommendations. http://www.environment.gov.au/system/files/resources/6d83bc2d-62ce-47c1-95b9-7d289d2a8285/files/oecd.pdf. Accessed 2 March 2014.

Office of the Auditor General of British Columbia. (2002). Managing contaminated sites on provincial lands. http://www.bcauditor.com/pubs/2002/report5/managing-contaminated-sites-bcprovincial-lands. Accessed 6 April 2014.

Olivari, A. (2014). Financial assurance for mine closure in Chile. Proceedings of Mine Closure Solutions, April 26–30, Brazil. http://www.mineclosuresolutions.com/wp-content/uploads/2014/05/Olivari-Alfonso-Financial-assurance-for-mine-closure-in-Chile.pdf. Accessed 25 July 2014.

Paramita D., Asri, Y. (2011). Portrait of reclamation and post mining closure in Indonesia: case study coal mining in Samarinda, East Kalimantan. Indonesian Centre for Environmental Law, Jakarta.

Pearman, G. (2009). *101 Things to do with a Hole in the Ground.* Eden Project. Bodelva, Cornwell: UK.

Peck P., Balkau, F., Kadyrzhanova, I., Daussa, R. (2005). Mining for closure: policies and guidelines for sustainable mining practice and closure of mines. United Nations Development Program (UNDP), Organisation for security and cooperation in Europe (OSCE), North Atlantic Treaty Organisation (NATO). http://enrin.grida.no/environment-and-security/mining_src.pdf. Accessed 15 April 2014.

Poulin, R., Jacques, M. (2009). Environmental bonding in mining – comparative analysis of regulatory requirements. Proceedings of the International seminar on environmental issues in the mining industry. *EnviroMine*, 30 September – 2 October. Santiago, Chile.

Queensland Audit Office. (2014). Environmental regulation of the resources and waste industries. Report 15: 2013-14. https://www.qao.qld.gov.au/files/file/Reportspercent20andpercent20publications/Reportspercent20topercent20Parliamentpercent202013-14/RTP15Environmentalregulation oftheresourcesandwasteindustries.pdf. Accessed 23 April 2014.

Queensland Audit Office. (2013). Results of audit: state public sector entities for 2012-13. https://www.qao.qld.gov.au/files/file/Reportspercent20andpercent20publications/Reportspercent20topercent20Parliamentpercent202013-14/RtP11Statepublicsectorentities2012-13.pdf. Accessed 23 February 2014.

Queensland Government. (2006). New $3.4M water treatment plant opens at Mount Morgan - Media Statement. Minister for Natural Resources, Mines and Water. http://statements.qld.gov.au/Statement/Id/45237. Accessed 23 April 2014.

Rittel, H., Webber, M. (1973). Dilemmas in a general theory of planning. *Policy Sciences*, 4(2), 155-169.

Sassoon, M. (2009). Financial surety – guidelines for the implementation of financial surety for mine closure. The World Bank, Mining industry for Development. http://siteresources.worldbank.org/INTOGMC/Resources/7_eifd_financial_surety.pdf. Accessed 19 March 2014.

Standing Council for Energy and Resources. (2013). Multiple land use framework and background document. http://www.scer.gov.au/workstreams/land-access/mluf/. Accessed 23 March 2014.

Surface Mining Control and Reclamation Act of 1977. http://www.osmre.gov/lrg/docs/SMCRA.pdf. Accessed 23 April 2014.

Tasmanian Government, Environmental Protection Agency. (n.d.). Mt Lyell acid drainage remediation. http://epa.tas.gov.au/epa/mt-lyell-acid-drainage-remediation. Accessed 14 December 2014.

The Equator Principles Association. (2013). The Equator Principles 2013. http://www.equator-principles.com/resources/equator_principles_III.pdf. Accessed 12 August 2014.

Thomson, B., Pavey, C., Reardon, T. (2005). Recovery plan for cave-dwelling bats, *Rhinolophus philippinensis, Hipposideros semoni and Taphozous troughtoni 2001–2005*. http://www.environment.gov.au/system/files/resources/25a608d2-0755-48b2-972a-37c30ee38ac5/files/cave-dwelling-bats.pdf. Accessed 3 March 2014.

Unger, C., Lechner, A., Kenway, J., et al. (2015). A jurisdictional maturity model for risk management, accountability and continual improvement of abandoned mine remediation programs. *Resources Policy*, 43, 1-10.

Unger, C., Lechner, A., Wilson, I. (2013). Prevention of negative mining legacies – a mine rehabilitation perspective on legislative changes in Queensland. AusIMM Bulletin, February.

Unger, C., Lechner, A. M., Glenn, V., Edraki, M., Mulligan, D. (2012). Mapping and prioritising rehabilitation of abandoned mines in Australia. Life of Mine 2012: Maximising Rehabilitation Outcomes: 10 - 12 July 2012. Brisbane, Australia. Australasian Institute of Mining and Metallurgy.

Unger, C. (2012a). Managing mining legacies forum, 16-17 July 2012. http://www.cmlr.uq.edu.au/Portals/0/MMLF/MMLpercent20forumpercent20report.pdf.

Unger, C. (2009). Leading practice abandoned mine rehabilitation and post-mining land use, in Austria, Germany, England and Canada. http://www.churchilltrust.com.au/media/fellows/UNGER_Corinne_2009.pdf.

Unger, C., Laurencont T., Keliher, L., et al. (2003). Rehabilitation plan for the Mount Morgan mine site. Queensland Department of Natural Resources & Mines. http://www.cmlr.uq.edu.au/Portals/0/MMLF/LOMpercent20Paperpercent20Ungerpercent20etpercent20alpercent20Julypercent202012.pdf

United Nations. (1987). Our Common Future: Report of the World Commission on Environment and Development, http://conspect.nl/pdf/Our_Common_Future-Brundtland_Report_1987.pdf. Accessed 24 May 2014.

United Nations Environment Programme. (2001). Abandoned mine sites: problems, issues and options. http://www.cepal.org/drni/noticias/noticias/9/7029/ProgrammeUNEP.pdf. Accessed 24 May 2014.

Victoria Auditor General's Office. (2011). Managing Contaminated Sites. Victorian Auditor-General's Office. Melbourne. www.audit.vic.gov.au/publications/20111207-Contaminated-Sites/20111207-Contaminated-Sites.html. Accessed 1 February 2014.

van de Graaff, S., Unger, C., and Evans, R. (2012). AusIMM Abandoned Mines Survey Report. AusIMM and Centre for Social Responsibility in Mining. https://www.ausimm.com.au/content/docs/abandoned_mines_survey-final_report.pdf.

van Holthe, W. (2013). Mining management act passed. http://newsroom.nt.gov.au/mediaRelease/8347. Accessed: 23 April 2014.

Western Australian Auditor General. (2011). Ensuring compliance with conditions on mining. Report 8. https://audit.wa.gov.au/wp-content/uploads/2013/05/report2011_08.pdf. Accessed 30 April 2014.

Whitbread-Abrutat P., Kendle, A., Coppin, N. (2013). Lessons for the mining industry from non-mining landscape restoration experiences. Proceedings of Mine Closure.

Whitbread-Abrutat P. (2008). Mining legacy survey informing the background paper. Eden Project, February 2008. http://www.postmining.org/documents/surveysummaryreportFINAL25Feb08.pdf. Accessed 1 March 2014.

Worrall, R., Neil, D., Brereton, D., et al. (2009). Towards a sustainability criteria and indicators framework for legacy mine land. *Journal of Cleaner Production* 17(16), 1426-1434.

Part IV
Mitigating Risk

Chapter 21
Investment Decision-Making Tools for the Mining, Oil, Gas, and Infrastructure Sectors

Terry O'Callaghan and Vlado Vivoda

T. O'Callaghan (✉)
Centre for International Risk, School of Communication, International Studies and Languages, The University of South Australia, Adelaide, SA 5001, Australia
e-mail: terry.ocallaghan@unisa.edu.au

V. Vivoda
Sustainable Minerals Institute, The University of Queensland, St Lucia, QLD 4072, Australia
e-mail: v.vivoda@uq.edu.au

T. O'Callaghan, G. Graetz (eds.), *Mining in the Asia-Pacific*, The Political Economy of the Asia Pacific, DOI 10.1007/978-3-319-61395-6_21

Mining Company Investment Decision-Making Tool

Geological	Profitability	Strategic	Fiscal and monetary	Regulatory	Operational	Social and environmental (CSR)	Country factors	Host country public relations
Commercially exploitable mineral endowment	Measures of profitability (IRR, NPV, payback)	Successful extractive history in country/province (prior examples of successful ventures)	Ability to raise external financing	Favourable export/import policy (export on volume of minerals; imports on capital goods and equipment)	Ability to apply geological assessment techniques during feasibility studies	Availability of stringent environmental requirements	Political stability (central, provincial and local)	Stability of government's mining policy
Availability of geo-scientific data		Prior mining company experience in the country/region	Realistic and non-corrupt share market regulations	Favourable mining legislation	Ability to secure access to land	Ability to predetermine environment related obligations	National security (from domestic and external threats)	Pro/con FDI policy
			Ability to repatriate profits	Stability of exploration/mining terms (security of tenure)	Quality of transportation and other infrastructure (water, power, and utilities)	Obligations for mine closure and reclamation	Availability of political risk insurance	Investor friendly publications and presentations
			Favourable taxation regime (methods and level of tax levies)	Majority equity stake	Security and health risks for staff	Environmental impact assessment and requirements	Availability of international dispute-resolution mechanisms	Well-developed web presence

			Ability to predetermine tax liability	Majority management control	Favourable climate	Areas closed to mineral exploration due to social/ environmental issues (Forestry laws, native title)	Form of government (democracy, authoritarian, socialist)	Relations with home country
			Availability of tax incentives	Right to transfer ownership	Availability of skilled workforce	Ability to predetermine social obligations (social licence to operate)	Presence of strong populist leaders (anti-mining)	War and sabotage
			Stability of the fiscal regime	Size of exploration/ mining blocks	Locational factors	Power of anti-mining NGOs	Presence of lucrative domestic markets for minerals	Expropriation
			Ability to keep external accounts	Duration of exploration/ mining rights	Constraints on the use of expatriate staff	Ease with which to get local stakeholder support (informal)	Independent judiciary	Terrorism

(continued)

Geological	Profitability	Strategic	Fiscal and monetary	Regulatory	Operational	Social and environmental (CSR)	Country factors	Host country public relations
			Bilateral double taxation treaty with home country	Access to the regulatory	Restrictions on hiring and negotiating wages	Social infrastructure requirement (schools, hospital, roads)	Stability of local currency	
			Due diligence as a legal requirement in the host country	Mineral agreements to supplement the mining code	Availability of mining technology and expertise for hire	Effects on company reputation of having operations in the country	Level of FDI	
			Use of international accounting standards	Predictability of compliance/ enforcement mechanisms	Labour costs and strength of labour unions	Religious attitude to mining FDI	Level of economic development	
				Regulatory overlap/ duplication	Availability of mining specialist lawyers	Labour attitude to mining FDI	Level of corruption	
				Regulatory capture	Impact of small-scale mining on foreign mining operations	Media attitude to mining FDI	Level of decentralisation	

				Regulatory capacity (human and financial resources		Native title requirements	Civil liberties (freedom of speech, human rights)	
				Ability to form flexible operating structures (JVs)		Endangered species habitats located in mineral rich regions	Presence of large domestic mining companies	
				Existence of ‘one-stop shop’		Reputation risk	Signatory of voluntary codes of conduct	
				Clear and effective application and approval procedures			International border issues	
				Equal treatment of foreign and domestic mining companies				
				Requirements to process, refine and smelt all minerals in country				

(continued)

Geological	Profitability	Strategic	Fiscal and monetary	Regulatory	Operational	Social and environmental (CSR)	Country factors	Host country public relations
				Quality of mineral titles system (cadastral maps)				
				Local partner requirement				
				Continuous chain of title (grandfathering clauses)				

Oil and Gas Company Investment Decision-Making Tool

Geological	Profitability	Strategic	Fiscal and monetary	Regulatory	Operational	Social and environmental (CSR)	Country factors	Host country public relations
Commercially exploitable oil and gas endowment	Measures of profitability (IRR, NPV, payback)	Successful extractive history in country/ province (Prior examples of successful ventures)	Ability to raise external financing	Favourable export/ import policy (export on volume of crude oil and gas; imports on capital goods and equipment	Ability to apply geological assessment techniques during feasibility studies	Availability of stringent environmental requirements	Political stability (central, provincial and local)	Stability of government's oil and gas policy
Availability of geo-scientific data		Prior oil and gas company experience in the country/ region	Realistic FOREX regulations	Favourable oil and gas legislation	Ability to secure access to land	Ability to predetermine environment related obligations	National security (From domestic and external threats)	Pro/con FDI policy
Reserve/ production (R/P) ratio			Ability to repatriate profits	Stability of exploration/ production terms (security of tenure)	Quality of transportation (roads, pipelines) and other infrastructure (water, power, and utilities)	Obligations for oil well closure and reclamation	Availability of Political Risk Insurance	Investor friendly publications and presentations

(continued)

Geological	Profitability	Strategic	Fiscal and monetary	Regulatory	Operational	Social and environmental (CSR)	Country factors	Host country public relations
			Favourable taxation regime (methods and level of tax levies)	Majority equity stake	Security and health risks for staff	Environmental impact assessment and requirements	Availability of international dispute-resolution mechanisms	Well-developed web presence
			Ability to predetermine tax liability	Majority management control	Favourable climate	Areas closed to mineral exploration due to social/ environmental issues (Forestry laws, native title)	Form of government (democracy, authoritarian, socialist)	Relations with home country
			Availability of tax incentives	Right to transfer ownership	Availability of skilled workforce	Ability to predetermine social obligations (social licence to operate)	Presence of strong populist leaders (anti-oil)	
			Stability of the fiscal regime	Size of exploration/ production blocks	Locational factors (physical lay of the land, proximity to goods and services)	Power of anti-oil NGOs (Local and international)	Presence of lucrative domestic markets for products	

			Ability to keep external accounts	Duration of exploration/ production rights	Constraints on the use of expatriate staff	Ease with which to get local stakeholder support (informal)	Independent judiciary	
			Bilateral double taxation treaty with home country	Access to the regulator	Restrictions on hiring and negotiating wages	Social infrastructure requirement (schools, hospital, roads)	Stability of local currency	
			Due diligence as a legal requirement in the host country	Mineral agreements to supplement the oil and gas law	Availability of oil and gas technology and expertise for hire	Effects on company reputation of having operations in the country	Level of FDI	
			Use of international accounting standards	Predictability of compliance/ enforcement mechanisms	Labour costs and strength of labour unions	Religious attitude to oil and gas FDI	Level of economic development	
				Regulatory overlap/ duplication	Availability of oil and gas industry specialist lawyers	Labour attitude to oil and gas FDI	Level of corruption	

(continued)

Geological	Profitability	Strategic	Fiscal and monetary	Regulatory	Operational	Social and environmental (CSR)	Country factors	Host country public relations
				Regulatory capture		Media attitude to oil and gas FDI	Level of decentralisation	
				Regulatory capacity (human and financial resources)		Native title requirements	Civil liberties (freedom of speech, human rights)	
				Ability to form flexible operating structures (JVs)		Endangered species habitats located in mineral rich regions (land and water)	Presence of large domestic oil and gas companies	
				Existence of 'one-stop shop'			Signatory of voluntary codes of conduct	
				Clear and effective application and approval procedures			International border issues	
				Equal treatment of foreign and domestic oil and gas companies			War/sabotage	
				Requirements to process/refine oil and gas in country				

				Quality of oil and gas titles system				
				Local partner requirement				
				Continuous chain of title (grandfathering clauses)				

Infrastructure Investment Decision-Making Tool

Geological and geographical	Profitability	Strategic	Fiscal and monetary	Regulatory and legal	Operational	Social and environmental (CSR)	Country factors	Host country investment environment
Access to fossil and other fuels (non-renewable and renewable)	Measures of profitability (IRR, NPV, payback)	Successful engagement in country/region (prior examples of successful ventures)	Ability to raise external financing	Favourable import policy (imports on capital goods and equipment)	Ability to secure access to land	Availability of clear environmental requirements	Political stability and change (central, provincial and local)	Stability of government's infrastructure policy
Availability of geo-scientific data	Cost overruns	Prior infrastructure experience in the country/region (company specific)	Realistic and non-corrupt share market regulations	Favourable infrastructure legislation	Quality of transportation and existing infrastructure (water, power, and utilities)	Ability to predetermine environment related obligations	National security (from domestic and external threats)	Pro/con FDI policy
Unforeseen geological conditions i.e. Natural disasters			Ability to repatriate profits	Stability of project terms (security of tenure)	Security and health risks for staff	Environmental impact assessment and requirements	Availability of political risk insurance	Investor friendly publications and presentations
			Favourable taxation regime (methods and level of tax levies)	Potential for majority equity stake	Availability of skilled workforce	Areas closed to development due to social/environmental issues (Forestry laws, native title)	Availability of international dispute-resolution mechanisms	Well-developed web presence

			Ability to predetermine tax liability	Potential for majority management control	Locational factors (physical lay of the land, proximity to goods and services)	Ability to predetermine social obligations (social licence to operate)	Form of government (democracy, authoritarian, socialist)	Relations with home country
			Availability of tax incentives	Right to transfer ownership	Constraints on the use of expatriate staff	Power of anti-corporate NGOs (local and international)	Presence of strong populist leaders (anti-corporate)	Tendering processes
			Stability of the fiscal regime	Access to the regulator	Restrictions on hiring and negotiating wages	Ease with which to get local stakeholder support (informal)	Independent judiciary	Existence of a 'one-stop shop'
			Ability to keep external accounts	Lack of predictability of compliance/ enforcement mechanisms	Availability of technology and expertise for hire	Social infrastructure requirements (schools, hospital, roads)	Level of FDI	Publications in English

(continued)

Geological and geographical	Profitability	Strategic	Fiscal and monetary	Regulatory and legal	Operational	Social and environmental (CSR)	Country factors	Host country investment environment
			Bilateral double taxation treaty with home country	Regulatory overlap/ duplication	Strengths of labour unions	Reputation risk (from operating in host country)	Level of economic development	
			Due diligence as a legal requirement in the host country	Regulatory capture	Labour costs	Cultural attitude to FDI	Existence and level of corruption/ nepotism/ bribery	
			Use of international accounting standards	Regulatory capacity (human and financial resources)	Availability of specialist lawyers (in-host country)	Labour attitude to FDI	Level of decentralisation	
			Global financial fluctuations (exchange and interest rates/ inflation movements)	Ability to form flexible operating structures (JVs)	Potential breakdown in commercial partnership(s)	Media attitude to FDI	Civil liberties (freedom of speech, human rights)	

			Incompatible legal structures for project finance	Clear and effective application and approval procedures	Access to quality materials	Native title requirements	Presence of large domestic infrastructure companies	
			Import duties	Equal treatment of foreign and domestic infrastructure companies	Supply chain barriers	Endangered species habitats located in project location	Signatory of voluntary codes of conduct	
			Stability of local currency	Local partner requirement	Completion delays		International border disputes	
				Continuous chain of title (grandfathering clauses)	Failure to meet performance specifications		Expropriation risk	
							War/sabotage	
							Terrorism	
							Reputation risk	

MIX
Papier aus verantwortungsvollen Quellen
Paper from responsible sources
FSC® C105338

If you have any concerns about our products,
you can contact us on
ProductSafety@springernature.com

In case Publisher is established outside the EU,
the EU authorized representative is:
Springer Nature Customer Service Center GmbH
Europaplatz 3, 69115 Heidelberg, Germany

Printed by Libri Plureos GmbH
in Hamburg, Germany